Energy Options Impact on Regic

NATO Science for Peace and Security Series

This Series presents the results of scientific meetings supported under the NATO Programme: Science for Peace and Security (SPS).

The NATO SPS Programme supports meetings in the following Key Priority areas: (1) Defence Against Terrorism; (2) Countering other Threats to Security and (3) NATO, Partner and Mediterranean Dialogue Country Priorities. The types of meeting supported are generally "Advanced Study Institutes" and "Advanced Research Workshops". The NATO SPS Series collects together the results of these meetings. The meetings are co-organized by scientists from NATO countries and scientists from NATO's "Partner" or "Mediterranean Dialogue" countries. The observations and recommendations made at the meetings, as well as the contents of the volumes in the Series, reflect those of parti-cipants and contributors only; they should not necessarily be regarded as reflecting NATO views or policy.

Advanced Study Institutes (ASI) are high-level tutorial courses intended to convey the latest developments in a subject to an advanced-level audience

Advanced Research Workshops (ARW) are expert meetings where an intense but informal exchange of views at the frontiers of a subject aims at identifying directions for future action

Following a transformation of the programme in 2006 the Series has been re-named and re-organised. Recent volumes on topics not related to security, which result from meetings supported under the programme earlier, may be found in the NATO Science Series.

The Series is published by IOS Press, Amsterdam, and Springer, Dordrecht, in conjunction with the NATO Public Diplomacy Division.

Sub-Series

A.	Chemistry and Biology	Springer
B.	Physics and Biophysics	Springer
C.	Environmental Security	Springer
D.	Information and Communication Security	IOS Press
E.	Human and Societal Dynamics	IOS Press

http://www.nato.int/science
http://www.springer.com
http://www.iospress.nl

Series C: Environmental Security

Energy Options Impact on Regional Security

edited by

Frano Barbir
FESB, University of Split,
Split, Croatia

and

Sergio Ulgiati
University of Naples Parthenope
Naples, Italy

 Springer

Published in cooperation with NATO Public Diplomacy Division

Proceedings of the NATO Advanced Research Workshop on
Energy Options Impact on Regional Security
Split, Croatia
17–20 June 2009

Library of Congress Control Number: 2010935019

ISBN 978-90-481-9567-1 (PB)
ISBN 978-90-481-9564-0 (HB)
ISBN 978-90-481-9565-7 (e-book)

Published by Springer,
P.O. Box 17, 3300 AA Dordrecht, The Netherlands.

www.springer.com

Printed on acid-free paper

CONTENTS

PREFACE

Energy appears to be a fundamental driving force of economic and political strategies as well as planetary stability. Energy-related issues such as (1) the availability of new energy sources and viable technologies, (2) the disparity in access to energy sources, (3) the role of energy in our societies (energy societal metabolism), (4) the energy support to the life of our cities (where about half of world population is going to live very soon), and (5) the energy demand for food security all over the world, are "hot" problems that humans will have to face within the framework of sustainability (ecologically sound production and consumption patterns associated with socially acceptable life styles), in terms of policies, technological development and economic processes.

A coherent energy strategy is required, addressing both energy supply and demand, security of access, development problems, equity, market dynamics, by also taking into account the whole energy lifecycle including fuel production, transmission and distribution, energy conversion, and the impact on energy equipment manufacturers and the end-users of energy systems. Issues of energy efficiency and rebound effect must also be taken into proper account. In the short term, the aim should be to achieve higher energy efficiencies and increased supply from local energy sources, in particular renewable energy sources. In the long term, redesign of life styles, further increase of alternative energy sources and shift to new energy carriers such as hydrogen is expected to contribute to solve or alleviate the problems generate by declining availability of fossil fuels. Both points of view must include national accounting procedures that also consider resource depletion and environmental degradation, and questions concerning growth, carrying capacity, sustainability, and inter- and trans-generational equity.

Concerned scientists and policy-makers cannot disregard the need for increased awareness of the energy problem among stakeholders (population, business communities, energy companies) in order to lead to redesign of societal structures and metabolism towards low energy life style and attitudes. How this can be achieved is not an easy matter, since the present trend relies of past habits of low cost energy use and lack of sufficient awareness of environmental problems. It is evident that education must play a significant role in this regard.

Urgent issues of energy security call for increased reliance on local resources, with special focus to the role of agriculture for biomaterials and bioenergy supply as well as to development of cost effective solar & wind technologies for electricity production. The use of thermal solar modules and devices should also be explored in depth, since a large part of energy demand occurs locally in the form of low and medium temperature heat. Energy security also calls for diversification of import supplies both in terms of quality and nature of energy resources (oil, natural gas, coal, etc.) and areas from which resources come from.

This book addresses the above issues in light of the energy options that the European and Mediterranean countries are facing and attempts to identify what economic, environmental, societal and regional security issues that may come out of those options. It is a collection of papers presented at the Advanced Research Workshop entitled Energy Options Impact on Regional Security that was held 17–20 June 2009 at the Mediterranean Institute for Life Sciences in Split, Croatia, generously supported by the NATO Science for Peace and Security Programme.

The following are the key conclusions from the Workshop regarding energy options and their impact on regional security:

- Local policies are needed that integrate development and environmental protection (renewable energies, efficiency, sustainable tourism). However, effective global policies are also needed because global paradigms determine local possibilities.
- Plural assessment methods through spatial, time and hierarchical scales and levels are needed due to existence of legitimate (and sometimes irreducible) different points of view.
- Energy security is linked to social stability and development.
- Technological progress is important, but it is not likely to provide a final and stable solution to energy and environmental problems. Life style changes are urgently needed, as well as policies that align economic, energy, environmental and security goals.
- Some regional policies presented at the workshop sought energy security by investing in the rising-demand fossil-fuelled paradigm that is actually the source of the insecurity, and therefore not a viable solution.
- The EU strategies are not easily implemented at regional and local levels.
- Reliable and meaningful statistical data and their interpretation are identified as one of the crucial problems in making energy decisions.

Editors
Frano Barbir
Sergio Ulgiati

MULTI-METHOD AND MULTI-SCALE ANALYSIS OF ENERGY AND RESOURCE CONVERSION AND USE

SERGIO ULGIATI[1*], MARCO ASCIONE[1],
SILVIA BARGIGLI[1], FRANCESCO CHERUBINI[2],
MIRCO FEDERICI[3], PIER PAOLO FRANZESE[1],
MARCO RAUGEI[4], SILVIO VIGLIA[1],
AND AMALIA ZUCARO[1]
[1]*Department of Sciences for the Environment,
Parthenope University, Napoli, Italy*
[2]*Department of Energy and Process Engineering,
Norwegian University of Science and Technology (NTNU),
Trondheim, Norway*
[3]*Department of Chemistry, University of Siena, Siena, Italy*
[4]*Environmental Management Research Group - GiGa
(ESCI-UPF), Barcelona, Spain*

Abstract   Optimizing the performance of a given process requires that many different aspects are taken into account. Some of them, mostly of technical nature, relate to the local scale at which the process occurs. Other technological, economic and environmental aspects are likely to affect the dynamics of the larger space and time scales in which the process is embedded. These spatial and time scale effects require that a careful evaluation of the relation between the process and its 'surroundings' is performed, so that hidden consequences and possible sources of inefficiency and impact are clearly identified. In this work the authors summarise a number of studies in which they applied a multi-method and multi-scale approach in order to generate a comprehensive picture of the investigated systems/ processes. The benefits of such an integrated investigation approach are discussed.

Keywords: Life Cycle Assessment, integrated evaluation, emergy

* To whom correspondence should be addressed: Sergio Ulgiati, Department of Sciences for the Environment, Parthenope University, Napoli, Italy; E-mail: sergio.ulgiati@uniparthenope.it

1. Introduction

Life Cycle Thinking (LCT) techniques are used worldwide to assess material and energy flows to and from a production process. LCT methodologies are aimed at assessing the environmental impacts of a product (or service) from 'cradle to grave' or better 'from cradle to cradle', including recycling and reclamation of degraded environmental resources. More than a specific methodology, LCT is a cooperative effort performed by many investigators throughout the world (many working in the industrial sectors) to follow the fate of resources from initial extraction and processing of raw materials to final disposal. This effort is converging towards standard procedures and common frameworks, in order to make results comparable and reliable. SETAC (the International Society for Environmental Toxicology and Chemistry) developed a "code of practice" to be adopted as a commonly agreed procedure for reliable LCAs (SETAC 1993). The SETAC standardization has been followed by a robust effort of the International Organization for Standardization (ISO) to develop a very detailed investigation procedure for environmental management based on LCT, namely Life Cycle Assessment (International Standards ISO 14040/2006 – LCA Principles and framework and 14044/2006 – Requirements and guidelines, www.iso.org). The ISO documents suggest clear and standard procedures for the description of data categories, definitions of goals and scope, statements of functions and functional units, assessments of system boundaries, criteria for inclusions of inputs and outputs, data quality requirements, data collection, calculation procedures, validation of data, allocation of flows and releases, reuse and recycling, reporting of results. Nowhere in the ISO documents is preference given to a particular impact assessment method. In the opinion of the authors, LCA can be looked at as a standardized (and still to be improved) framework, where most of the methodologies already developed for technical and environmental investigations may be included and usefully contribute.

Finally, a European Platform on Life Cycle Assessment (http: //lct.jrc.ec.europa.eu/eplca) has been established by the European Commission in support of the implementation of the EU Thematic Strategies on the Prevention and Recycling of Waste and on the Sustainable Use of Natural Resources, the Integrated Product Policy (IPP) Communication and Sustainable Consumption and Production (SCP) Action Plan. The purpose is to improve the credibility, acceptance and practice of Life Cycle Assessment (LCA) in business and public authorities, by providing reference data and recommended methods for LCA studies.

1.1. FACTORS OF SCALE AND SYSTEM BOUNDARIES

It is self-evident that each evaluation can be performed at different space and time scales. Figure 1 shows the systems diagram of a generic thermal power plant, with fossil fuels used to produce electricity. The local scale only includes direct energy and mass inputs. The latter include plant components and buildings, discounted over the plant lifetime.

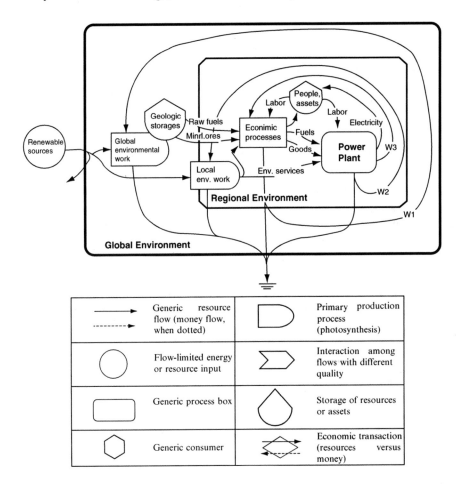

Figure 1. Systems diagram of a generic power plant, showing the convergence of matter and energy input flows from the larger to the local scale. Systems symbols are from Odum (1996)

As the scale is expanded to the regional level [whatever the regional area is, it should include the production process for machinery components (boiler, turbine, insulating materials, etc.) and plant building materials like concrete and steel], new mass and energy inputs must be accounted for.

If the scale is further expanded, the mass of raw minerals that must be excavated to manufacture the pure metals for plant components also contribute to all of the calculated performance indicators. At this larger scale, raw oil used up in the extraction and refining of minerals and oil itself must also be accounted for. The evaluation may therefore be carried out at three different scales (local, regional and global), each one characterized by well-specified processes (respectively, resource final use, manufacturing and transport of components, resource extraction and refining), so that inefficiencies at each scale may be easily identified and dealt with.

The larger the spatial scale, the larger the cost in terms of material and energy flows, i.e. the worse the related conversion efficiency and other impact indicators. In fact, if a process evaluation is performed at a small scale, its performance may not be well understood and may be over-estimated due to a lack of inclusion of some large scale impacts. Depending upon the goal of the investigation, a small-scale analysis may be sufficient to shed light on the process performance for technical or economic optimization purposes, while a large-scale overview is needed to investigate how the process interacts with other upstream and downstream processes as well as the biosphere as a whole. Defining the system boundary and clarifying at what time-scale an assessment is performed is therefore of paramount importance, even if the scale of the assessment is sometimes implicit in the context of the investigation. It is very important to be aware that a 'true' value of net energy return or other performance indicators does not exist. Each value of a given indicator is only 'true' at the scale at which it is calculated. When the same value is used at a different scale, it does not necessarily become false. It is, however, out of the right context, and therefore inapplicable and useless.

1.2. ACCOUNTING FOR TIME EMBODIED IN RESOURCES

Time is an important, but very often neglected, issue in any kind of evaluation. The simplest case is when we have inputs, whose lifetime exceeds the time frame of the analysis. It is easy to transform extended-lifetime inputs into annual flows by dividing by their lifetime (in years). Another and perhaps more important time scale is hidden in the resources used, i.e. the time nature takes to concentrate or produce a given resource (e.g. oil). A resource turnover time is often a good measure of its renewability. An effort to go beyond the concept of turnover time in resource evaluations is the introduction of emergy accounting procedures (Odum 1988, 1996; Brown and Ulgiati 2004), also included in the integrated approach discussed in the following of this paper.

2. Towards an Integrated Evaluation Approach

Environmental and socio-economic accounting procedures so far proposed by many authors were applied to different space–time windows of interest and were aimed at different investigations and policy goals (Szargut et al. 1988; Odum 1996; Herendeen 1998; Ayres and Masini 1998; Giampietro et al. 1998; Lozano and Valero 1993; Foran and Crane 1998; Finnveden and Moberg 2005, Ulgiati et al. 2006; Gasparatos et al. 2008). These Authors offered valuable insight towards understanding and describing important aspects of resource conversions and use. However, due to their different focus on specific scales, numeraires and questions, results from these methods are hardly comparable. Further effort is needed towards the development of LCA-like approaches that are capable to integrate different socio-economic and environmental points of view for a more comprehensive picture at different scales. The context and the goal of both the process and the investigation procedure are of paramount importance and likely to affect the results. For example, investigating only the behaviour of a single process and seeking maximization of one parameter (efficiency, production cost, jobs, etc.) is unlikely to provide sufficient insight for sustainable policy making. Instead, if suitable approaches are selected, applicable at different scales and designed in such a way as to complement each other, integration would be feasible. Each method may supply a piece of information about system performance at an appropriate scale, to which the others do not apply. Integration supplies an overall picture, characterized by an 'added value' that could not be achieved through each approach individually. The choice of the set of approaches is therefore of crucial importance. More-over, if the integration is carried out properly, the same set in input data (similar to an LCA inventory, but complemented with other typologies of input, e.g. free environmental flows, environmental services, socio-economic data such as labor and economic services, etc.) may serve to expand the focus of the evaluation beyond the common accounting of energy costs and environmental impacts.

2.1. ACCOUNTING FOR MATTER FLOWS

Quantifying input and output mass flows is a preliminary step. We need to assess not only the amount of input materials to the local process, but to the greatest possible extent the amount of outputs (products, co-products, and emissions) from the process. In addition, when we expand our scale of investigation, we realize that each flow of matter supplied to a process has been extracted and processed elsewhere. Additional matter is moved from place to place, processed and then disposed of to supply each input to the process. Sometimes a huge amount of rock must be excavated per unit of metal or other chemical actually delivered to the user. Most of this rock is then returned to the mine site and the site reclaimed, but its stability is lost and several chemical compounds become soluble with rainwater and may affect the environment in unexpected ways. Accounting for the materials directly and indirectly involved in the whole process chain has been suggested as a measure of environmental disturbance by the process itself (Schmidt-Bleek 1993; Hinterberger and Stiller 1998; Bargigli et al. 2005). A quantitative measure can be provided by means of Material Intensity Factors (MIPS, Material Input Per unit of Service) calculated for several categories of input matter, namely abiotic, biotic, water, and air (Ritthoff et al. 2003). The resulting MAterial Intensity Analysis (MAIA) method is aimed at evaluating the environmental disturbance associated with the withdrawal or diversion of material flows from their natural ecosystemic pathways. In this method, appropriate material intensity factors (g/unit) are multiplied by each input, respectively, accounting for the total amount of abiotic matter, water, air and biotic matter that is directly or indirectly required in order to provide that very same input to the system. The resulting material intensities (MIs) of the individual inputs are then separately summed together for each environmental compartment (again: abiotic matter, water, air and biotic matter), and assigned to the system's output as a quantitative measure of its cumulative environmental burden from that compartment.

2.2. ACCOUNTING FOR HEAT FLOWS

First-law heat accounting is very often believed to be a good measure of energy cost and system efficiency. The energy invested in the overall production process is no longer available to the user of the product. It has been used up and is not contained in the final product itself. The actual energy content (measured as combustion enthalpy, H.H.V. – Higher Heating Value, L.H.V. – Lower Heating Value) of the product differs from the total input energy because of conversion losses occurred in many steps leading to the final product. Energy analysts refer to the total energy required in the form of crude oil equivalent as to "embodied energy". The Embodied Energy Analysis method (Slesser 1974; Herendeen 1998) deals with the gross (direct and indirect) energy requirement of the analysed system, and offers useful insight on the first-law energy efficiency of the analysed system on the global scale, taking into consideration all the employed commercial energy supplies. In this method, all the material and energy inputs to the analysed system are multiplied by appropriate oil equivalent factors (g/unit), and the cumulative embodied energy requirement of the system's output is then computed as the sum of the individual oil equivalents of the inputs, which can be converted to energy units by multiplying by the standard calorific value of 1 g of oil (41,860 J/g). The chosen cumulative indicator is the so-called "gross energy requirement" (GER), expressing the total commercial energy requirement of one unit of output in terms of equivalent joules of oil.

Quantifying the total energy invested into a process allows an estimate of the total amount of primary energy invested and, as a consequence, the extent of the depletion of nonrenewable energy resources caused by the process. If the evaluation deals with an energy conversion process (e.g. conversion of oil to electricity or conversion of wind into electricity), the comparison of the energy output to the energy invested provides a measure of the performance named EROI (Energy Return on Investment): the higher the EROI the higher the energy benefit from the process.

2.3. ASSESSING USER-SIDE RESOURCE QUALITY. THE EXERGY APPROACH

Not all forms of energy are equivalent with respect to their ability to produce useful work. While heat is conserved, its ability to support a trans-formation process must decrease according to the second law of thermo-dynamics (increasing entropy). This is very often neglected when calculating efficiency based only on input and output heat flows (first-law efficiency) and leads to an avoidable waste of still usable energy and to erroneous efficiency estimates. The same need for quality assessment applies to every kind of resource supplied or released in a process. The ability of resources to supply useful work or to support a further transformation process must be taken into account and offers opportunities for inside-the-process optimiz-ation procedures, recycle of still usable flows, and downstream allocation of usable resource flows to another process.

The ability of driving a transformation process and, as a special case, producing mechanical work, may be quantified by means of the exergy concept. According to Szargut et al. (1988) exergy is "the amount of work obtainable when some matter is brought to a state of thermodynamic equili-brium with the common components of the natural surroundings by means of reversible processes, involving interaction only with the abovementioned components of nature". Chemical exergy is the most significant free energy source in most processes. Szargut et al. (1988) calculated chemical exergy as the Gibbs free energy relative to average physical and chemical para-meters of the environment.

By definition, the exergy (ability of doing reversible work) is not conserved in a process: the total exergy of inputs equals the total exergy of outputs (including waste products) plus all the exergy losses due to irreversibility. Quantifying the exergy losses due to irreversibility (which depends on deviations from an ideal, reversible case) for a process offers a way to calculate how much of the resource and economic cost of a product can be ascribed to the irreversibility affecting the specific technological device that is used as well as to figure out possible process improvements and optimization procedures aimed at decreasing exergy losses in the form of waste materials and heat. Exergy losses due to irreversibilities in a process are very often referred to as "destruction of exergy."

The exergy evaluation method is not routinely used in LCA assessments, where it could be instead adopted and implemented for the assessment of the second law efficiency of the process, to be considered as another important LCA parameter. Each input to the process can be accounted for in terms of its exergy content. The ratio of the exergy content of the output to the sum of the input exergies is a measure of the maximum conversion efficiency attainable in theoretical reversible conditions. Exergy has also sometimes been suggested as an ecological metric to gauge ecosystem health and stability (Fath and Cabezas 2004; Jørgensen 2005), but the authors have made the choice to stick to a strictly thermodynamic evaluation of the systems under study, leaving the evaluation of direct and indirect ecosystem disturbance to the Emergy method, as well as to other "downstream" impact categories.

2.4. ASSESSING DONOR-SIDE RESOURCE QUALITY. THE EMERGY APPROACH

The same product may be generated via different production pathways and with different resource demand, depending on the technology used and other factors, such as boundary conditions that may vary from case to case and process irreversibility. In its turn, a given resource may require a larger environmental work than others for its production from nature. As a development of these ideas, Odum (1988, 1996) introduced the concept of *emergy*, i.e. "the total amount of available energy (exergy) *of one kind* (usually *solar*) that is directly or indirectly required to make a given product or to support a given flow". We may therefore have an oil emergy, a coal emergy, a solar emergy, etc. according to the specific goal and scale of the process. In some way, this concept of embodiment supports the idea that something has a value according to what was invested into making it. This way of accounting for required inputs over a hierarchy of levels might be called a "donor system of value", while exergy analysis and economic evaluation are "receiver systems of value", i.e. something has a value according to its usefulness to the end user. Solar emergy was therefore suggested as a measure of the total environmental support to all kinds of processes in the biosphere, including economies. Flows that are not from solar source (like deep heat and gravitational potential) are expressed as solar equivalent energy by means of suitable transformation coefficients (Odum 1996). The Emergy Accounting method focuses on the environmental performance of the system on the global scale, but this time also taking into account all the free environmental inputs such as sunlight, wind, rain, as well as the indirect environmental support embodied in human labour and services, which are not usually included in traditional embodied energy analyses. Moreover, the accounting is extended back in time to include the environmental work needed for resource formation.

The amount of input emergy dissipated per unit output exergy is called *solar transformity*. The latter can be considered a "quality" factor which functions as a measure of the intensity of biosphere support to the product under study. The total solar emergy of a given item may be calculated as: (solar emergy) = (exergy of the item) * (solar transformity). Solar emergy is usually measured in solar emergy joules (seJ), while the unit for solar transformity is solar emergy joules per joule of product (seJ/J). Sometimes emergy per unit mass of product or emergy per unit of currency are also used (seJ/g, seJ/$, etc). In so doing, all kinds of flows to a system are expressed in the same unit (seJ of solar emergy) and have a built-in quality factor to account for the conversion of input flows through the biosphere hierarchy. The specific emergy or transformity of a system's output is calculated as the sum of the total emergy embodied in the necessary inputs to the system, respectively, divided by the output mass or exergy. The total emergy requirement thus calculated can be interpreted as an indication of the total appropriation of environmental services by the analysed human activity. In particular, while the total *non-renewable* emergy input to the system under study provides a quantitative estimate of global non-renewable resource depletion, the total *renewable* emergy requirement is a measure of all the natural exchange-pool resources that are diverted from their natural pathways, and that can therefore no longer provide their natural ecosystemic functions.

Values of transformities are available in the scientific literature on emergy. When a large set of transformities is available, other natural and economic processes can be evaluated by calculating input flows, throughput flows, storages within the system, and final products in emergy units. As a result of this procedure, a set of indices and ratios suitable for policymaking

(Ulgiati et al. 1995; Ulgiati and Brown 1998; Brown and Ulgiati 1999) can be calculated. Such indicators expand the evaluation process to the larger space and time scales of the biosphere. While the emergy approach is unlikely to be of practical use in making decisions about the price of food at the grocery store or the way a process should be improved to maximize exergy efficiency at the local scale, its ability to link local processes to the global dynamics of the biosphere provides a valuable tool for adapting human driven processes to the oscillations and rates of natural processes, towards sustainable patterns of human economies. LCA procedures do not include the emergy method among the tools available to the analyst. Instead, emergy (a measure of the demand for environmental support by a process) could be easily included as an upstream, broad-focus, impact category, thus providing a comprehensive measure of cost at the scale of biosphere.

2.5. ACCOUNTING FOR AIRBORNE AND WATERBORNE EMISSIONS

Airborne and waterborne emissions are a crucial issue in the evaluation of a process impacts. They can be measured directly or estimated indirectly according to existing databases or stoichiometric parameters. In our investigations we used indirect emission factors from published databases (Corinair 2007; CML2 baseline 2000, among others) which calculate the potential environmental damage of airborne, liquid and solid emissions by means of appropriate equivalence factors to selected reference compounds for each impact category. The impact potential of the analysed system for each category is calculated by multiplying all emissions by their respective impact equivalence factors, and then summing the results. The CML2 method was preferred among other similar methods for its versatility and completeness. The impact categories analysed in the case studies are:

- *Global warming potential*, expressed in grams of CO_2 equivalent per gram of product
- *Acidification potential*, expressed in grams of SO_2 equivalent per gram of product
- *Eutrophication potential*, expressed in grams of $PO4^{3-}$ equivalent per gram of product
- *Tropospheric ozone and photosmog formation potential*, in grams of ethene equivalent per gram of product
- *Stratospheric ozone depletion potential*, in grams of CFC-11 equivalent per gram of product
- *Ecotoxicity potential*, in grams of 1,4-dichlorobenzene equivalent per gram of product (this category is then sub-divided into freshwater, soil and sea water ecotoxicity potentials)

Within the framework of this downstream approach, the possibility for an update of the specific equivalence factors remains open for the future, as is usually the case for any equivalence factor. Similarly, the inclusion of further impact categories (e.g. radioactivity), in order to meet the specific requirements of the analysed case study, is also theoretically possible.

3.   Results from Selected Case Studies

The case studies presented in this Section are the results of recent investigations performed by the authors by means of the multi-method and multi-scale approach described in the previous sections. Since results refer to the past years 2001–2009, some of them might be no longer representative of the technological progresses achieved by far, specially for those technologies that are undergoing a very fast development (e.g. photovoltaic and fuel cells). They are shown in order to support the understanding of the integrated evaluation approach discussed in this paper. The joint use of complementary methods, points of view and numeraires is the basis for an evaluation framework named SUMMA (Sustainability Multi-method Multi-scale Approach; Ulgiati et al. 2006) that allows the generation of a large set of performance indicators to be used for technological improvement, investment choices, resource and environmental policy. Figure 2 shows a schematic overview of how the framework is applied: The analysed system or process is treated as a 'black box', and a thorough inventory of all the input and output flows is firstly performed on its local scale. It is important to underline that this inventory forms the common basis for all subsequent impact assessments, which are carried out in parallel, thus ensuring the maximum consistency of the input data and inherent assumptions. Each individual assessment method is applied according to its own set of rules. The 'upstream' methods are concerned with the inputs, and account for the depletion of environmental resources, while the 'downstream' methods are applied to the outputs, and look at the environmental consequences of the emissions. The calculated impact indicators are then interpreted within a comparative framework, in which the results of each method are set up against each other and contribute to providing a comprehensive picture on which conclusions can be drawn. Results reflect the specific characteristics of each case study evaluated and do not claim to be generalisable. They are only presented here to illustrate what can be obtained by means of an integrated approach, not to support or counter the feasibility of a specific technology or process. For this to be done, the set of case studies should be increased, in order to rely on a representative sample of indicators.

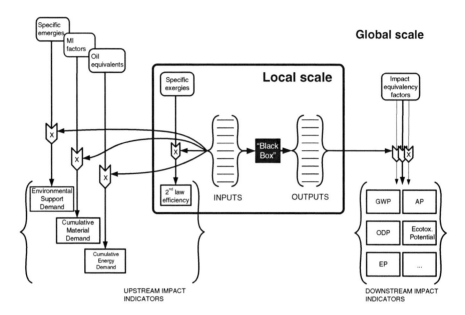

Figure 2. Flow diagram of the multimethod multiscale approach. The system is treated as a black box. Input and output flows are multiplied by specific exergy, energy, matter, emergy and emission factors to yield estimates of upstream and downstream impacts on resource and environmental dynamics (After Ulgiati et al. 2006, modif.)

3.1. COGENERATION OF HEAT AND ELECTRICITY

Table 1 compares the results obtained from the application of the approach to a selection of different power plant for electricity and heat generation. The power plants investigated are all powered by natural gas, but differ by technology (internal combustion engine, gas turbine, steam turbine, combined cycle, fuel cells) and size, ranging from a power of 0.5– (MCFC plant) 1,200 MWe (NGCC plant). Their technical characteristics and the full details of the evaluations performed are provided in Raugei et al. (2005), Bargigli et al. (2008, 2009). Comparison is made possible by the use of intensive indicators (i.e. indicators per unit of output or indicators of efficiency), that are not dependent on the size of the system.

TABLE 1. Performance indicators of selected cogeneration electricity production processes

Process/ product Indicator	ICE (internal combustion engine)	MCFC[a] (molten carbonate fuel cells)	Hybrid[a] (MCFC + GT100)	Gas turbine (TURBEC GT 600)	STGT (steam turbine + gas turbine)	NGCC (Natural Gas Combined Cycle)
Material resource depletion						
MI_{abiot} (g/kWhe)	1,030	264	190	640	276	146
MI_{water} (g/kWhe)	3,530	1,144	870	1,890	916	875
MI_{air} (g/kWhe)	–	878	–	–	5,003	2,655
Energy resource depletion and First Law efficiency						
GER of electricity (10^6 J/kWhe)	9.00	10.30	7.83	11.78	13.80	7.35
Oil equiv of electricity (g/kWhe)	215.10	246.00	187.00	281.40	331.00	176.00
Electric Energy efficiency	0.40	0.35	0.46	0.20	0.26	0.49
Cogeneration energy efficiency	0.82	0.72	0.72	0.74	0.75	0.71
Exergy, Second Law efficiency						
Cogeneration Exergy efficiency	0.66	0.61	0.62	0.55	0.45	0.60
Emergy, demand for environmental support						
Transformity (10^5 seJ/J), with services	11.10	2.64	2.38	11.10	4.01	1.70
EYR	1.02	–	1.00	1.01	–	–
ELR	63	–	1,896	73	–	–
Climate change						
GWP (CO_2-equiv, g/kWhe)	921	583	493	788	750	398
Acidification (SO_2-equiv, g/kWhe)	–	0.33	–	–	0.62	0.54

Source of data: Raugei et al. 2005, Bargigli et al. 2008, 2009.
[a] Data for MCFC systems are from pilot scale production. As such, the corresponding results presented here should not be considered to be fully representative of the current state of the art.

A pictorial view of results if provided by Fig. 3, where five different typologies of plants selected from Table 1 are compared according to the values of their contribution to selected impact categories or performance indicators. Absolute values are reported in Table 1, while instead the values in Fig. 3 are adjusted and normalized for better visualization of results. It clearly appears that each plant shows a good performance concerning some impact categories, while ranks lower in other categories. In most cases, the higher the value of the indicator the worse is the performance. Since efficiencies have an opposite meaning (the higher the better), we reported their inverse values in the diagram. In so doing, the larger is the area identified by the indicators for each given plant on the radar diagram, the larger is the potential impact of that system in relative terms. According to the results shown in Fig. 3, the best performance is shown by the cogeneration NGCC plant while ICE plant ranks lower (it should be acknowledged, however, that data for MCFC systems are from pilot scale production (first-of-a-kind). As such, the corresponding results presented here – although already satisfactory – should not be considered to be fully representative of the current state of the art, and they should not be included in the comparison if the emphasis is placed on identifying the most recommendable option for the near future).

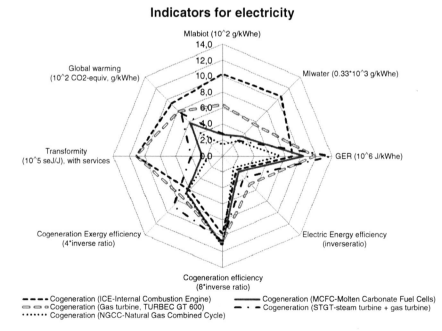

Figure 3. Radar diagram comparing the performance of selected typologies of cogeneration power plants, according to data from Table 1

3.2. PHOTOVOLTAIC ELECTRICITY

Photovoltaic (PV) energy generation devices have experienced a sharp, almost 10-fold increase in production in less than a single decade (Jaeger-Waldau 2008). Since the early 2000s, new PV technologies have begun to be employed commercially alongside the more traditional Si-based systems, among which in particular cadmium telluride (CdTe) PV, which now represents 5% of the global PV market (EPIA 2008).

Three types of innovative PV modules were selected to be analysed in the original study performed by some of the authors (PVACCEPT 2005; Raugei et al. 2007):

1. Micro-perforated, semi-transparent poly-crystalline silicon modules: in this technology the photoactive material is a layer of high-purity silicon, which is 'doped' on its opposite sides by introducing into its lattice structure a small number of atoms of the third and fifth chemical groups, respectively (e.g. boron and phosphorous). This procedure effectively turns the Si layer into an electrical P/N junction, characterised by a suitable bandgap energy. The single Si cells are then electrically connected in series and sandwiched between a transparent glass pane and a rear cover to form the finished module.
2. Cadmium telluride (CdTe) thin film modules: in this technology the photoactive P/N junction is made up of two semiconductor compounds, CdTe and CdS, which are directly deposited in extremely thin layers on a treated transparent glass pane by means of a vapour transport deposition process. Series connection of adjacent P/N junctions is achieved by means of a repeated automated scribing process, and then a second protective glass pane is added on top to form the finished module.
3. Copper indium diselenide (CIS) thin film modules: this technology shares many similarities with the previous one (CdTe), the main difference between the two consisting of the chemical compound used for the P part of the heterojunction, i.e. $CuInSe_2$ instead of CdTe.

Resulting performance indicators from that original study are shown in Table 2 (absolute values) and diagrammed in Fig. 4 (values adjusted for better visualization).

A sensitivity analysis allowed to identify the input flows that more likely affect the performance of each device and explore how suggested changes may translate into better performance. According to these results, CdTe modules could be identified as the ones that performed best.

It is however of paramount importance to underline that the results presented here only apply to the early pre-production (in the case of CIS) and pilot production (in the case of CdTe) modules available at the time

when the original study was performed (2001–2004). Thin film PV technologies, and CdTe PV in particular, have since undergone staggeringly fast
progress (e.g. CdTe module efficiency has increased from below 8% to
almost 11%), and even comparatively more mature c-Si technologies have
advanced considerably (e.g. c-Si wafer thickness has generally been reduced
by almost 30%). As a consequence of such progress, recent studies have
proven that, for instance, the Global Warming Potential of CdTe PV
modules is now down to only slightly over 10 g CO_2-eq/kWhe (Fthenakis
et al. 2009). Similar order-of-magnitude changes may be expected to have
occurred in other impact indicators, too, thereby making the numerical
results presented in Table 2 and Fig. 4 devoid of any practical relevance
with respect to the actual environmental performance of current-production
PV modules.

TABLE 2. Performance indicators of selected thin-film photovoltaic modules

Process/product Indicator	Poly-Si – poly- cristalline silicon module	CdTe – cadmium telluride module[a]	CIS – copper indium selenide module[a]
Material resource depletion			
MI_{abiot} (g/kWhe)	140	70	150
MI_{water} (g/kWhe)	530	300	630
Energy resource depletion and first law			
efficiency			
GER of electricity (10^6J/kWhe)	1.0	0.30	1.1
Oil equiv of electricity (g/kWhe)	23.9	7.2	26.3
Exergy, second law efficiency			
Exergy efficiency	10.5%	6.0%	8.2%
Emergy, demand for environmental support			
Transformity (10^4seJ/J), without services	2.80	1.50	3.0
Climate change			
Global warming (CO_2-equiv, g/kWhe)	52	19	70
Acidification (SO_2-equiv, g/kWhe)	0.44	0.11	0.37

Source: PVACCEPT 2005, Raugei et al. 2007. Assumptions for all systems: lifetime = 20 years;
irradiation = 1,700 kWh/(m² year); Performance ratio = 75%; efficiencies = 14% (c-Si), 8% (CdTe),
11% (CIS).

[a] Data for CdTe and CIS modules are from pilot scale production. PV systems, and those based on thin
films in particular, have been proven to have improved considerably in the last few years; therefore, the
corresponding results presented here should *not* be considered to be representative of the current state of
the art, and are only presented for the purposes of illustrating the adopted methodology.

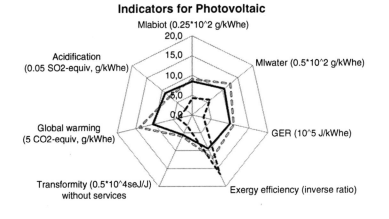

Figure 4. Radar diagram illustrating the relative performance of three typologies of photovoltaic devices, based on data from Table 2

The value of the comparison remains, though, *from a purely methodological point of view*, in that it illustrates the importance of going beyond the limited set of the most commonly employed indicators, and extending the analysis to also include non-conventional metrics of environmental impact.

Results for CdTe and CIS modules are from pilot scale production. PV systems, and those based on thin films in particular, have been proven to have improved considerably in the last few years; therefore, the corresponding results presented here should not be considered to be representative of the current state of the art, and are only presented for the purposes of illustrating the adopted methodology.

3.3. ALTERNATIVE FUELS AND BIOFUELS

Fuels and biofuels alternative to the commonly used fossil gasoline and diesel were explored and results presented in Table 3 (absolute values). A more detailed discussion of each process investigated (syngas, hydrogen, biofuels) as well as of the meaning of the results obtained is provided in Bargigli et al. (2004) and Giampietro and Ulgiati (2005).

TABLE 3. Performance indicators of selected fuels and biofuels

Process/product Indicator	Syngas (from coal gasification)	Hydrogen (from steam reforming of natural gas)	Hydrogen from water electrolysis (thermo-electricity)	Bioethanol (from corn)	Biodiesel (from sunflower)	Methanol (from wood)
Material resource depletion						
MI_{abiot} (g/g)	30.90	3.60	12.20	7.45	13.97	–
MI_{water} (g/g)	40.80	10.60	71.80	4,811	2,853	–
MI_{air} (g/g)	8.80	21.30	165.10	–	–	–
MI_{biot} (g/g)	–	–	–	0.35	0.79	–
Energy resource depletion and First Law efficiency						
GER (10^4 J/g)	2.20	18.80	42.90	2.51	3.43	0.47
Oil equiv (g_{oil}/g)	0.48	4.48	10.20	0.60	0.82	0.11
Energy Return on Investment (EROI)	0.76	0.64	0.28	1.15	0.98	1.10
Exergy, Second Law efficiency						
Exergy efficiency	0.75	0.71	0.27	–	–	–
Emergy, demand for environmental support						
Transformity (10^4 seJ/J)	9.56	12.30	36.60	18.90	23.10	26.60
EYR	–	–	–	1.24	1.09	2.35
ELR	–	–	–	10.90	25.90	2.10
Climate change						
Global warming (CO_2-equiv, g/g)	5.20	9.50	33.70	2.02	3.21	1.54
Solid emissions (g/unit)	3.00	–	–	–	–	–

Source of data: Bargigli et al. 2004, Giampietro and Ulgiati 2005.

Figure 5 shows the systems diagram of the industrial conversion of corn into ethanol (Sciubba and Ulgiati 2005). The main process steps are indicated and the evaluation was performed step by step, in order to be able to identify the parts of the process that are crucial for the final performance. The diagram indicates the main input flows (matter, energy and labor) that support each step of the process; these flows are then listed in an inventory table and converted into embodied matter, embodied energy, exergy, emergy and emission flows according to the approach described in Section 2. Similarly, Fig. 6a, b shows the systems diagram of two methods for hydrogen production (from steam reforming of natural gas and from water electrolysis powered by fossil generated electricity).

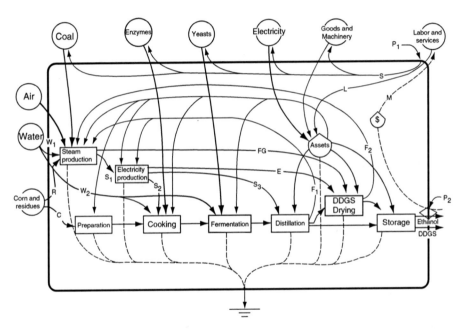

Figure 5. Energy systems diagram of industrial conversion of corn into bioethanol and animal feedstock (DDGS, Distilled Dry Grains with Solubles). The diagram shows the matter and energy input flows from outside to each step of the process as well as flows exchanged among system's components (From Sciubba and Ulgiati 2005)

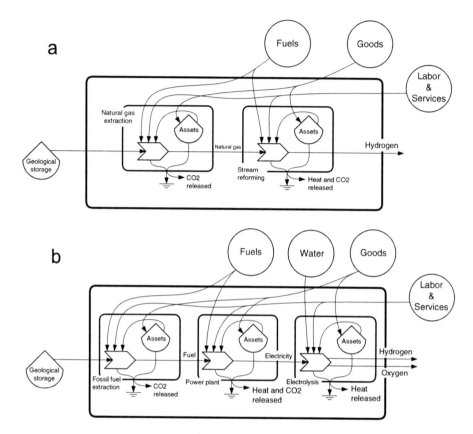

Figure 6. Systems diagram of hydrogen production from natural gas via steam reforming (a) and via water electrolysis (b) (From Brown and Ulgiati 2004, modif.; systems symbols are described in Fig. 1)

The final results for a selection of the investigated fuels are compared in the radar diagram in Fig. 7. As with the previous radar diagrams, values are adjusted for better visualization and a larger area indicates a worse performance. That is identifies the system that is characterized by the largest potential impact. In the case of Fig. 7, hydrogen from steam reforming shows the best relative performance. It should be noted here the large water demand related to the production of biodiesel from sunflower, mainly for the agricultural step. Such a result makes it apparent that even when most performance parameters are very good (e.g. EROI, GWP, etc), the global result may be easily worsened by one parameter only. However, this also provides a way to identify the process bottlenecks and try to remove them.

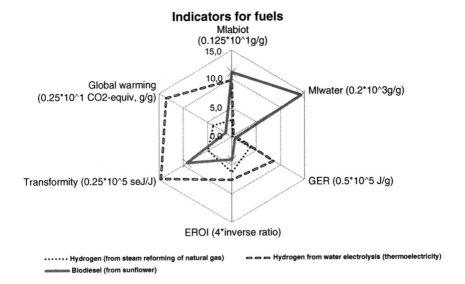

Figure 7. Radar diagram of three selected fuel production processes (Data from Table 3) showing the relative global impact (depending on the selected set of performance indicators)

3.4. TRANSPORTATION SYSTEMS

A selection of transportation modalities at local and national scales in Italy was investigated by Federici et al. (2003, 2008, 2009). These authors compared local transportation systems at urban level and national highway, railway and air systems, for passenger and commodity transport. A selection of their results is shown in Table 4 (absolute values), where values are, as always in this paper, presented as intensive indicators, per unit of service delivered (per passenger or ton transported per km, p-km and t-km), in order to allow a fair comparison. In particular, the authors focus on two different typologies of railway transport (regular intercity trains and high speed trains). As a result of their study, the authors point out the huge importance of infrastructure (concrete and steel for viaducts and tunnels, road asphalt and other building materials, as well as special steel for machinery) and discuss how infrastructure, average occupancy and speed affect the final performance indicators. Figure 8 suggests that the railway passenger transport is characterized by the lowest relative impact.

TABLE 4. Performance indicators of selected transportation modalities

Process/ product Indicator	Highway passenger transportation by car (p-km)	Highway commodity transportation by truck (t-km)	Railway passenger transportation by intercity train (p-km)	High speed railway, passenger transportation by HS train (p-km)	Air passenger transportation by A320 aircraft (p-km)
Material resource depletion					
MI_{abiot} (g/unit)	530	600	770	1200	800
Energy resource depletion and first law efficiency					
GER (MJ/unit)	1.87	1.25	0.69	1.23	2.20
Oil equiv (g_{oil}/unit)	44.70	29.90	16.50	29.40	52.60
Emergy, demand for environmental support					
Specific emergy (10^{11} seJ/unit)	1.74	1.08	1.10	1.41	1.28
Climate change					
Global warming (CO_2-equiv, g/unit)	134	90	49	88	158
Acidification (SO_2-equiv, g/unit)	1.69	0.80	0.39	1.53	n.a.

Source of data: Federici et al. 2003, 2008, 2009.

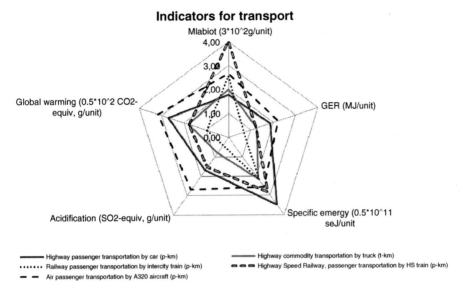

Figure 8. Radar diagram of selected transportation modalities (Data from Table 4, normalised for comparison)

3.5. WASTE MANAGEMENT

Four different urban waste management processes related to the city of Rome, Italy, were investigated by Cherubini et al. (2008, 2009). The study focused on (1) landfilling, (2) landfilling with biogas recovery, (3) conversion to biogas and Refuse Derived Fuel (RDF, for electricity production), and finally (4) direct incineration and electricity production. Calculated indicators are shown in Table 5. All steps were accounted for, including preliminary sorting of recoverable materials, collection and transport, landfilling of combustion ash and biogas digestion process. The energy generated from waste biomass was credited to the process, in so decreasing the global energy cost of management and the global emissions from the whole cycle. Performance indicators relative to the electricity generated are also shown in Table 5. In one case, (conversion to biogas and RDF) the process delivers a non-negligible amount of net energy, so that the net emissions (= actual process emissions minus avoided emissions due to the energy delivered) are negative.

The radar diagram in Fig. 9 allows a relative comparison of the three processes characterised by energy recovery, suggesting that landfilling with biogas recovery and direct incineration (the most commonly used tech- nologies) are also the ones characterized by the higher global impact.

TABLE 5. Performance indicators of urban waste management, Roma, Italy

Process/ product / Indicator	Landfilling	Landfilling with biogas recovery[a]	Sorting and conversion to biogas and RDF[a]	Direct incineration[a]
Material resource depletion				
MI_{abiot} (g/g_{waste})	0.24	0.24	0.30	0.36
MI_{abiot} (g/kWhe)	–	1,899	334	552
MI_{water} (g/unit)	0.03	0.02	2.09	1.04
MI_{abiot} (g/kWhe)	–	0.82	2,398	1,578
Energy resource depletion and first law efficiency				
GER (kJ/g_{waste})	0.05	2.15	9.71	9.52
GER (J/kWhe)	–	2.67E+07	1.38E+07	1.60E+07
$GER_{oil\ equiv}$ (g_{oil}/g_{waste})	0.001	0.05	0.23	0.23
$GER_{oil\ equiv}$ (g_{oil}/kWhe)	–	637.84	329.67	382.23
Energy efficiency (%)	–	13	52	22
Emergy, demand for environmental support				
Specific emergy (10^8 seJ/g_{waste})	1.58	1.54	1.22	1.83
Specific emergy (seJ/kWhe)	–	5.36E+05	2.28E+04	8.66E+04
Climate change				
Global warming (CO_2-equiv, g/g_{waste})	1.31	0.59	−0.23	0.15
Acidification, total emissions (SO_2-equiv, mg/ g_{waste})	0.37	0.13	−0.30	0.53

Source: Cherubini et al. 2008, 2009
[a] Followed by conversion to electricity

3.6. URBAN SYSTEMS

The metabolic patterns of an urban system (Rome, Italy) were investigated from 1962 to 2002 by Ascione et al. (2008, 2009). Figure 10 shows a systems diagram of the city and its surrounding environment (natural areas, agriculture) and infrastructure. The investigation took into account all the matter, energy, and emergy flows supporting the urban system over 40 years of growth and development, with the aim of ascertaining the total cost of supporting the urban system, its population, its economic activity and generation of GDP.

Waste Management

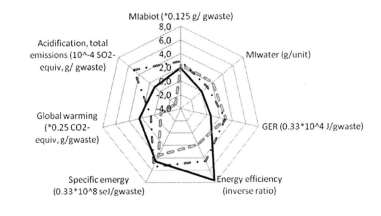

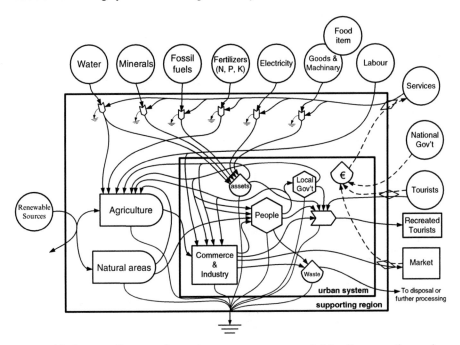

Figure 9. Radar diagram of three selected waste management options, based on data from Table 5. Landfilling option without biogas recovery was not included

Figure 10. Systems diagram of an urban ecosystem surrounded by its supporting region. Systems symbols are described in Fig. 1

More than 20 indicators of performance and sustainability were calculated in this urban case study and their trends compared and discussed. The above waste management case study (Section 3.5) was part of such an investigation effort, with focus on a specific sector (waste) that plays an important role in the whole urban metabolism. Results, shown in Table 6, point out a continuous growth of resource demand by the urban system. Unfortunately, such a demand was mainly demand for nonrenewable flows from outside the system, thus making it largely dependent on imports.

TABLE 6. Trend of urban metabolism, Roma, Italy (1962–2002)

Indicator \ Year	1962	1972	1982	1992	2002
Material resource depletion					
MI_{abiot} (10^7g/person)	1.72	2.93	2.82	3.87	4.50
MI_{abiot} (10^3g/€)	82.70	40.60	5.98	2.30	1.99
MI_{water} (10^8g/person)	3.19	4.91	5.65	6.82	8.16
MI_{water} (10^4g/€)	154	68.20	12.00	4.05	3.61
Energy resource depletion and energy efficiency					
GER per person (10^{10}J/person)	6.34	11.40	14.80	19.70	27.30
GER per unit currency (10^7J/€)	30.50	15.80	3.14	1.17	1.21
Oil equiv (10^6g/person)	1.52	2.72	3.54	4.72	6.51
Oil equiv (10^2g/€)	73.00	37.80	7.51	2.80	2.88
Emergy, demand for environmental support					
Specific emergy (10^{16}seJ/person)	2.61	3.53	3.92	6.36	5.45
Specific emergy (10^{12}seJ/€)	126	49.10	8.33	3.78	2.41
EYR	1.05	1.03	1.02	1.01	1.02
ELR	40.85	61.94	52.38	94.73	64.47
Ecological footprint					
Area per person (ha/person)	2.13	2.12	2.81	2.84	3.60
Area per unit GDP (ha/€)	0.0103	0.0030	0.0006	0.0002	0.0002
Climate change					
Global warming (10^6CO_2-equiv, g/person)	4.62	8.42	11.00	14.40	20.00
Global warming (10^2CO_2-equiv, g/€)	223	117	23.40	8.57	8.83
Acidification (10^4SO_2-equiv, g/person)	1.30	2.54	3.08	4.00	5.67
Acidification (SO_2-equiv, g/€)	62.70	35.30	6.54	2.38	2.51

Source: Ascione et al. 2008, 2009

The radar diagram in Fig. 11 was drawn by normalising each perform-ance index relative to the value of the same index in the year 1962. The figure shows very clearly the impressive expansion of the urban system's impact over time, due to both increase of population and increase in demand for resources. The expansion seems to have been accelerating in the last decades.

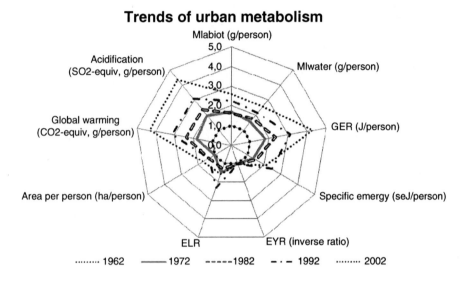

Figure 11. Radar diagram showing the trend over time of urban metabolism performance indicators. Values from Table 6 were normalised with reference to the year 1962

3.7. AGRICULTURE

A case study about the agricultural sector of Campania region in Southern Italy was performed and its performance assessed over a time span of 20 years (Ulgiati et al. 2008). As with the previous case studies, an inventory of the main input and output flows served as a basis for assessing the direct and cumulative support by the economic system as well as by the environ-ment. The systems diagram of the process is shown in Fig. 12, where all input flows from the environment and the economy are indicated together with system's components and internal exchanges of matter and energy.

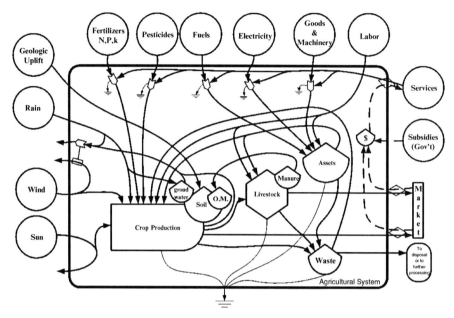

Figure 12. Systems diagram of a generic agricultural system, showing the main driving forces, system's components and internal interactions. Systems symbols are described in Fig. 1

The calculated performance indicators are listed in Table 7 and graphically shown in the radar diagram of Fig. 13. The performances of the years 1993, 2002 and 2006 are referred to the year 1985, so that the expansion of the areas related to each investigated year suggests a higher relative impact over time. Such a trend is due, depending on the years, to: changed mix of crops, changed machinery use, changed climate conditions (mainly less rainfall), changed market value of products. It clearly appears that the agricultural system is increasingly impacting whatever is the performance indicator considered. The star-like shape is due to the fact that some composite indicators grow faster than others in the investigated years, which may depend on either numerator (energy used, amount of emissions, etc) or denominator (total product mass, product economic value, etc).

4. Discussion

The discussion about each specific case study was already provided in the original studies referred to, published by the authors. Here we would like to point out the usefulness of an integrated approach as well as how it can be used for process improvement.

Table 7. Trend of the agricultural sector of Campania region, Italy (1985-2006).

Indicator	1985	1993	2002	2006
Material resource depletion				
MI_{abiot}(g/g $_{d.m.}$)	0.30	0.24	0.60	0.57
MI_{abiot}(10^3g/€)	2.31	1.84	1.28	1.30
MI_{water}(10 g/g $_{d.m.}$)	3.84	1.89	3.23	2.37
MI_{water}(10^5g/€)	2.98	1.43	0.69	0.54
Energy resource depletion and energy efficiency				
GER per unit mass (10^3J/ $g_{d.m.}$)	1.24	1.32	2.80	2.97
GER per unit currency (10^6J/ €)	9.65	9.97	6.00	6.82
Oil equiv (g/$g_{d.m.}$)	0.03	0.03	0.07	0.07
Oil equiv (10^2g/ €)	2.30	2.38	1.43	1.63
EROI ($10^{1)}$	1.34	1.26	0.58	0.53
Emergy, demand for environmental support				
Specific emergy with L&S (10^8seJ/ $g_{d.m.}$)	4.16	4.30	10.11	11.49
Specific emergy with L&S (10^{12}seJ/€)	3.24	3.25	2.17	2.63
EYR (with L&S)	1.30	1.19	1.16	1.14
ELR (with L&S)	3.71	5.81	6.81	7.75
Climate change				
Global warming (CO_2-equiv, g/$g_{d.m.}$)	0.10	0.10	0.20	0.21
Global warming ($10^2 CO_2$-equiv, g/€)	7.54	7.61	4.35	4.91
Acidification (10^{-2}SO2-equiv, g/gd.m.)	0.03	0.03	0.06	0.06
Acidification (SO_2-equiv, g/€)	2.36	2.50	1.34	1.43

Source of data: Ulgiati et al., 2008.

4.1. THE 'ADDED VALUE'

Quantifying direct and indirect flows of matter and energy to and from a system permits the construction of a detailed picture of the process itself as well as of its relationship with the surrounding environment. Processing these data in order to calculate performance indicators and material and energetic intensities makes it possible to compare the process output to other products of competing processes. Results may differ depending on the goal, the boundaries, the time scale, and the technology and may suggest different optimization procedures. If the analyst is able to provide com-prehensive results as well as to explain divergences at the appropriate scales of the investigation, a process can be more easily understood. Conclusions are also reinforced and are more likely to be acceptable for research, application and policy strategies.

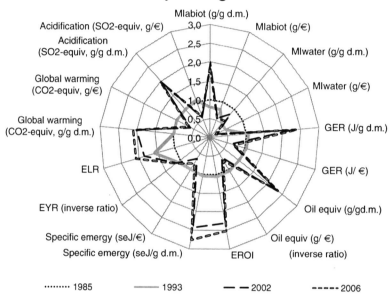

Figure 13. Radar diagram showing the trend over time of the performance of the agriculture of Campania region (Italy). Values from Table 7 are normalised with reference to the year 1985

Assessing a process performance on different scales offers an effective way to refine the analysis and improve the process. Results from the simultaneous application of a multiple set of methods yield consistent and comparable performance indicators and call for a twofold optimization pattern:

1. Upstream: trying to decrease the use of or replace those input flows which affect the material, energy and environmental support demands more heavily
2. Downstream: trying to decrease the use or avoid misuse of the investigated product, in order to negatively affect the input demand by controlling the end of the life cycle chain

As already pointed out in the introduction, it was the authors' explicit choice not to provide a means of combining the results from the different upstream and downstream methods into one single "super-indicator", since this is contrary to the fundamental idea that separate indicators provide a much more comprehensive environmental profile, the interpretation of which should be left to the analyst. The latter should in fact clarify to the maximum possible extent the meaning and all the possible implications of the single results, highlighting the often inevitable trade-offs that all strategic choices entail, rather than simply attach to them one numerical tag, which would inevitably conceal much of the valuable detail of the study.

Last but not least, since the approach is based on a single common inventory of all the system's inputs and outputs, a systematic sensitivity analysis can simultaneously be performed on all calculated data and indicators, simply by allowing for variable cells for all input quantities as well as for the associated impact coefficients in the spreadsheet-based calculation procedures. Such an analysis is invaluable in order to estimate the actual reliability of the impact assessment itself, accounting for the inevitable uncertainties and variability in the input data and/or impact coefficients, as well as to single out which are the most critical key points of the analysed process in the light of the different assessment methods.

4.2. WEIGHTING FACTORS

Tables 1–7 list a selection of the calculated performance indicators. In general, many more indicators are generated by an integrated approach, but listing all of them is not necessary for the purpose of the present paper. If the focus is on the specific behaviour of the system from a single point of view (e.g. energy consumption), the analyst can refer to the calculation procedure that leads to that specific indicator and look carefully at the options available to minimize such an impact. Each choice can be tested in the calculation procedure (either implemented within an excel platform or a commercial software) and its consequences can be simulated, being aware that the improvement of the value of one indicator may lead to the worsening of another one. Since inventory data are linked to all the calculation procedure of all the indicators, this is a relatively easy task.

If the goal is providing a global picture of a process impact, then a selection of many indicators is needed in order to have a comprehensive evaluation across space and time scales. Since many indicators are generated by the same calculation procedure (e.g. Gross Energy Requirement, energy efficiency, EROI; or EYR, ELR, transformity) the analyst might select them according to his/her experience or according to the specific target of the investigation (e.g. assessing the environmental sustainability of the process). Once indicators are chosen, they can be normalized and diagrammed in such a way (histograms, radar, lines) that their values can be compared to a reference value or year, or simply looked at together, in order to provide a global picture of the impact. This is the case, for example, of Figs. 3, 4, 7–9, where several systems are compared each other and the normalization is performed in such a way that the area within each curve indicates a measure of the impact relative to the other processes investigated for comparison. Instead, Figs. 11 and 13 show the behaviour of an urban system and an agricultural sector over time, by referring to the global performance in a given year. In this case, the larger the area the larger the variation of the impact over time, i.e. the diagram assesses how fast is the change for each individual impact and globally.

All the radar diagrams put a selection of indicators on each axis. Such a choice might lead to the erroneous assumption that their impacts are the same. This is not true, of course. Each indicator refers to a specific impact category in LCA as well as other evaluation methods (e.g. contribution to global warming, demand for environmental support, water footprint). Assigning weighting factors to the whole assessment is not easy nor there is any agreement in the scientific community in this regard, so that such a step is not mandatory in the ISO norms (ISO 14040/2006; ISO 14044/2006) that codify the LCA approach. In fact, weighting and grouping is explicitly discouraged by ISO 14044/2006 for all studies intended for public disclosure. Our choice of putting all the indicators on the same importance basis does not affect the final understanding of the impact, because assigning a different weighting factor to any indicator in Figs. 3, 4, 7–9 would change the values of all the investigated processes, but would leave the ranking unchanged. Same would happen with Figs. 11 and 13. The shapes and areas in the diagram would be affected by choosing weighting factors, but relative ranking would not and therefore the same relative impacts would be suggested by the diagram. Some evaluation methods suggest weighting factors and scores (Eco-Indicator 1999), but the debate about the opportunity of such a choice is still open. The way data and indicators are used in the approach presented in the present paper does not require weighting factors to be applied.

4.3. OPTIMIZATION PROCEDURES

The ultimate goal of any investigation about a process is to generate a clear picture of the crucial steps as well as crucial input and output flows, i.e. those steps and those flows that affect more heavily the process performance.

In so doing it is possible to focus on these steps and flows, to understand how important are they in the global economy of the investigate process, and to suggest changes capable to lead to an improved performance. Some steps may be replaced by alternative patterns, some flows may be decreased by means of more efficient machinery or sub-processes, and finally some flows may simply be avoided without any important consequence for the final product. Suggesting an optimization procedure is not an easy task. Indicators are the result of a calculation procedure where the inventory data are multiplied by intensity factors specific of each given method (e.g. oil equivalent factors, transformity, global warming potential, etc). Therefore, when a performance indicator (e.g. the Acidification Potential for a coal powered plant) is not satisfactory, the analyst goes back to the calculation procedure in order to identify the input items that are responsible of the largest contributions to that impact category and may suggest to decrease their amount by applying technological changes to the process (e.g. use of de-sulphurized fuel). After the suggested changes have been implemented (or their adoption has been simulated) in the process, the analyst will recalculate the indicator under consideration and will assess the extent of the performance improvement. However, it is very likely that the suggested change affects other impact categories and, due to the reliance on the same set of input data, the improvement in one category might translate into a worse performance in another category (e.g. fuel de-sulphurization requires an additional technological process and increased energy input and generates additional waste to dispose of).

5. Conclusion

Investigating a system performance is by itself a very difficult task, due to the complexity of the problems that are always involved. When a simplified model is adopted, this is certainly a way to address part of the problem at the cost of leaving unsolved another part of it. Depending upon the goal of the investigation, this is sometimes a useful procedure. However, investigators very often run the risk of neglecting the complexity of the problem and taking their model as reality. As a consequence, they assign a value to a process product according to the results of their simplified investigation. The outcome of this evaluation process is then used in other subsequent evaluations and translated into economic and policy actions. In so doing, the complexity is lost: reality does not fit the model and the planned policy fails or is inadequate.

An integrated approach is therefore suggested, to overcome the limits of individual methods and generate the added value of a comprehensive picture for the process steps, the process as a whole, the local scale and global scale environmental interactions, as well as the thermodynamic process performance. Evaluating comparable alternatives, when specific answers regarding different possible uses of resources in the space-time frame of interest are sought, necessarily requires the adoption of a multi-criteria approach. It must be realised that in virtually all cases there is no single 'optimal' solution to all problems. Only an analysis based on several complementary approaches can highlight the inevitable trade-offs that reside in alternative scenarios, and thus enable a wiser selection of the option embodying the best compromise in the light of the existing economic, technological and environmental conditions.

Acknowledgement

The Authors at Parthenope University gratefully acknowledge the financial support received from the European Commission within the seventh Framework Programme, Project no.217213, SMILE – Synergies in Multi-scale Inter-Linkages of Eco-social systems. Socioeconomic Sciences and Humanities (SSH) Collaborative Project FP7-SSH-2007-1

References

Ascione, M., Campanella, L., Cherubini, F., Bargigli, S., and Ulgiati, S., 2008. Investigating an urban system's resource use. How investigation methods and spatial scale affect the picture. ChemSusChem, 1: 450–462.

Ascione, M., Campanella, L., Cherubini, F., and Ulgiati, S., 2009. Driving forces of urban growth and development. An emergy-based assessment of the city of Rome, Italy. Landscape and Urban Planning 93: 238–249.

Ayres, R.U. and Masini, A., 1998. Waste Exergy as a Measure of Potential Harm. In: Advances in Energy Studies. Energy Flows in Ecology and Economy. Ulgiati S., Brown M.T., Giampietro M., Herendeen R.A., and Mayumi K. (Eds). Musis Publisher, Roma, Italy; pp. 113–128.

Bargigli, S., Raugei, M., and Ulgiati, S., 2005. Mass flow analysis and mass-based indicators. In: Handbook of Ecological Indicators for Assessment of Ecosystem Health. Sven E. Jorgensen, Robert Costanza, and Fu-Liu Xu (Eds). CRC Press, Boca Raton, FL; pp. 353–378.

Bargigli, S., Raugei, M., and Ulgiati, S., 2004. Comparison of thermodynamic and environ-mental indexes of natural gas, syngas and hydrogen production processes. Energy – The International Journal 29(12–15): 2145–2159.

Bargigli, S., Cigolotti, V., Moreno, A., Pierini, D., and Ulgiati, S., 2008. An Emergy Comparison of Different Alternatives for the Cogeneration of Heat and Electricity. In: Emergy Synthesis. Theory and Applications of Emergy Methodology – 5. Brown, M.T., Campbell, D., Comar, V., Huang, S.L., Rydberg, T., Tilley, D.R., and Ulgiati, S., (Ed). Center for Environmental Policy, University of Florida, Gainesville, FL, forthcoming.

Bargigli, S., Pierini, D., Cigolotti, V., Moreno, A., Iacobone, F., and Ulgiati, S., 2009. Cogeneration of heat and electricity. An LCA comparison of gas turbine, internal combustion engine and MCFC/GT hybrid system alternatives. Journal of Fuel Cell Science and Technology 7(1), published online 11 November 2009: http://link.aip.org/link/?FCT/7/011019

Brown, M.T. and Ulgiati, S., 2004. Emergy Analysis and Environmental Accounting. In: Encyclopedia of Energy, C. Cleveland (Ed). Academic, Elsevier, Oxford, UK, pp. 329–354.

Cherubini, F., Bargigli, S., and Ulgiati, S., 2008. Life cycle assessment of urban waste management: Energy performances and environmental impacts. The case of Rome, Italy. Waste Management 28(12): 2552–2564.

Cherubini, F., Bargigli, S., and Ulgiati, S., 2009. Life Cycle Assessment (LCA) of Waste Management Strategies: Landfilling, sorting plant and incineration. Energy – The International Journal 34(12): 2116–2123.

CML2 baseline 2000 (Centre of Environmental Science, Leiden University, NL; http://www.pre.nl/simapro/impact_assessment_methods.htm

CORINAIR, 2007. EMEP/CORINAIR Emission Inventory Guidebook – 2007; http://www.eea.europa.eu/publications/EMEPCORINAIR5/

Finnveden, G., and Moberg, A., 2005. Environmental systems analysis tools: an overview. Journal of Cleaner Production 13: 1165–1173.

Foran, B. and Crane, D., 1998. The OzECCO Embodied Energy Model of Australia's Physical Economy. In: Advances in Energy Studies. Energy Flows in Ecology and Economy. Ulgiati S., Brown M.T., Giampietro M., Herendeen R.A., and Mayumi K. (Eds). Musis Publisher, Roma, Italy; pp. 579–596.

Giampietro, M. and Ulgiati, S., 2005. Integrated assessment of large scale biofuel production. Critical Reviews in Plant Sciences, 24: 365–384.

Giampietro, M., Mayumi, K., and Pastore, G., 1998. A Dynamic Model of Socio-economic Systems described as Adaptive Dissipative Holarchies. In: Advances in Energy Studies. Energy Flows in Ecology and Economy. Ulgiati S., Brown M.T., Giampietro M., Herendeen R.A., and Mayumi K. (Eds). Musis Publisher, Roma, Italy; pp. 167–190.

European Photovoltaic Industry Association / Greenpeace (2008) "Solar Generation V – Solar electricity for over one billion people and two million jobs by 2020". The Netherlands / Belgium. http://www.epia.org/

Federici, M., Ulgiati, S., Verdesca, D., and Basosi, R., 2003. Efficiency and sustainability indicators for passenger and commodities transportation systems. The Case of Siena, Italy. Ecological Indicators 3(3): 155–169.

Federici, M., Ulgiati, S., and Basosi, R., 2008. A thermodynamic, environmental and material flow analysis of the Italian highway and railway transport systems. Energy – The International Journal 33: 760–775.

Federici, M., Ulgiati, S., Basosi, R., 2009. Air versus terrestrial transport modalities: An energy and environmental comparison. Energy 34(10): 1493–1503.

Fthenakis, V.M., Held, M., Kim, H.C., Raugei, M., 2009. Update of Energy Payback Times and Environmental Impacts of Photovoltaics. Presented at 24th European Photovoltaic Solar Energy Conference and Exhibition, Hamburg, Germany.

Gasparatos, A., El-Haram, M., and Horner, M., 2008. A critical review of reductionist approaches for assessing the progress towards sustainability. Environmental Impact Assessment Review 28: 286–311.

Herendeen, R.A., 1998. Embodied Energy, embodied everything...now what? In: Advances in Energy Studies. Energy Flows in Ecology and Economy. Ulgiati S., Brown M.T., Giampietro M., Herendeen R.A., and Mayumi K. (Eds). Musis Publisher, Roma, Italy; pp. 13–48.

Hinterberger, F. and Stiller H., 1998. Energy and Material Flows. In: Advances in Energy Studies. Energy Flows in Ecology and Economy. Ulgiati S., Brown M.T., Giampietro M., Herendeen R.A., and Mayumi K. (Eds). Musis Publisher, Roma, Italy; pp. 275–286.

ISO, 14040, 2006. Environmental management – Life cycle assessment – Principles and framework. International Standard Organization, June 2006, Brussels. www.iso.org

ISO, 14044, 2006. Environmental management – Life cycle assessment – Requirements and guidelines. International Standard Organization, June 2006, Brussels. www.iso.org.

Jaeger-Waldau, J., (2008) PV Status Report 2008 – Research, Solar Cell Production and Market Implementation of Photovoltaics. European Commission Joint Research Centre Technical Notes. EUR_23604EN_2008. http://sunbird.jrc.it/refsys/pdf/PV%20Report% 202008.pdf

Jørgensen, S.E., 2005. Towards a thermodynamics of biological systems. International Journal of Ecodynamics. 1: 9–27.

Lozano, M.A. and Valero A., 1993. Theory of the exergetic cost. Energy 18(9): 939–960.

Odum, H.T., 1988. Self organization, transformity and information. Science 242: 1132–1139.

Odum, H.T., 1996. Environmental Accounting. Emergy and Environmental Decision Making. Wiley, New York.

PVACCEPT, 2005. Report to the EU Commission of the results of the project IPS-2000-0090, 5th Framework Programme. Action "Promotion of innovation and encouragement of SME participation" http://www.pvaccept.de

Raugei, M., Bargigli, S., and Ulgiati, S., 2007. Life cycle assessment and energy pay-back time of advanced photovoltaic modules: CdTe and CIS compared to poly-Si. Energy – The International Journal, 32(8):1310–1318.

Ritthoff, M., Rohn, H., and Liedtke, C., 2003. Calculating MIPS. Resource Productivity of Products and Sources. Wuppertal Institut for Climate, Environment and Energy – Science Centre North Rhine-Westphalia. ISBN 3-929944-56-1e. Pp.53.

Schmidt-Bleek, F., 1993. MIPS re-visited. Fresenius Environmental Bulletin 2: 407–412.

Sciubba, E. and Ulgiati, S., 2005. Emergy and exergy analyses: complementary methods or irreducible ideological options? Energy – The International Journal 30(10): 1953–1988.

SETAC, Society of Environmental Chemistry and Toxicology, 1993. Life Cycle Assessment: A Code of Practice. www.setac.org

Slesser, M., 1974. Energy analysis workshop on methodology and conventions. Report IFIAS No. 89. Stockholm: International Federation of Institutes for Advanced Study.

Szargut, J., Morris, D.R., and Steward, F.R., 1988. Exergy Analysis of Thermal, Chemical and Metallurgical Processes. Hemisphere Publishing Corporation, London.

Ulgiati, S., Raugei, M., and Bargigli, S., 2006. Overcoming the inadequacy of single-criterion approaches to Life Cycle Assessment. Ecological Modelling, 190: 432–442.

Ulgiati, S., Zucaro, A. and Franzese, P. P., 2008. Matter, Energy and Emergy Assessment in the agricultural sectors of the Campania region. Constrains, bottlenecks and perspectives. In: Proceedings of the 6th Biennial International Workshop Advances in Energy Studies. Towards a holistic approach based on science and humanity, 978-3-85125-018-3. OeH TU Graz, Graz, Austria: pp. 550–560.

USING THE MUSIASEM APPROACH TO STUDY METABOLIC PATTERNS OF MODERN SOCIETIES

MARIO GIAMPIETRO*, ALEVGUL H. SORMAN,
AND GONZALO GAMBOA
*Institute of Environmental Science and Technology (ICTA);
Universitat Autònoma de Barcelona (UAB) 08193 Bellaterra,
Spain*

Abstract   This paper presents examples of application of the MuSIASEM approach (Multi-Scale Integrated Analysis of Societal and Ecosystem Metabolism). The text is organized as follows: Section 1 briefly discusses, using practical examples, the theoretical challenge implied by the quantitative analysis of complex metabolic systems. Complex metabolic systems are organized over multiple hierarchical levels, therefore, they require the adoption of different dimensions and multiple scales of analysis. This challenge has to be explicitly addressed by those performing quantitative analysis; Section 2 discusses two key characteristics to be considered when studying the evolution in time of the metabolic pattern of modern societies: the implications associated with changes in demographic structures (Section 2.1); and, the need of providing an integrated analysis of the structural change of socio-economic systems (Section 2.2). Such an analysis can be obtained by integrating the two functional/structural parts: (i) the part in charge for the production; and (ii) the part in charge for the consumption of goods and services. Section 3 provides an example of an integrated analysis of the evolution in time of the metabolic patterns of European Countries (data from an ongoing European Project – SMILE). This analysis shows clearly the problem generated by the use of data referring to the societal level (characteristics of whole countries). Looking only at the characteristics of the black-box, neglecting the differences of key parts operating inside the black-box, can lead to erroneous interpretation of data (e.g. the theory of Environmental Kuznets Curves).

---
* To whom correspondence should be addressed: Mario Giampietro, Institute of Environmental Science and Technology (ICTA); Universitat Autònoma de Barcelona (UAB) 08193 Bellaterra, Spain; E-mail: Mario.Giampietro@uab.cat

F. Barbir and S. Ulgiati (eds.), *Energy Options Impact on Regional Security*,
DOI 10.1007/978-90-481-9565-7_2, © Springer Science + Business Media B.V. 2010

Keywords: Multi-scale integrated analysis, societal metabolism, ecosystem metabolism, comparative analysis of European countries, environmental Kuznets curves, demographic changes

1.　The Problems with Existing Quantitative Sustainability Analyses

1.1.　THE CHALLENGE OF MULTI-LEVEL ANALYSIS

In this section we provide a few practical examples of a systemic challenge faced by quantitative analysis, when applied to the analysis of complex systems (e.g. the sustainability of societies).

By definition a complex phenomenon is a phenomenon which can and should be perceived and represented using simultaneously several narratives, dimensions and scales of analysis (Ahl and Allen 1996; Allen et al. 2001; Funtowicz et al. 1999; O' Connor et al. 1996; Rosen 1977, 1986, 2000; Simon 1962, 1976).

Therefore, complexity represents a challenge for the generation of quantitative indicators, since, in order to be able to generate an accurate quantitative representation of an event, scientists must focus only on a limited subset of these possible scales and dimensions (Rosen 1977, 2000; Giampietro 2003; Giampietro et al. 2006). Any quantitative representation of a complex system – which must be based on the use of only a chosen dimension and a chosen scale at the time – entails a dramatic "reduction" in the set of possible useful perceptions and representations. This is to say that quantitative analysis entails an important epistemological trade-off: in order to gain a robust and accurate quantitative representation of one aspect of the investigated system (perceived by using just one dimension and one scale) the scientist must accept to lose the possibility of perceiving and representing other aspects, which would be relevant for other analysts having other interests and therefore purposes.

This is to say that when dealing with a complex issue such as the sustainability of modern societies, the power and strength of quantitative analysis may entail also a potential weakness due to the excessive reliance on reductionism, which they require. This point is beautifully explained by Box (1979), under the heading "all models are wrong, but some are useful" he discusses the usefulness of quantitative models as follows: *"For such a model there is no need to ask the question "is the model true?". If "truth" is to be the "whole truth" the answer must be "No". The only question of interest is "Is the model illuminating and useful?""* (pp. 202–203). This consideration points directly at the fact that the pre-analytical definition of the purpose of the analysis – why are we doing this analysis in the first place?, whose problems are addressed by this analysis? – will determine the usefulness and pertinence of the quantitative results.

In the rest of this section we illustrate a few practical examples of typologies of problems associated with this predicament:

1.1.1. *Example #1 – Scale issues matter: it is essential to establish interlinkages between events taking place at different hierarchical levels*

The example given in Fig. 1 refers to a multi-level reading of the score of a tennis match. In this example Player A won the match after winning three sets at the tie-breaker (with a score of 7-6 games in each one of the three sets). On the contrary Player B won only two sets with a score of 6-3 and 6-2. Let's now imagine that a scientist, trying to discover the rules of tennis, wants to find out the winner, looking at this quantitative description (score). If he decides to use an index based on the accounting of the number of games won, he would get a completely wrong picture of the result of the match. In fact, Player B (who lost the match) won 30 games versus the 26 won by Player A.

Explanation of causalities requires the choice of the right scale

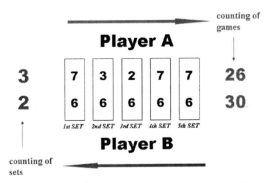

Figure 1. The multi-scale accounting of the score in tennis

This example may seem trivial, but it points at a very dangerous pitfall: one should never rely on statistical information gathered at a given hier- archical level – e.g. the household level, the sub-economic sector level, the whole society level – without having first seen and understood the "big picture" – the meaning of the relative numbers within the investigated system. It is the hierarchical structure of relations across levels, which provides the meaning of the data gathered at different levels.

The example given in Fig. 1 illustrates that it is not always true that by working more in details, with more accurate measurements one can get a more reliable picture (or explanation!) of a given situation. In fact, let's imagine that two different scientists trying to study this "unknown game" – one counting the "number of games won" and the other counting the "number of sets won" – get into a public scientific controversy about the determination of who is the winner of this match. Let's also imagine that in order to solve this controversy the scientific community would follow the traditional recipe of reductionism: gathering more accurate data, coming from a more detailed study. This decision may lead to a more accurate analysis of the recorded tape of the match. For example, a new team of scientists can keep record of the individual points won by the two players within each game. This additional check based on an additional dose of reductionism would simply add confusion to confusion. In a tennis match, the number of points won, does not necessarily map either in the number of sets or the number of games won by the two players. There is no escape from the necessity of having first of all a clear understanding of the meaning of the numerical relations found when observing a complex system operating across different levels of organization.

In the example given in Fig. 1 the three relevant rules to be considered for gaining understanding are: (i) how the winning of points within a game translates into the winning of games; (ii) how the winning of games in a set translates into the winning of sets; and (iii) how the winning of sets in a match translates into the winning of a match. In technical jargon this entails to develop a grammar capable of addressing the issue of scale. The problem is "how to scale" the results of the quantitative analysis performed at one level to the next one. In fact, within different levels, one should expect to find different rules. That is, we cannot understand the emergent behavior of the whole if we are not able to establish first effective interlinkages between the various representations of events referring to different hierarchical levels.

This problem can be described using various concepts such as: non-linear behavior, emergent properties, the need of using hierarchy theory. To make things more challenging one should notice that determining the winner of the match – using a variable defined on the highest hierarchical level "number of sets won" – is not the only information which may be relevant when studying a tennis match. If we are also interested in the duration of the match, then the variable we have to consider is the "number of points played" (defined at the lowest hierarchical level). Finally, by looking at the games won within the various sets one can get an idea of the typology of the match. For example, a score of 6-0, 6-0, 6-0 indicates a overwhelming triumph in a match played over five set; whereas, a score of a 7-6, 6-7, 7-6 indicates a very tight win in a match played over three sets. This to say, that when compressing the information gathered about a complex system to just a number – a single indicator – referring to just one of the hierarchical levels, we are losing a lot of potential information. For this reason, it would be wise to keep as much as possible the gathered information organized over different "variables" – categories – referring to different hierarchical levels.

1.1.2. *Example #2 – assessments per capita (using a variable at higher level) miss important qualitative differences between societies (= the importance of demographic variables – referring to a lower level)*

Economic development entails both quantitative and qualitative socio-economic changes, which can be missed when adopting quantitative assessment "per capita" (e.g. GDP per capita, energy consumption per capita, number of teachers per capita).

For example, the very same assessment of 'per 1,000 people' (equivalent to a 'per capita' assessment) can imply quite different supplies of work hours per year into the economy depending on the demographic and social structure of society.

As shown in Fig. 2, in 1999, Italian population supplied 680,000 h of work to the economy per 1,000 people, while Chinese population supplied 1,650,000 h of work per 1,000 people (2.46 times more!). In China, 1 out of every 5 h of human activity was allocated to paid work, while in Italy this was only 1 out of every 13 h (Table 1).

This difference can be easily explained: more than 60% of the Italian population is not economically active, including children, students and elderly (retired). The human activity associated with this part of the population is therefore not used in the production of goods and services but allocated to consumption. Furthermore the active population, the 40% of the population included in the work force, works less than 20% of its available time (yearly work load per person of 1,700 h).

The economic implications of this qualitative difference (per 1,000 people) on economic variable are illustrated in Table 1. If we imagine that these two countries were operating at the same level of GDP per capita – e.g. assuming a common value of 20,000 €/person/year – this difference in work supply would imply that in order to be able to generate the same level of GDP per capita, the amount of GDP produced per hour of labour in Paid Work in Italy would have to be 2.4 times higher than in China (29.4 €/h versus 12.1 €/h). This qualitative difference is at the root of one of the trade-offs linked to progress that will be discussed later on and it is completely missed if we adopt indicator of economic performance based on the "per capita" basis.

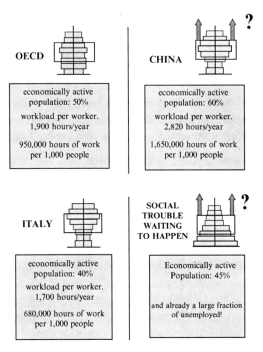

Figure 2. Relation between demographic structure and supply of work hours at the level of society (Adapted from Giampietro 2009)

TABLE 1. Allocation of human activity (HA in hours) to paid work (PW) and household (HH) sectors for Italy and China in 1999 (per 1,000 people per year); THA = 1,000 people × 8,760 h/year, and THA = HA_{PW} + HA_{HH}

	Italy	China
Total Human Activity (THA) in hours/year	8,760,000	8,760,000
Paid Work sector (HA_{PW}) in hours/year	680,000	1,650,000
Household sector (HA_{HH}) in hours/year	8,080,000	7,110,000
Ratio Paid Work/Total Human Activity (HA_{PW}/THA)	1/13	1/5
Hypothetical level of GDP (20,000 €/year per capita)	20,000,000	20,000,000
Flow of added value to be generated in Paid Work! (€/h)	29.4	12.1

1.1.3. *Example #3 – assessments of any variable at higher-level (the whole society seen as a black box) must be combined with assessments of variables at lower-levels (qualitative differences within economies) by looking inside the black box*

The I = PAT relation – introduced by Elhrlich in the 1980s – can be used to explain the main point we want to make in this example. The four terms of this relation are:

(I) standing for the Impact on the environment; which is determined by the combination of other three terms: (P) Population; (A) Affluence; and (T) Technology. Within this relation (T) Technology is individuated as the key factor that makes it possible to decouple economic growth and environmental degradation. According to the traditional gospel about the positive effect technical progress (e.g. the Environmental Kuznets' Curve hypothesis), improvements in Technology can effectively counteract the effects of increasing population (P) and affluence (A). That is, even though these two factors have the effect of increasing the amount of goods and services which have to be produced and consumed in a given society, technical progress, by improving the performance of technology (T), can reduce the impact per unit of goods and services produced and consumed by society.

Can we check the validity of this hypothesis using empirical data? Again, in order to be able to answer this question it is crucial to be careful in handling in a wise way the set of possible assessments of performance referring to different hierarchical levels. Let's start by comparing the characteristics of three European countries (Spain, the UK and Germany) adopting the rationale of "I=PAT". This comparison is given in Table 2.

TABLE 2. Indicators relevant for the I=PAT relation and the "black-box level" (level n)

	U.K.	Spain	Germany
I – CO_2 emissions p.c. (ton/year)	352	897	558
P – Population (millions)	42.3	82.5	50.1
A – GDP per capita (€/year)	17,900	26,800	27,000
T – CO_2 emission intensity (kg/€)	0.46	0.41	0.35

Looking at this dataset it seems that the data back up the hypothesis of the Environmental Kuznets curves. That is, the Affluence A (estimated in this case using the proxy GDP p.c.) seems to explain the differences in emission intensity (estimated using the proxy CO_2 emission per unit of GDP). UK with a higher GDP per capita than Spain has a lower energy intensity of its economy. According to this hypothesis the variable Technology

(T) is explaining this difference, since, according to this analytical frame-work Technology is "better" where the GDP is higher.

But how robust is such an analysis if we check the same data set across different hierarchical levels? To test the robustness of this result we can use a multilevel system of accounting proposed by the MuSIASEM approach. In this way we can "open up" the black box and move down the analysis through three hierarchical levels:

- **Level n:** the "whole society"
- **Level $n - 1$:** the "Paid Work sector"
- **Level $n - 2$:** individual compartments within the economy (e.g. socio-economic sectors)

After opening the black-box we look for benchmark values referring to the proxy variables chosen to characterize the semantic categories A and T at level $n - 2$. That is, in this way we can look for indicators of performance referring to the level $n - 2$.

This characterization can be done both in economic terms (e.g. *extensive variable*: sectorial GDP; *intensive variable*: pace of added value generated per hour of labor) and in biophysical terms (e.g. *extensive variable*: energy consumption per sector; *intensive variable*: exosomatic energy spent per hour of labor in the sector). In this way, we can generate a different – richer – view of the key characteristics (expressed at a lower hierarchical level) generating the overall level of CO_2 emission per capita and of the overall energy intensity (MJ of primary energy consumption/€ of GDP) – measured as aggregated value for the whole country.

An example of what we see after opening the black boxes is given in Fig. 3.

After moving to a lower hierarchical level (that is, the "Paid Work sector" inside the black box [level $n - 1$]), we can check for differences and similarities in sub-sectors of the Paid Work sector (i.e. socio-economic sectors at level $n - 2$) between the three considered countries. The three sub-sectors considered in this example are:

(i)   **AG** = agriculture
(ii) **PS** = Productive Sector (Building and Manufacturing, plus Energy and Mining)
(iii) **SG** = Service and Government

For these three sub-sectors we can now compare benchmark values referring to both economic, demographic and energy related variables:

- Economic variables: Sectoral Gross Domestic Product per hour of Human Activity – GDPi/HAi – measured in €/h (data from Eurostat EEA)

- Demographic variables: Human Activity in the Paid Work sector (HA_{PW}) and Human Activity in the Production sector (HA_{PS}), Human Activity in the Service and Government sector (HA_{SG}), Human Activity in the Agricultural sector (HA_{AG}) (data from ILO)
- Energy related variables: Sectoral Energy Throughput per hour of Human Activity in the sector i – ET_i/HA_i – measured in MJ/hour (data from IEA)

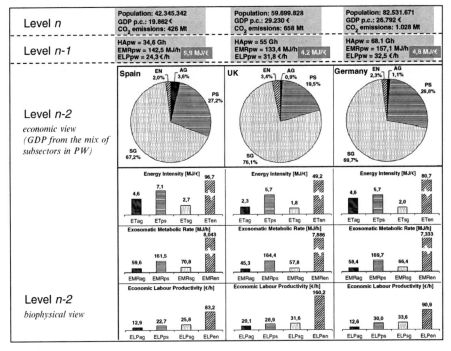

Figure 3. Opening the "black-box": what is behind the "I=PAT" relation (Data source: Eurostat)

In order to be able to read the system across levels, the MuSIASEM approach uses assessment of energy and added value flows "per hour of human activity" instead of using values "per capita/per year". Then, as soon as one looks at the integrated characterization (based simultaneously on economic and biophysical data) across the levels given in Fig. 3, one can make the following observations:

1. The differences in the aggregate value of CO2 emission intensity (Table 1) have very little to do with the values of individual proxy variables used to characterize the semantic categories I, A, T at the level n − 2. That is, the lower level of CO2 emission of the UK (Table 1) is not about a more efficient production of steel or construction than in Germany or Spain.

Rather the differences in this value are more related to the different composition of the Paid Work sector in these three countries (UK does not produce the same level of steel and construction, and it has to rely more on import for its internal consumption).

2. The different economic performances of these three countries depend on their different socioeconomic structures (i.e. different characteristics of the sub-compartments of the whole), that is by the relative importance of the different economic sectors. UK gets a large fraction of her GDP from the service and the financial sector.

Looking at this example we can conclude that the ratio MJ/€ measuring the energy intensity (which is often erroneously labelled as a proxy of Technology in the analysis of Environmental Kuznet Curves) either of a whole economy (at the level n) or of an economic sector (at the level n − 1) does not have a meaningful external referent. In fact, this emergent property of the whole can be explained by both: (i) differences in technology over the various sectors (detectable only at level n − 2 and level n − 3); and (ii) differences in the relative profile of sectoral GDPs.

When using the chosen label (Technology) these values could (mis)lead us to think that Spain is using worse technology than UK and Germany. In fact, when looking at the amount of fossil energy consumption per hour of work (a proxy of the amount of technical devices used per worker) of the Industry, Building and Manufacturing sector (PS) the three countries present very similar values: 161.5 MJ/h in Spain, 164.4 MJ/h in the UK and 169.7 MJ/h in Germany.

In conclusion looking at the dataset presented in Figure 3 we can say that the differences of values in energy intensity (or CO2 emissions) found at the level of the whole economy do not necessary imply better or worse technology.

This point will be illustrated in detail in Section 2 when describing the pattern of energy metabolism of these three countries across levels.

1.1.4. *Example #4 – Multiple-scale [short-term vs long-term view]*
      and Multiple-dimensions [steady-state analysis vs evolutionary
      analysis]

Demographic changes provide an easy example of lag time dynamics that imply a predictable trade-off in economic performance. As observed in Fig. 2 a wave of individuals moving across age class will determine a change in the performance of an economy, which can imply a different effect when considering short-term versus long-term.

This effect is illustrated in Fig. 4. What is bad in the short term (in 1980) a high dependency ratio, associated with a baby-boom in the 1970s, will become a very positive situation in the year 2000 (after 20 years), providing an incredible high fraction of working population. Needless to say that we can expect that the situation will revert again to bad in another 20 years when the dependency ratio will go up again because of the massive ageing of the population. Therefore, the example in Fig. 4 shows the existence of predictable pattern of qualitative (structural) change in time, which will determine a contrasting performance at different points in time (short term vs long term).

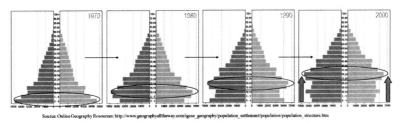

Source: Online Geography Resources: http://www.geographyalltheway.com/igcse_geography/population_settlement/population/population_structure.htm

Figure 4. A view of the changes in demographic structure in China 1970–2000

This contrasting effect of demographic variables is obviously extremely relevant for sustainability analysis, and it can only be analyzed by adopting an evolutionary narrative of sustainability. That is, when looking at existing indicators of performance and demographic dynamics one can guess a situation of instability. For example, we can develop an analysis that says: when the economy of China will reach a dependency ratio similar to that of Italy now, it is very unlikely that it will be able to generate a flow of added value of 29.4 €/h of labour, since the large population would make quite unlikely the accumulation of a huge amount of capital per capita (required by this performance). However, such a model can only predict a reason for instability (lack of viability) due to demographic changes. It cannot predict if this instability will translate into wars, riots, massive emigration, or rather in a positive transformation determining a new form of organization of social activities, which will make it possible for the society to work in a desirable way, in presence of a much higher level of dependency ratio. Such a model can only point at the existence of critical bottlenecks and at forced transformations to be expected in the future.

2.   Key Characteristics of the Pattern of Societal Metabolism

2.1.   THE IMPLICATIONS OF CHANGES IN DEMOGRAPHIC STRUCTURE

It is well known that an increase in material standard of living translates into a longer life expectancy. The relation between an increase in material standard of living and adjustment in demographic characteristics of a society is well known (e.g. the theory of demographic transition, the work of 1993 Nobel Prize in Economics Robert Fogel on the implications of demographic changes). Very shortly, we can see in Fig. 5 the differences in population structure between a pre-industrial society, a developing country and two developed countries. (Giampietro and Mayumi 2000a).

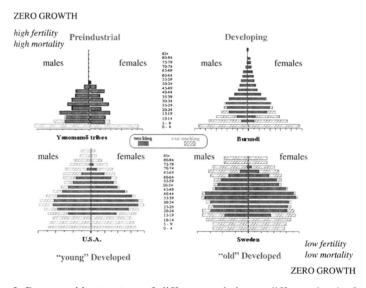

Figure 5. Demographic structure of different societies at different level of economic development

When performing the same analysis within OECD countries, we can still observe a clear difference in population structure between countries at different levels of economic development. As illustrated in Fig. 6 Mexico and Turkey do have at the moment a population structure belonging to the "pyramid type" (associated with developing country), whereas richer countries, such as Sweden and Japan have a population structure belonging to the "mummy type" (associated with developed countries).

However, it is possible to see, from the projections of population structure, that in the year 2050 Mexico and Turkey, because of their economic development, will get into the "mummy type" shape as illustrated in Fig. 6 by the white bars describing the size of each age class. Obviously, these projections of demographic changes are based on the steady-state assumption of continuous growth of 3% of the global economy.

We saw before that population changes may entail a non linear change in the feasibility of the dynamic equilibrium between: (i) the requirement – what is consumed by the whole economy; and (ii) the supply – what can be supplied by the specialized compartments of society in charge for the production of goods and services. In particular the MuSIASEM approach can focus on this dynamic budget in terms of congruence over the flows that are produced and consumed per hour of human activity.

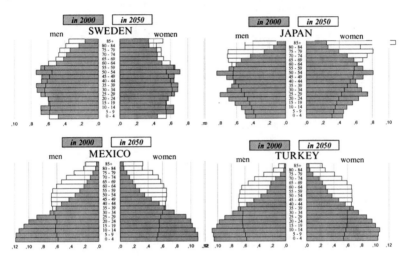

Figure 6. Demographic structure of different societies in OECD countries

As illustrated in Fig. 7 we can see that there is a standard breakdown in the "expected" profile of human activity across different compartments of the economy, defined at different levels of organization. As illustrated in this example this implies that the amount of hours of human activity per capita available for each specialized task is very limited. In the example of Fig. 7 the task considered is producing energy carriers, but the same constraint applies to other tasks such as producing food, mining, generating an adequate supply of water, activity of doctors, teachers, etc.

In more general terms, we can use the same approach, used by ecologist to study the structural/functional organization of ecosystems, to study the set of internal constraints affecting the metabolism of a modern society. This implies studying the forced relation between what **can be** produced per hour by the different compartments of the economy (on the production side) and what **is required** by the various compartments of the economy (on the consumption side).

The trade-off between economic development and economic competitiveness can be now explained by the systemic change in internal relations which is associated with economic development.

Economic development entails an integrated set of changes in social variables: (i) longer life expectancy = larger dependency ratio; (ii) subsidies to unemployed people = longer periods of unemployment since the unemployed can wait for a desirable job offer; (iii) longer periods of morbidity in the work force; (iv) better level of secondary education = further reducing the economically active population within the work force; (v) smaller work load per year and paid leaves = reducing the actual work supply of the economically active population. The final result of this combined set of changes is an expansion of the size of the Household Sector – the hours of human activity allocated in consuming, by performing activities outside the Paid Work sector. The consequence of economic development is therefore a dramatic reduction of the ratio HA$_{PW}$/THA – the hours of work in the Paid Work sector versus the Total Hours of Activity of the society.

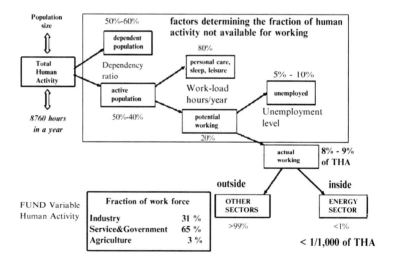

Figure 7. The standard dendrogram of splits of hours of human activity over different compartments of a developed economy (From Giampietro and Mayumi 2009)

The implications of this mechanism are clear and can be visualized as in Fig. 8, where we can see (on the upper right quadrant – whole society level) that the value of *GDP per hour at the level of the whole country (Spain)* – US$1.8 /h – is determined by: (i) the value of the total GDP (US$611 billion) on the horizontal axis; and (ii) the value of 344 Gh (the hours of human activity in a year of 39 million people in Spain). Then on the left upper quadrant we can see the reduction in the value of THA (Total Human Activity) due to the social factors discussed before, leaving only 23 Gh (a mere 7% of THA) available for the Paid Work sector.

This entails that at a given value of the ratio HA_{PW}/THA there is a forced relation between *the level of GDP per hour of the whole society* and the rate of *production of GDP per hour in the Paid Work sector* (ELP_{PW}).

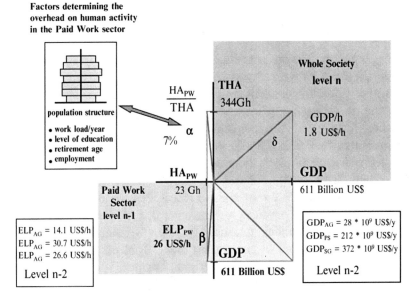

Figure 8. The forced relation between the level of GDP of the whole and the rate of production of GDP per hour of the Paid Work sector (Spain 1999)

Using the label indicated in Fig. 8 we can write the following relation:

$$\text{GDP per hour}_{(\text{societal level})} = ELP_{PW} \times HA_{PW}/THA$$

That is, *IF* ageing and other social changes are continuously reducing the ratio HA_{PW}/THA *THEN* either the GDP per hour (which is the same as GDP per capita!) will be reduced in the same proportion, or the economy must be able to continuously increase the value of ELP_{PW} = rate of generation of added value per hour of work in the Paid Work sector.

As it will be discussed later on, a continuous increase in ELP_{PW} can only be obtained by continuously re-adjusting the activities across economic sectors with the goal of increasing continuously the amount of added value of the products and services produced and consumed per hour of work in PW. One of the most popular solutions to this challenge among developed countries is to: (i) stop producing goods and rather importing them; (ii) use financial leverage to boost the amount of debt in the national economies. That is in developed countries ageing imposes a massive switch from industrial economy to the "bubbles economy", as predicted more than 50 years ago by Soddy in his seminal book: "Wealth, Virtual Wealth and Debt".

3. The Continuous Restructuring of the Socio-economic Metabolic Pattern

3.1. THE METABOLIC PATTERN OF SOCIETY ACROSS LEVELS

Another important change associated with economic development is that not only the ratio between "working hours of human activity" and "non-working hours of human activity" is continuously reduced, but also that within the shrinking compartment of working hours in Paid Work, an increasing fraction of the available work must be allocated into the compartments of services. That is, not only in a developed country per each hour allocated in Paid Work there are 12 h allocated in final consumption, but also (and this is even more scaring) more than 60% of the working time in Paid Work is allocated in the services. Put in another way, when considering the flow of products we consume in a modern society, per each hour allocated in producing products (in the PS compartment) there are more than 25 h allocated in consuming them. That is, the more a society increases its level of consumption per capita, the more it reduces the amount of human activity allocated in producing goods. In biophysical terms, this entails the existence of the same phenomenon found for the activities generating added value illustrated in Fig. 8 (the higher the GDP the lower the fraction of human activity in Paid Work, the higher the need of getting a high ELP).

To study the consequences of these structural changes we can analyze the metabolic pattern of a modern society based on the fund flow model proposed by Georgescu–Roegen using the graphical representation proposed by H.T. Odum. An example of this representation is given in Fig. 9 (references to the theoretical approach followed to develop this graph are given at the beginning of Section 3). This application of the MuSIASEM deal with the metabolic pattern of a modern society divided into four compartments: (i) HH – Household on the consumption side; and on the production side: (ii) PS* – Productive Sector minus the energy sector; (iii) ES – the energy sector; (iv) SG – Service and Government. For each compartment we have a characterization based on three different variables:

(1) an extensive variable for the Fund Human Activity (hours of human activity in the compartment per year); (2) an extensive variable for the Flow (MJ of exosomatic energy per year); and (3) an intensive variable obtained dividing the flow by the fund (MJ/h of human activity).

The network of flows over this graph must guarantee the congruence across levels of these three variables. The average values for society (in the upper-right box) for these three variables, must be compatible with the triplet of values written for each elements at the lower hierarchical level.

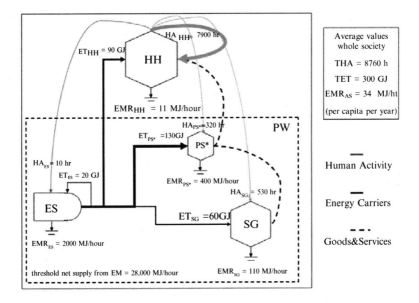

Figure 9. The forced relation between parts and whole in terms of relative size of funds and flows (the typical metabolic pattern) in a modern society (From Giampietro and Mayumi 2009)

3.1.1. *The dynamic budget between requirement of human activity and supply of human activity between the Household (HH) and the Paid Work (PW) sector*

Depending on the characteristics of household types and the given profile of distribution of the population of households over the set of types, we can calculate the overall flow of products and services required by the HH sector (on the side of final consumption) – aggregate at the **level** $n-1$ – and at the same time the supply of hours of work that the HH sector provides to the PW sector.

An example of this analysis, applied to Catalonia, is given in Fig. 10. The overall amount of Total Human Activity (THA) per year in Catalonia is 58.8 Gh (the total activity of 6.7 million people × 8,760 h/year).

Recalling the relation of congruence across contiguous levels – THA (level n) = HA_{HH} + HA_{PW} (the two compartments defined at level n − 1) – we can say that, when considering the level n − 1, this fund of human activity (58.8 Gh at the level n) is divided between hours of human activity in the Household sector (HA_{HH} = 52.1 Gh) and hours of human activity within the Paid Work (HA_{PW} = 6.6 Gh). According to this analysis we can also say that there is a reciprocal supply/requirement: (A) the HH sector is supplying the amount of working hours required by the PW sector to perform its activities; (B) the PW sector is supplying the amount of goods and services required by the HH sector to perform its activities.

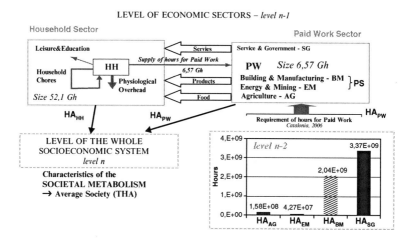

Figure 10. The dynamic budget of hours of paid work (requirement vs supply) between the Household Sector and the Paid Work sector – Catalonia – 2007

The multilevel analysis described in Fig. 10 can be used to understand the constraints affecting the viability of the dynamic budget associated with this pattern of metabolism. When representing the characteristics at the *level n* we obtain the standard indicators referring to the country as a whole, people, employment, GDP, flows of goods and services. But if we move across hierarchical levels, to the *level n − 1* – by addressing the distinction between the production side and the consumption side of the economy – then we can see that the division of THA (Total Human Activity) between consumption (HA_{HH}) and (HA_{PW}) determines a bridge between social indicators and economic performance determined by the ratio HA_{PW}/THA (see discussion in Section 2.1). If we move down another level – the red box on lower right corner of Fig. 10 – at the *level n* − 2, the level of the economic subsectors, we address the existence of another relation of congruence over subsectors: in fact PW (Paid Work) can be expressed as the sum of four sectors: (i) AG (AGriculture); (ii) EM (Energy and Mining); (iii) BM (Building and Manufacturing); and (iv) SG (Service and Government)

$$HA_{PW} = HA_{AG} + HA_{EM} + HA_{BM} + HA_{SG}$$

In semantic terms this means that the available work supply – the 6.57 Gh supplied by the household sector – must be divided among these four sectors, which are competing for the available supply of working hours. This new constraints can be calculated also using a bottom-up direction. That is, given: (i) the overall requirement of goods and services expressed by society; and (ii) the technical coefficients of the various activities carried out in the different subsectors, we can determine the requirement of jobs for each sector.

With the analytical scheme provided in Fig. 10 we can discuss of the implication of an increase in the **Bio-Economic Pressure (BEP)** associated with economic development. The terms BEP indicates an increase in the requirement of goods and services at the very moment in which the supply of work is decreasing, and a dramatic increase in the requirement of services is experienced in PW.

Looking at the relation of congruence illustrated in Fig. 10, there are very few options for a developed society facing a continuous increase in BEP. The options are:

- To externalize the requirement of work for products by importing, rather than producing, labour intensive goods (e.g. from China)
- To import workers from elsewhere (immigration) for those activities which cannot be externalized - e.g. low paid services such as cleaning, cooking, assistance of elderly
- To use the financial leverage – i.e. spreading debts across different hierarchical levels to increase as much as possible the generation of virtual added value per hour of work.

As a matter of fact, these three options represent the exact picture of the strategy of development which has been followed by developed countries in the last decade.

4. Using an Integrated Analysis of the Metabolic Pattern of European Countries to Check the Hypothesis of Environmental Kuznet Curves

A theoretical presentation of the MuSIASEM approach has been given in the previous NATO workshop (Giampietro 2008) and it is available in several publications (Giampietro and Mayumi 2000a, b, 2009; Giampietro 1997, 2000, 2001; Giampietro et al. 1997, 2001, 2006a, b; Giampietro and Ramos-Martin 2005; Ramos-Martin et al. 2007).

4.1. AN INTERNATIONAL COMPARISON OF EU METABOLIC PATTERNS

In this section we want to show that the change in the pattern of metabolism of European societies (associated with a decrease in energy intensity) in the last decade can be easily explained by a change in the profile of productive activities of the economy. This means that by using a favourable terms of trade and relying as much as possible on financial transactions, the EU has been able to externalize those economic activities more intensive in terms of: (i) consumption of natural resources (including energy); and (ii) environmental impact; to developing countries. In this way, it has been possible to move the shrinking EU work force on those activities in the service sectors and in the productive sector, which were providing more added value per unit of energy. This has required covering: (i) the internal demand of services on low paid jobs with immigrants; and (ii) the internal demand of labor for intensive and resource intensive products with imported goods, outsourcing work in the services whenever possible.

To illustrate this point, let's start by looking at the change in time of the profile of energy intensity of EU countries. To do so, we use a plane having on the two axes: (i) energy consumption per capita (vertical axis); and (ii) GDP per capita (horizontal axis). This graph is shown in Fig. 11. It should be noted, that in order to be able, later on to study these changes across levels, we provide these two values in MJ/hours and €/h. To obtain the equivalent value of MJ or € per person per year one has only to multiply the value found on the axes by 8,760 (hours per capita per year).

A few notes about this graph: (i) the original point indicating the starting position of each country in 1992 is indicated by a label identifying the various countries; (ii) the position of the country is represented year after year to visualize the change over the historic series (from 1992 to 2005); (iii) the relative size of the country (indicated by the size of the disk) is reflecting their population size.

Looking at the trajectories of European countries one can note two different slopes:

Slope #1 – refers to a simultaneous increase in time of the values on both axes – that is, over this period countries like Portugal, Greece, Spain, Italy did increased their GDP but they also consumed more energy

Slope #2 – indicates an increase in time of the value of GDP faster than the increase in energy consumption – that is, over this period countries like Germany, France, the Netherlands (and other northern European countries not shown in this figure) have been increasing the rate of generation of GDP quicker than their increase in consumption of energy. The analysis of this trend (a reduction of energy intensity of the countries with higher levels of GDP) led to the formulation of the Environmental Kuznet Curve hypothesis – that is energy intensity is reduced with economic growth, because of the adoption of better technology.

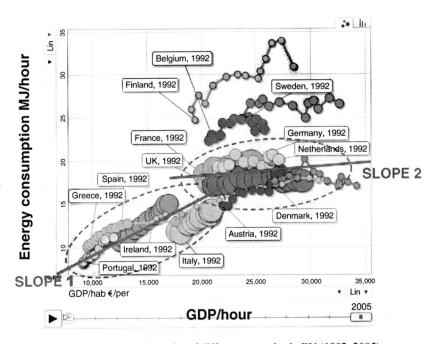

Figure 11. Energy intensity of different countries in EU (1992–2005)

We claim here that the robustness of this hypothesis is very doubtful. In relation to this claim, it can be very instructive to analyze the trend expressed by the various countries (represented in Fig. 11) at the level of the whole society, after opening the black box, by looking at the internal changes of energy intensity at the hierarchical level of individual subsectors (level n − 2) – this is done below.

The differences found among countries in Fig. 11 can be explained by making hypotheses based on common sense and our general knowledge of the various countries. For example, we can hypothesize that the higher energy intensity of Belgium in relation to EU average is a legacy of its heavy industry based on coal in the 1960s. In the same way, we can explain the higher energy intensity of northern countries – such as Finland and Sweden – with the higher consumption of energy associated with the expression of the same function expressed by the other European countries in a colder environment at a much lower density of population. However, by opening the black box and comparing the performance of each sector and subsector in terms of benchmarks, at lower hierarchical levels, we can actually check the validity of this type of hypotheses.

Let's start this exercise from the comparison of the evolution in time of the energy intensity of the three countries compared in Table 2 and Fig. 3 – Spain, Germany and UK.

The four graphs shown in Fig. 12 represent: (A) an integrated analysis – since it addresses both economic (€/h) and biophysical (MJ/h) variables and in indirect way demographic variables – the difference in size between the whole (average society) and the part PW (the ratio HA_{PW}/THA); and (B) a

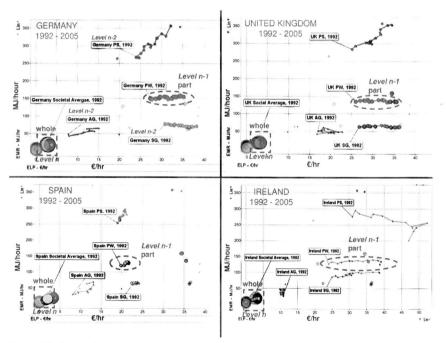

Figure 12. Comparing the metabolic pattern of Spain, Germany and UK using simultaneously three hierarchical level

multi-scale analysis – since it provides a simultaneous representation at: (i) the *level n* is the value for the whole country; (ii) *level n*–1 is the value for the productive side of the economy: the PW sector; and (iii) *level n*–2 is the value for each one of the three sub-sectors of the productive sector (PS, SG, AG). That is, it is a multi-scale integrated analysis of the metabolic pattern of these countries.

The trajectory of change for each one of the representation across scale can be seen by looking at the "movement" of the disks in time (the shape of the "worms" made of the same type of disks). Let's now see more in detail the information given by this graph (Fig. 12):

* The characteristics of the whole – level n – are represented by:
(A) the size of the disk of the country, in terms of Total Human Activity – Population × 8,760 h/year) – the average value for the whole country is in a red broken line box
(B) The consumption of energy per hour (the value of the variable on the vertical axis)
(C) The GDP per hour (the value of the variable on the horizontal axis)

* The characteristics of the Paid Work sector (a part) – level n–1 – are represented by:
(A) The size of the disk of the PW sector, in terms of HA_{PW} – defined as the number of workers in the PW sector x average hours of work/year) – the average value for the paid work sector (the production side) is in a red broken line ellipsoid
(B) The consumption of energy per hour (the value of the variable on the vertical axis)
(C) The added value per hour (the value of the variable on the horizontal axis)

* The characteristics of the sub sectors – level n–2 – in this example we consider three sectors: SG (Service and Government); PS (Productive Sectors); AG (Agricultural sector), which are represented by:
(A) The size of the disk of the sub-sector, in terms of HA_i – defined as the number of workers in the sector *i* times the average hours of work/year) – in the figure the average value for the various sub-sectors is represented by the disks moving in time
(B) The consumption of energy per hour (the value of the variable on the vertical axis)
(C) The added value per hour (the value of the variable on the horizontal axis)

4.1.1. *Moving from level n (societal average) to level n−1 (PW sector)*

We can immediately see from this multi-level integrated analysis of the metabolism of a country, that the energy intensity of the country – the ratio (MJ/hour)/(GDP per hour) of the whole at the *level n* (the disk in the square box with broken line) is different from that of the productive part of the economy – PW at the *level n−1* (the disk in the ellipsoid with broken line). This difference depends on the relation between the characteristics of the Household Sector and the Paid Work Sector (HA_{PW}/THA and the energy consumption of the Household Sector – these two values reflect socio-economic variables describing the material standard of living in the compartment of final consumption – the whole set of relation is illustrated in Fig. 13). We can recall here the analysis in Fig. 10 referring only to the flow of $/h – the difference between the pace of metabolism of the whole – GDP per hour of human activity – and the amount of GDP generated per hour of work in PW – ELP_{PW} – depends on the fraction HA_{PW}/THA.

4.1.2. *Moving from level n−1 (PW sector) to level n−2 (subsectors of PW)*

We can immediately see from this multi-level analysis of the metabolism of a country, that changes in the characteristics of the PW sector (the energy intensity of the productive part of the economy) cannot be related directly to changes in the technical coefficients of lower level compartments. Rather they reflect: (1) the differences in energy intensity *which are typical of the different subsectors making up the PW sector*; (2) the relative size in percentage of sectoral GDP of the three subsectors in determining the overall value for PW. The combination of these two factors will determine the intensity of the PW sector.

4.2. PUTTING THE VARIOUS PIECES OF THE PUZZLE TOGETHER

An overview of the various pieces of the puzzle defining an overall value of energy intensity of a country is given in Fig. 13. The figure clearly shows that using a simple ratio TET (Total Energy Throughput – the energy consumption of a society) over GDP, a choice often done in econometric analysis, as an indicator of performance of economies, in reality has the effect of generating a number which does not carry any meaning. With this we mean that the ratio "Energy consumption"/"GDP" does not have an external referent. When using this indicator TET/GDP if we analyze the energy intensity of Finland and El Salvador we find the same value: 12.5 MJ/US$.

On the other hand, we claim that if we use rather than the direct ratio TET/GDP a ratio between the two paces of metabolism of a country: (i) expressed in GDP per hour (flow of added value generated in the GDP per person per year); and (ii) MJ/h (aggregate amount of energy consumed per person per year at the level of the whole society), then we are dealing with benchmark values (external referents).

Since this distinction can result not clear to the reader we provide an explanatory example in Fig. 13. When calculating energy intensity of an economy using the ratio between two metabolic paces (GDP/THA and TET/THA) we deal with two benchmark values, which in turn can be related to other expected benchmark values. For example (i) a value of GDP of $2/h ($16,000/year p.c.) or an energy consumption of 20 MJ/h are typical of a developed country (in blue on the left of Fig. 13); as well as (ii) a value of GDP of $0.4/h ($3,500/year p.c.) or an energy consumption of 4 MJ/h are typical of a developing country (in red on the left of Fig. 13).

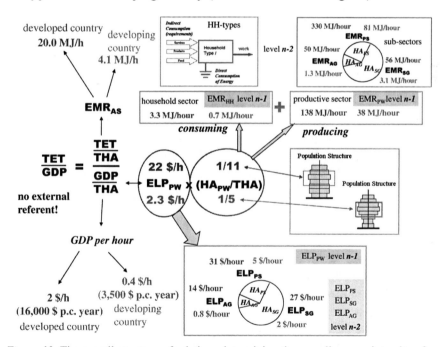

Figure 13. The overall structure of relations determining the overall energy intensity of an economy

If we perform a multi-level analysis of the relations over the values of different benchmarks, we can find the existence of *an integrated set of bench-mark values* (interlinkages!) referring to the intensity of flows per hour both for *Added Value* and *Total Energy Throughput* across hierarchical levels:

a. At the *level n* (whole countries) as indicated on the left of the figure - for developed (in blue) and developing countries (in red)

b. At the *level n−1* for the average production of GDP per hour of human activity in PW – the value of ELP_{PW} – given in the middle of the picture for developed (in blue) and developing countries (in red)

c. At the *level n −1* for the average ratio between hours of working activity/total hours of human activity (the ratio HA_{PW}/THA discussed above)

d. At the *level n−2* for the average value of flows of both €/h and MJ/h typical of sub-sectors of the economy. In Fig. 13 the value of ELP_i are given in the pie in the box on the lower-right corner – for developed (in blue) and developing countries (in red)

e. Both at the *level n−1* in relation to the flow of energy consumption per hour – in the division in boxes on the upper-right corner – the expected relation between the fraction of TET going into the sectors of production and consumption

f. At the *level n−2* in relation to the flow of energy consumption per hour – in the boxes on the upper-right corner – within the productive sector for individual subsectors – for developed (in blue) and developing countries (in red)

The values of these benchmark values have to be changed *in an integrated way* to generate the overall value of GDP/THA or TET/THA. On the contrary, the two hypothetical countries represented in Fig. 13 are consistently different across the different benchmarks of their metabolic pattern across levels, but they have the same energy intensity – 10 MJ/$ – when using the ratio TET/GDP [*the choice of standard econometric analysis!*].

Looking at this complex set of relations it is easy to understand that changes in technical coefficients (= better technology) can only change the value of one of these benchmarks in just one of the lower level elements. But any change in lower level elements will only imply a re-adjustment over the various benchmarks across different levels. There is no guarantee that a change at the lower level will arrive to affect the emergent property of the whole in a predictable way. As a matter of fact, in order to be able to distinguish the effect generated by technological changes one should move down to another hierarchical level – moving at *level n−3* and even at *level n−4* – to individuate a compartment expressing an homogeneous set of activities in which improvements in technical coefficients can make a difference in terms of generation of €/h and MJ/h.

Let's now observe – in Fig. 14 – a comparison between the four countries considered in Fig 12. This comparison is made by focusing only on the characteristics of the subsectors defined at the *level n−2*.

In this figure the same type of graph is used to compare – over the three chosen subsectors – the differences between the four countries considered in Fig. 12. From this figure it is easy to see that the differences *across typologies of sectors* over the three countries (e.g. energy intensity of PS sector versus energy intensity of AG) are much higher than the differences, *within the same sector over the three countries, due to gradients in technological performance.*

Exploring the metabolic pattern at level n-2

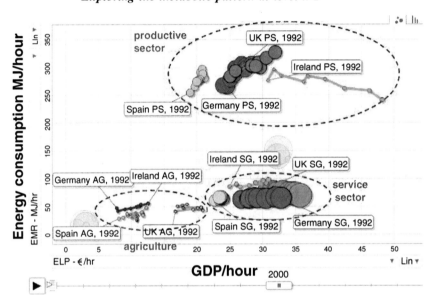

Figure 14. Characteristics of the metabolic pattern of sub-compartments of the economy of Germany, UK, Spain and Ireland (level n − 2)

In particular, we can notice that: (i) the PS sector is much more energy intensive than the others; and (ii) the energy consumption per hour of work in the PS sector is increasing in time everywhere (but in Ireland…). On the contrary the intensity of the service sector is much lower and decreasing in time.

Therefore, we can observe two important points: (i) technical changes are not generating a reduction of energy intensity in the PS sector; (ii) the overall reduction of energy intensity of PW is due to a progressive reduction of the weight of the PS sector in determining the average in PW. In fact, we know that both in terms of hours of working time in PW and in terms of relative proportion of the sectoral GDP the SG sector is continuously increasing its share in the PW sector. Again this confirms that developed societies change their metabolic pattern allocating more time and energy to the final consumption sector (HH) and within the PW sector they are allocating more working time and energy to the Service and Government sector.

As a matter of fact, when comparing the trend of change in energy intensity – as observed at the *level n*−1 (the characteristics of the PW sector, the side of production of the economy) and the characteristics of the SG sector at the *level n*−2 – as illustrated in Fig. 15 – we can immediately observe that it is the slope describing the direction of changes of the SG service which is determining the overall slope describing the direction of change of PW. However, it should be noted that this post-industrialization of developed economies depends on the phenomenon of globalization. That is, someone else must produce the energy and resource intensive products consumed but not produced by post-industrial countries.

Comparing the metabolic pattern across level n-1 and level n-2

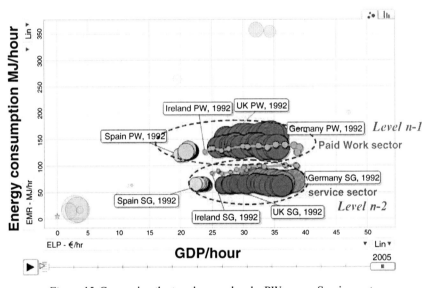

Figure 15. Comparing the trend across levels: PW versus Service sector

4.3. THE LESSON TO BE LEARNED FROM THIS INTEGRATED ANALYSIS OF METABOLIC PATTERN OF EU COUNTRIES

In his seminal book "Wealth, virtual wealth and debt" the Nobel Prize Frederick Soddy, in 1926, made the point that it is very dangerous to rely only on economic accounting to develop indicator of development of modern societies. In his book he explains that the very idea associated with money is that society will have to provide to the money holder either a product or a service of an equivalent price. Money "per se" should be associated with a debt that the society has with those holding money. Therefore, tracking flows of money means tracking flows of debt, this means that those studying monetary flows are studying only a reflection (as in the Plato metaphor of the cave) of real wealth. Real wealth is the biophysical process of *production and consumption* of actual goods and services which is behind the flows of money. The generation of 'real wealth', therefore, is necessarily affected by the biophysical constraints associated with thermodynamic transformations.

In spite of these unavoidable biophysical constraints, debts, such as loans, make it possible to spend money now that refers to goods and products which have not yet been generated. The money that is spent now is based on the promise – e.g. made by someone that buys a house – that she/he will pay back the credit. However, the buyers can pay back the debt only if they will be able to earn money. To do that, they will have to be able to take part in economic activities which will be performed in the future earning either wages or profits. Here the discrepancy between real economy and virtual economy indicated by Soddy becomes relevant. If the pie keeps expanding, then these economic activities will keep increasing, also in biophysical terms. But when the wealth is generated by stock depletion (extracting and consuming oil and other non-renewable resources) we will end up in the paradox indicated by Soddy: the virtual economy based on the accounting of monetary flows will perceive as economic growth any massive increase of debt – which many will confuse with an increase of wealth – at the very moment in which the resources are depleted – the real economy is reducing its original endowment of wealth. The relevance of his analysis to the situation that is experienced in these years in Europe and in the world is clear.

As soon as we start looking at the structural and functional organization required to be able to generate the processes of production and consumption of a given set of goods and services, then we can immediately see the existence of trade-offs in globalization and specialization. A high level of specialization (taking advantage of what is called in economics as "comparative advantages") implies a strong dependence on the assumption about the viability and possibility of importing and exporting. This is at the basis of the modern economic civilization based on market economy. However, this process of specialization reached in the last decade a scale and a degree of openness, which was never experienced before at the level of the whole planet. Some small countries can afford to take this risk (e.g. like Ireland, since they are protected by the size of EU), but how risky is a total reliance on imports for a big macro-economic entity such as the EU?

As a matter of fact both UK and Ireland that got the maximum advantage in terms of reduction of energy intensity from the process of externalization are now the most affected by the global crisis. We can recall here the famous sentence of Mahatma Gandhi about the possibility of having a major economic development in India following the example set by developed countries: "If it took Britain to use the resources of half the world to be where it is, how many worlds would India need?"

Looking at the EU context, an excessive reliance on imports and externalization could imply an increasing fragility in the case of a big world crisis (in the case nobody else is capable of exporting and importing) or a sudden contraction of trade due to various reasons – increasing costs of transportation or inability of guaranteeing safe transportation routes. Another problem may be represented by future bottlenecks in the production of critical imports due to shortage of natural resources. Finally, in the long term, all these risks will be further increased by the increasing internal demand expected to grow within developing countries. That is, what if China in 20 years will be able to express and internal demand so high which will absorb the majority of its production (the very same production which is exported at the moment), and because of this fact, will stop exporting cheap to developed countries?

References

Ahl, V. and Allen, T.F.H. 1996. *Hierarchy Theory*. Columbia University Press, New York.

Allen, T.F.H., Tainter, J.A., Pires, J.C. and Hoekstra, T.W. 2001. Dragnet Ecology, "Just the facts Ma'am": the privilege of science in a post-modern world. *Bioscience* 51:475–485.

Box, G.E.P. 1979. Robustness in the Strategy of Scientific Model Building. In: R.L. Launer and G.N. Wilkinson (Editors) *Robustness in Statistics*, Academic, New York, pp 201–236.

Funtowicz, S., Martinez-Alier, J., Munda, G., Ravetz, J. 1999. Information tools for environmental policy under conditions of complexity, *European Environmental Agency, Experts' Corner*, Environmental Issues Series, No. 9.

Giampietro, M. 1997. 'Linking technology, natural resources, and the socioeconomic structure of human society: A theoretical model'. In: L. Freese (Editor) *Advances in Human Ecology*, vol 6, JAI Press, Greenwich, CT, pp 75–130.

Giampietro, M. (guest editor) 2000. Societal metabolism, part 1: Introduction of the analytical tool in theory, examples, and validation of basic assumptions, *Population and Environment*, special issue, vol 22, no 2.

Giampietro, M. (guest editor) 2001. Societal metabolism, part 2: Specific applications to case studies, *Population and Environment*, special issue, vol 22, no 3.

Giampietro, M. 2003. *Multi-Scale Integrated Analysis of Agro-ecosystems*. CRC Press, Boca Raton, FL, 472 pp.

Giampietro, M. 2008. 'Studying the "addiction to oil" of developed societies using the approach of Multi-Scale Integrated Analysis of Societal Metabolism (MSIASM)'. In: F. Barbir and S. Ulgiati (Editors) *Sustainable Energy Production and Consumption and Environmental Costing*, NATO Science for Peace and Security Series: C–Environmental Security, Springer, The Netherlands.

Giampietro, M. 2009. The future of agriculture: GMOs and the agonizing paradigm of industrial agriculture. In: A. Guimaraes Pereira and S. Funtowicz (Editors) *Science for Policy: Challenges and Opportunities*, Oxford University Press, New Delhi.

Giampietro, M. and Mayumi, K. 2000a. 'Multiple-scale integrated assessment of societal metabolism: Introducing the approach', *Population and Environment*, vol 22, no 2, pp 109–153.

Giampietro, M. and Mayumi, K. 2000b. 'Multiple-scales integrated assessments of societal metabolism: Integrating biophysical and economic representations across scales', *Population and Environment*, vol 22, no 2, pp 155–210.

Giampietro, M. and Mayumi, K. 2009. *The Biofuel Delusion: The Fallacy of Large-Scale Agro-Biofuel Production*, Earthscan, London 320 pp.

Giampietro, M. and Ramos-Martin, J. 2005. 'Multi-scale integrated analysis of sustainability: a methodological tool to improve the quality of the narratives', *International Journal of Global Environmental Issues*, vol 5, no. 3/4, pp 119–141.

Giampietro, M., Bukkens, S.G.F. and Pimentel, D. 1997. 'The link between resources, technology and standard of living: Examples and applications', in L. Freese (ed), *Advances in Human Ecology*, vol. 6, JAI Press, Greenwich, CT, pp 129–199.

Giampietro, M., Mayumi, K. and Bukkens, S.G.F. 2001. 'Multiple-scale integrated assessment of societal metabolism: an analytical tool to study development and sustainability', *Environment, Development and Sustainability*, vol 3, no 4, pp 275–307.

Giampietro, M., Allen, T.F.H. and Mayumi, K. 2006. The epistemological predicament associated with purposive quantitative analysis *Ecological Complexity*, vol 3, no 4, pp 307–327.

Giampietro, M., Mayumi, K., and Ramos-Martin J. 2006a. 'Can biofuels replace fossil energy fuels? A multi-scale integrated analysis based on the concept of societal and ecosystem metabolism: Part 1', *International Journal of Transdisciplinary Research*, vol 1, no 1, pp 51–87, www.ijtr.org/

Giampietro, M., Mayumi, K., and Ramos-Martin J. 2006b. 'How serious is the addiction to oil of developed society? A multi-scale integrated analysis based on the concept of societal and ecosystem metabolism: Part 2', *International Journal of Transdisciplinary Research*, vol 2, no 1, pp 42–92.

O' Connor, M., Faucheux, S., Froger, G., Funtowicz, S.O., Munda, G., 1996. Emergent complexity and procedural rationality: Post-normal science for sustainability. In: R. Costanza, O. Segura, J. Martinez-Alier (Editors) *Getting Down to Earth: Practical Applications of Ecological Economics*, Island Press/ISEE, Washington, DC, pp 223–248.

Ramos-Martin, J., Giampietro, M. and Mayumi, K. 2007. 'On China's exosomatic energy metabolism: an application of multi-scale integrated analysis of societal metabolism (MSIASM)', *Ecological Economics*, vol 63, no 1, pp 174–191.

Rosen, R., 1977. Complexity as a system property. *International Journal of General Systems*, vol 3, pp 227–232.

Rosen, R. 1986. *Anticipatory Systems: Philosophical, Mathematical and Methodological Foundations*, Pergamon Press, New York.

Rosen, R. 2000. *Essays on Life Itself*, Columbia University Press, New York, 361 pp.
Simon, H. A. 1962. The architecture of complexity. *Proceedings of the American Philosophical Society*, vol 106, pp 467–482.
Simon, H.A. 1976. From substantive to procedural rationality. In: J.S. Latsis (Editor) *Methods and Appraisal in Economics*, Cambridge University Press, Cambridge.
Soddy, F. 1926. *Wealth, Virtual Wealth and Debt*, George Allen & Unwin, London.

SEVEN POLICY SWITCHES FOR GLOBAL SECURITY

JAMES GREYSON[*]
BlindSpot, Lewes, UK

Abstract   Everyone desires a secure life. Yet the security of more and more regions is undermined by unreliable and unequal availability of basics such as energy, water, food, natural resources, funds, co-operation, trust and hope for the future. Shocks such as the credit crunch, infectious diseases, climate instability and ecological collapses are converging towards a 'planet crunch' where security would become a fond memory. Traditional policy-making, that manages problems separately and incrementally, offers only the illusion of protection against impending unaffordable and irreversible shocks affecting all people. Future security anywhere requires all facets of security everywhere. This 'global security' ambition can be sought with a new era of policy-making that encompasses the indivisibility, scale and urgency of all planet crunch issues. This paper offers a selection of seven simple 'policy switches' (or 'leverage points' in complex systems). Each policy switch offers an expanded vision of people's role on Earth and a whole-system change to implement it. Together the switches define a practical strategy for global security, for a serious attempt at revival of co-operation, ecosystems and prosperity. The proposed policy switches are: (1) The strategy of aiming to reduce problems can be switched to reversing them with 'positive development'. Less bad is not good enough. (2) Education can inspire a culture of joined-up thinking and engagement by switching from predetermined to curiosity-led learning. (3) Economic growth can be switched from consuming the basis for further growth to building it by correcting markets with 'precycling insurance'. (4) Rapid global disarmament can be launched by switching from Gross Domestic Product to 'Gross Peaceful Product', that omits weapons-related transactions. (5) Exploitive commodification of the Earth's surface can be switched to guardianship by international treaty that interprets ownership in terms of responsibility to future generations. (6) Surplus accumulations of financial wealth, which would be wiped out by the planet crunch, can be switched by the wealthy

---
[*] To whom correspondence should be addressed: James Greyson, BlindSpot, PO Box 140, Lewes BN7 9DS, UK, E-mail: james.greyson@blindspot.org.uk

F. Barbir and S. Ulgiati (eds.), *Energy Options Impact on Regional Security*,
DOI 10.1007/978-90-481-9565-7_3, © Springer Science + Business Media B.V. 2010

into investments that sustain all forms of wealth. (7) Global financial stability can be regained by switching money creation from the private sector to central public authorities and local currencies.

Keywords: Energy, climate, regional security, global security, credit crunch, planet crunch, revival, policy switches, vision, positive development, joined-up thinking, systems thinking, curiosity-led education, precycling insurance, Gross Peaceful Product, ownership, guardianship, wealth, philanthropy, money creation, money tree, local currencies

1. Security means Global Security

1.1. SECURITY INCLUDES NON-MILITARY THREATS

Security in the modern world means far more than military security and radical new non-combative solutions are needed to cope with new security challenges. This was highlighted by Professor Sir Brian Heap (2009), former UK Representative on the NATO Science Committee on the occasion of NATO's 60th anniversary: "security includes non-military threats arising from incompetent governance, corruption, organised crime, insecure borders, ethnic and religious conflict, proliferation of weapons of mass destruction, shortage of natural resources and, of course, terrorism." Traditional narrow concepts of security are obsolete. The future security of individuals, regions and nations requires a broad 'global security' vision that encompasses rapid effective solutions to all major economic, social and ecological challenges.

1.2. GLOBAL SECURITY IS NOT HAPPENING

The ultimate threat to global security is not to be found among the long list of specific challenges; rather it is that the moment of opportunity for effective action passes whilst humanity is otherwise occupied. According to UNEP executive director Achim Steiner, "political efforts to curb pollution, protect forests and avert climate change have proven totally inadequate" (Reuters 2008). Numerous other challenges are worsening, including insecure and unequal availability of energy, water, food, natural resources, funds, co-operation, trust and hope for the future. More than 50 states are already 'fragile' (United Nations University 2008a) and all others depend upon complexities, energy dependence and running costs that cannot be sustained. The credit crunch is becoming a 'planet crunch' (Greyson 2009b) of mounting instabilities and multiple converging shocks that threaten everyone.

1.3. THE AGE OF STUPID

A movie documentary released in March 2009 called 'The Age of Stupid' (Armstrong 2009) asked how humanity knew that climate change could make life unlivable and yet was entirely ineffective at solving it. This applies equally to every other planet crunch issue. Albert Einstein counselled, "We cannot solve today's problems with the same kind of thinking we used when we created them." Problems that are divided up to suit society's specialisms may appear more manageable yet global problems are indivisibly joined-up and codependent (Greyson 2008). Reductionism hasn't worked and 'solutions' devised within geographical, sectoral, organisational or symptomatic policy silos are inadequate, futile and illusory. The opportunity of seeking global security as a whole, for everyone, is neglected. The imperative of joined-up thinking and joined-up policy-making is a world-wide blindspot.

1.4. HOW TO FAIL

Failure to achieve global security will mean the collapse of modern civilisation, just as surely as the fate of past civilisations that were not sustainable. The timescale of failure is not predictable but likely to be abrupt due to troubled global systems having positive feedbacks (problems causing further problems) and 'tipping point' thresholds that trigger cascading shocks. Failure is assured by continuing to seek each facet of security separately and by managing problems with patchwork policies (Greyson 2008) devised in policy silos. The pursuit of ever more complex, expensive and tough controls on a barrage of worsening symptoms neglects the underlying circumstances that continue to cause those symptoms. Initiatives are considered practical and viable if they fit the same world-views and assumptions that perpetuate the problems. Security has been sought where it is ultimately unavailable, within financial, geographical and organisational 'bubbles' where some goals are temporarily met for some people.

1.5. THE PLANET CRUNCH PROCEEDS UNCHALLENGED

The potential loss of all that is valued is more than our minds can admit. Psychological self-defense diverts attention away from the overwhelming reality of the planet crunch and towards theatrical debates about special-interest topics such as emissions, weapons or economic growth. Roles that are played include defending and opposing the status quo, announcing tokenistic 'breakthroughs', promoting 'white elephant' investments, denial, fatalism and distraction by trivia. Newspapers and TV publicise and review the performances. Researchers provide data and ideas for scripts. Policy-makers produce, direct, act and applaud, doing whatever it takes to 'keep the show on the road'. Each successive shock becomes the new hot topic. The flourishing of drama at this evolutionary crisis point should not be mistaken for an effective response. Every converging threat of the planet crunch proceeds unhindered by any plausible challenge.

1.6. HOW NOT TO FAIL

As Einstein suggested, we can think differently to solve today's problems. The planet crunch is paradoxically less overwhelming and more manage-able when approached as a whole. Churchman (1979) prescribed a 'systems approach' where "...no problem can be solved simply on its own basis." Anyone can practise joined-up or 'systems' thinking. This can start with awareness that the selective targeting of intellect and compassion is the source of problems, not the solution. The relentless complexity of the planet crunch can be managed at 'leverage points' "where a small shift in one thing can produce big changes in everything" (Meadows 1999). Leverage points are here described as 'policy switches' to emphasise that the purpose is practical not academic. Carefully-designed policy switches are ideal for situations where rapid progress is necessary with big issues that have previously resisted meaningful progress, including all facets of global security. Obsolete paradigms can be changed rather than just constrained or accommodated.

1.7. MAKING THE SWITCHES

The switches do not neatly match up with symptomatic problems; for example there is no particular switch for climate change, energy, population, poverty or health. Each of these can be tackled by making all the policy switches and by the further actions that would then become viable. Policy-makers have previously been asked to do the impossible – to solve problems within policy silos. Thus the biggest problems remain unsolved. Now politics can demonstrate its relevance to people's lives by collaboration to create the circumstances for global security. The policy switches have the effect of aligning the self-interest of individuals and institutions with the shared imperative of a world that works, so the incentives are to do more rather than to resist change. The potential speed of an international response is illustrated not by the historical glacial pace of agreement on matters such as climate and disarmament but by the comparatively instant international action to bail-out the financial sector.

2. First Policy Switch: From a Strategy of Reducing Problems to Reversing them

2.1. LESS BAD IS NOT GOOD ENOUGH

The default strategy during decades of persistent global problems has been incrementalism – planning for 'less bad'. However, less bad has proven not to be good enough. For example, incremental planning to cut waste has produced net increases in waste and incremental planning to cut emissions of greenhouse gases (GHG) has produced net increases in emissions. The consequent continuing global loss of resources and rise in GHG concentrations is removing the potential for future security. Incrementalism was plausible when gradual long-term problems seemed to require gradual long-term solutions. Unfortunately this didn't solve any global issue. Today's critical problems invite immediate switching to another strategy on an entirely new scale of ambition and effectiveness.

2.2. THE ILLUSION OF PROGRESS

The urge to advance is so fundamental that there is a tendency to imagine progress and development even when it isn't happening. The looming planet crunch reveals a civilisation that has lost its way, where self-interest is misdirected to make things worse for everyone rather than better. *Is it still progress when a billion people go to sleep hungry? When ecosystems are exploited to the point of collapse? When debts outpace incomes? When nations seek peace and security behind walls of weapons? When accumulating waste gases re-approach the inhospitable atmosphere of the primordial past?* The planet crunch is progress in reverse, with systematic losses of financial, societal and ecological stability that undermine any realistic prospect of security in any region. Civilisation can proceed only with a new understanding of what it means to develop.

2.3. WHICH WAY FOR GROWTH?

Economic growth, the increase in income of nations, is the political icon of progress and development. Positive growth means more economic activity and growing tax revenues for government. However, the inventor of national income statistics, Simon Kuznets (1934) was the first to point out that growth was not designed to measure progress: "The welfare of a nation can scarcely be inferred from a measurement of national income...". Growth provides no protection against running an economy that systematically removes the potential for future growth and progress. The innumerable consequences of the planet crunch are expensive to cause, mitigate and adapt to, and all that spending contributes to economic growth but not progress. If the economy is like a vehicle then growth displays the changing speed but says absolutely nothing about which way it's going.

2.4.   NO GROWTH IS NO ANSWER

There are three possible strategies for future growth. The **first** is less-bad growth, funded by rising ecological and financial debt; 'greener', 'cleaner', 'responsible' adjustments to today's activities. This is the default incremental strategy that is still promoted on the world stage despite its record of reinforcing rather than challenging conventional paradigms. The **second** strategy sees planetary destruction as the only possible outcome of continued growth; it sees the failures of markets but not their potential. It calls for the goal of economic growth to be abandoned and for markets to be constrained with centralised caps (or fixed limits) on resources and emissions (Jackson 2009). However, growth ignores resources, emissions and destruction; it is interested solely in the added-up financial value of economic activity. Just as a bad diet cannot be corrected by limiting the grocery bill, no-growth limits on economic activity would do nothing to inspire the necessary flourishing of valuable new patterns of activity. A scarcity mentality is no answer to the world's growing scarcities.

2.5.   AIM TO REVERSE NOT REDUCE PROBLEMS

Any future for growth requires a **third** strategy. People are not inherently destructive and economic activity need not remain dependent on exploiting people, planet and the potential of the future. The economic vehicle need not remain stuck in reverse, making reverse progress. Janis Birkeland (2008) offers the third strategy option of 'positive development': "The view that negative impacts are an inevitable consequence of development has blinded us to the obvious. We could design development to increase the size, health and resilience of natural systems, while improving human health and life quality." This strategy is applicable to every planet crunch issue. For example, international climate talks have pursued a less-bad strategy of lower emissions (flows to atmosphere) when the crucial target (Hansen 2008) is lower concentrations (stocks in atmosphere), which are potentially achievable by positive development. All global problems must be reversed, not just worsened less fast.

2.6. THE PRIMARY LIMIT IS IMAGINATION

Anyone whose car is drifting backwards towards a cliff edge knows that the strategy for success is not to go slower or more steadily, but to change into forward gear and accelerate away safely. Humanity is speeding towards economic, social and ecological cliff-edges so why is attention absorbed by the decoy strategies of less-bad and no-growth? Less-bad is an appealing strategy for those focused on awareness and political will as limiting factors. The aim of reducing damage can be widely agreed and endlessly debated, with all participants 'doing what they can'. Those who focus on the limits of nature's capacity to accommodate human activity are attracted to the no-growth vision of tough government-enforced boundaries to contain unsustainable aspirations. Positive development offers both groups the opportunity to unite society and markets in achieving far more than just limiting further damage. This strategy is limited not by politics nor by nature, but by imagination.

2.7. REAL LASTING VALUE

Positive development goes further than not making things worse. It invites attention to the neglected stockpiles of financial debt (personal, corporate and national), ecological debt (such as lost nature and surplus concentrations of GHG) and social debt (such as overpopulation, surplus concentrations of weapons, habits of conflict and surplus concentrations of wealth). These combined 'debts' reveal the extent of civilisation's self-harm and must be promptly 'paid back' to ensure any form of future security. All this activity should not be viewed as a cost but as investments in the future that also boost current economic growth. Illusory progress and invented financial value can be replaced with real lasting value. The following policy switches can, if all are implemented soon enough, rapidly institute positive development world-wide.

3. Second Policy Switch: From Predetermined to Curiosity-Led Learning

3.1. ESCAPING THE OLD IDEAS

The remaining policy switches could enable positive development to become the defining vision of a new era of co-operation and abundance. Or the moment of opportunity could pass unnoticed amidst a predictable escalation of chaos. Which outcome depends not on any technical obstacles that shape the bounds of possibility but on the frame of mind of those who encounter the opportunity. The opportunity will be either dismissed as too hard or pursued into practice depending on the balance between the comfortable familiarity of old habits of thought and openness to the new. John Maynard Keynes (1936) prefaced his General Theory, "The ideas ... are extremely simple and should be obvious. The difficulty lies not in the new ideas, but in escaping the old ones."

3.2. CURIOSITY KILLED THE CATASTROPHE

The future is currently predetermined along 'tram-lines' of unreliable assumptions that are taken for granted rather than actively considered. The innately flexible human mind must be enabled to respond with sufficient creativity for the future to be instead determined consciously. This can be achieved by attending to the missing ingredient in both education and policy-making – curiosity. Lack of curiosity about the availability of options allows the planet crunch to proceed whereas a blossoming of curiosity would enable the rigorous, creative, joined-up thinking needed to elude impending catastrophes. "When we experience curiosity we are willing to leave the familiar and routine", according to psychologist Todd Kashdan (2009). Today's predicament requires an immediate awakening of curiosity on a planetary scale.

3.3. TEACHING DISENGAGEMENT

Any society that values its ability to face the future can allow its learning and education to be led by curiosity rather than the delivery of 'right answers'. "Kids start out creative but we lose it at school" was a 9-year old girl's comment recorded during the author's work in the UK government's flagship creativity project for government-funded schools (Greyson 2009a). Habits of creative thought cannot be cultivated by assuming that inquisitive young minds must be moulded into established patterns of thinking. In modern centrally-planned education, knowledge is chopped into lesson-sized chunks, pre-packaged and fed to children. Success is measured by children's acquiescence in first 'swallowing' and then 'bringing up' facts and skills when probed with tests. Politicians then wonder why so many people feel alienated and disengaged.

3.4. IN PURSUIT OF KNOWLEDGE

George Bernard Shaw long ago paraphrased the necessary switch; "what we want to see is the child in pursuit of knowledge and not knowledge in pursuit of the child." Schools that have made this switch, such as Lewes New School in England (Kettles 2009), simply allow learning to follow the curiosities of the class, which range broadly and deeply across the entire curriculum and the possibilities of our time. The UK's 5 year long Nuffield Review (Pring et al. 2009) of secondary education and training explains that teaching should be "an engagement of minds" not "delivery of a curriculum devised elsewhere for transmission to the learners". As role models for the fascination of discovery, teachers can facilitate an endless flow of learning. Children who experience this system get the same (or better) skills as other children, including strong personal and social problem-solving skills, but they do not get pre-packaged thinking.

3.5. AN ERA OF THINKING BIG

If education were to make this switch, society would be instituting a culture of creativity and innovation. The quality of ideas would rise along with the quality of participation in decision-making at all levels. Governments would find that engagement rather than control tackles disruptiveness both in the classroom and in society. The herd thinking that underlies both the credit crunch and the planet crunch would be perpetuated no longer. Any nation instilling a culture of curiosity would gain such a competitive advantage that all nations would be galvanised to follow. People everywhere could echo the words of another primary school student reflecting on the author's sessions in her school: "I learnt that I could have big ideas" (Greyson 2009a).

3.6. JOINED-UP SOCIETIES

Getting unstuck mentally is a change of mind and need not be a struggle. By contrast the mind that lacks flexibility is in a state of continual struggle that is expressed in real-life stresses and struggles. Conflict, crime, terrorism, anti-social behaviour and many other ailments are products of a compartmentalised, fatalistic, 'us versus them' world-view that excludes wider possibilities and perspectives. Fragmented thinking brings fragmented societies. A spread of curiosity, from governments switching educational models, to newspapers reporting it, to parents and children experiencing it could spark a cultural renaissance where populations surprise themselves with what they can achieve co-operatively.

4. Third Policy Switch: From Consuming to Building the Basis for Economic Growth

4.1. LINEAR ECONOMICS

Four decades ago the economist Kenneth Boulding (1966) wrote about the "reckless, exploitative and violent behaviour" associated with the mythical possibility of endless frontiers available to be claimed and fouled. He poetically called this the 'cowboy economy', though today it is commonly called the 'linear economy', to envisage the default economic vision of a conveyor belt of resources becoming wastes (Leonard 2008). All forms of wealth and security including; climate stability, co-operation, trust, bio-diversity, ecosystem services, resource availability, soil fertility, air and water purity, health, sharing and democratic accountability are depleted by the systemic error of running a linear economy. Linear economics consumes the basis for future growth so what is now growing fastest is unproductive activity, inactivity and instabilities. The credit crunch marks the withdrawal of faith in growth-as-usual and any reliable revival of growth and prosperity requires a switch of vision.

4.2. CIRCULAR ECONOMICS

Boulding envisaged the economy taking part in a "cyclical ecological system which is capable of continuous reproduction of material form even though it cannot escape having inputs of energy." This is not academic: China's 11th 5 year plan for 2006–2010 established a national goal of circular economics, "It is an overall, urgent and long-term strategic task for China to vigorously develop the circular economy" (Zhou 2006). The future for growth is circular economics where more economic activity would mean a faster pace of change away from waste-making and towards looking after the world and all its inhabitants. This would preserve and regenerate material value, co-operation and natural capital instead of losing it, so growth would work to build the basis for more growth. Today this may appear idealistic. Yet if circular economics was already practiced, and people were accustomed to prosperity based on resource security, then any proposal to adopt an exploitive self-defeating vision would be laughable.

4.3. PRECYCLING

Economic dependence on waste is perpetuated by managing waste primarily as an addiction to disposal, "how can we get rid of all this junk?" The 'waste hierarchy' (reduce, reuse, recycle, then dispose) that has been available since 1975 (European Union 2008) is commonly quoted but in practice the bulk of effort and funding provides for continuing long-term disposal to ecosystems (by landfill, waste-burning and pollution). The waste hierarchy is being used backwards and no nation has yet attempted to create the incentives for an economy that grows from the work done to end waste dumping and implement circular economics. This is achievable with the concept of 'precycling' (O'Rorke 1988) originally used for public waste education. Precycling is applicable throughout an economy (Greyson 2007) and may be understood as *action taken to prepare for current resources to become future resources*. The 'pre' prefix emphasises that this cannot be arranged after something becomes waste; it must be done beforehand. The scope of action extends far beyond recycling, to creating the economic, social and ecological conditions for all resources to remain of use to people or nature.

4.4. PRECYCLING INSURANCE

A simple economic tool is available to switch from linear to circular economics and from dumping waste to dumping the habit of wasting. This tool internalises diverse externalities efficiently within markets by paying the price of preventing problems instead of the larger or unaffordable price of not preventing them. Precycling insurance is an extension of the EU WEEE Directive's 'recycling insurance' (European Union 2002) from just recycling to all forms of preventing all products becoming waste in any ecosystem. This allows a single economic instrument to work with the issues at every stage of product life-cycles. Significant producers would be obliged to consider the risk of their products ending up as waste in ecosystems and to retain responsibility for insuring against that risk. Suitable design principles for precycling insurance have been fully outlined in the NATO Science Programme (Greyson 2008).

4.5. 'LIFE INSURANCE' FOR PRODUCTS AND PLANET

Precycling insurance is a form of regulation to be set-up in every nation but not centrally planned. The volume of regulation can be cut but its effective-ness drastically boosted. For example, emissions can be cut rapidly with no need for any further ineffectual negotiations about capping. Unlike taxes, the premiums from precycling insurance would not be handled by govern-ments (whose role would be to legislate, monitor and ensure full public transparency). Unlike conventional insurance, the premiums would not be collected up and then paid out following (potentially irrecoverable) planet crunch shocks. Premiums would be distributed by insurers and invested preventively throughout society, to cut the risk of resources being lost as wastes. Support would be provided for the dialogue, understanding, participation, capabilities, designs, efficiencies, facilities and ecological productivity needed to return used matter as new resources for people and for nature. Today's resources would feed tomorrow's economy.

4.6. A FREE MARKET IN HARMONY WITH NATURE

Precycling insurance would switch the power of markets to reversing the planet crunch. The speed and scale of change would exceed the expectations of all who are accustomed to ineffectual controls designed to make markets less-bad. All market participants (such as buyers, sellers, investors and governments) would adapt their decisions to the new incentives, profiting by addressing actual needs rather than superficial consumerist wants. Producers would remain free to choose how to meet customers' needs without waste, and even free to continue making wasteful products, in competition with other producers cutting their costs (including precycling insurance costs) by cutting their product's waste risk. Economic growth would no longer be a competitive scramble between people rushing to acquire and discard ever more resources from an every-shrinking stock. The economy would prosper in harmony, rather than in conflict, with nature.

4.7. SHRINKING MATERIAL AND ENERGY DEMANDS

The material requirements of today's linear economy would rapidly shrink since the new incentives would lead to the most needs being met with the least materials moved the least distance and then regenerated rather than dumped. The energy requirements of today's linear economy would rapidly shrink since a smaller material flow with higher quality materials closer to where they are needed requires less energy to process. For example, a factor 10 improvement in resource productivity would dampen energy requirements by up to 80% (Schmidt-Bleek 2008), putting renewables within easy reach world-wide and putting waste-making energy sources (such as new coal-fired plants, nuclear, food or forest-consuming biofuels and mixed-waste incineration) back on the shelf. Shrinking energy dependence is the key to energy security, economic recovery, climate restabilisation and prevention of conflict over diminishing non-renewable resources. The resource and energy efficiency of circular economics makes it realistic to plan the necessary reductions in GHG concentrations (ie net-negative emissions).

5. Fourth Policy Switch: Reversing an Unintended Incentive for Conflict

5.1. MILITARY SECURITY CAN'T BE BOUGHT

Global military spending in 2008 is estimated at $1464 billion (SIPRI 2009), an increase of 45% since 1999. Yet the acquisition of military security primarily by military spending is an unobtainable illusion. "One only need consider the enormous expenditures the United States has made to counter the threat posed by improvised explosive devices (IED). The United States has spent literally billions to counter these crude, inexpensive, and extraordinarily effective devices. If one were to multiply this ratio against a global enemy, it becomes unexecutable." (US Joint Forces Command 2008). Military personnel and nations can be protected not with expanded budgets for more weapons but with an expanded vision.

5.2. SECURITY MEANS GLOBAL SECURITY

A 2007 speech by the NATO Secretary General (de Hoop Scheffer 2007) set out a new preventive strategy for global security, "...our prevailing security paradigm has shifted. And the new paradigm can be summed up in just one word: engagement. We need to address the issues where they emerge, before they end up on your and my doorstep... NATO must be prepared to address security challenges at their source, whenever and wherever they arise." The old paradigm of high dependence on armed engagement with problems that were previously not prevented is a strategy for bankruptcy in a world of unwinnable conflicts. Effective engagement with the sources of conflict requires a new set of policies that works beyond traditional 'us versus them' security analysis. It requires unprecedented investment of intellect, compassion and money in every facet of security for all people. The new security paradigm must be global security.

5.3. ESCAPING CYCLES OF CONFLICT

If circular economics is implemented before planet crunch issues become irreversible then many of the current threats to international security would fade. However investment in a secure future would still be starved by the massive global spending on weapons. Rising weapons spending feeds a cycle of ever-stronger cultural dependence upon force and ever more suspicion between communities. Funds spent on threat and counter-threat are lost to the new security paradigm. The level of weapons spending reveals the degree of dependence on the old security paradigm and the use of weapons reveals the extent of failure to engage preventively with today's problems. All the campaigns on arms control, and all the diplomatic attention to the proliferation of conflicts have been unable to break the global cycle of conflict. However a simple macro-economic correction is available to directly switch security paradigms and to implement a replacement cycle of self-reinforcing peace and security.

5.4. SHOULD GROWTH INVITE GROWING CONFLICT?

The national income statistics used to measure economic growth include a perverse incentive in favour of greater economic and military dependence on weapons. Gross Domestic Product (GDP) currently includes weapons-related transactions so nations with high dependence on weapons gain higher GDP and higher political status despite accumulating armaments indicating poorer prospects. Politicians aspiring to boost economic growth cannot ask the wider economy to earn more but they can choose to spend more on weapons and indulge in military adventures. Spending on weapons feeds a cycle of conflict as other nations feel obliged to respond to defend themselves. Individual security efforts add up to a collective absence of security. The vast investments in preparing for the worst are unavailable for global security, thus ensuring worst-case outcomes.

5.5. GROWTH AS A 'WEAPON' FOR PEACE AND SECURITY

A replacement cycle of global disarmament and global security can be invited by switching from Gross Domestic Product to 'Gross Peaceful Product' (Greyson 2008) that simply omits weapons-related transactions. Nations can implement GPP as a diplomatic statement of intent to build a more secure world, and as a badge of peace. GPP provides a replacement benchmark for the economic growth of all nations in which higher GPP and higher growth more accurately indicates improved future prospects. Nations adopting Gross Peaceful Product would be rewarded for lower reliance on weapons with comparatively higher economic growth. The new security paradigm of global security would become real by being funded. A self-reinforcing cycle of less weapons spending, less armed threat and more co-operation would be instituted internationally.

5.6. GPP BY INTERNATIONAL AGREEMENT

GPP is the simplest of all proposals for GDP adjustment and requires no estimates or predictions of the costs of damage, yet it provides decision-makers with a powerful incentive. GPP can be presented on the world stage by any nation or any international body desiring peaceful international relations. It provides the means to implement the long-awaited commitment in Chapter 26 of the United Nations Charter (UN 1945) where member nations agreed "to promote the establishment and maintenance of inter-national peace and security with the least diversion for armaments of the world's human and economic resources". In a time of recession, the savings from a shrinking global dependence upon weapons spending would release scarce public funds to boost growth by lowering the tax burden and stimulating spending on productive approaches to security.

5.7. SPREADING A CULTURE OF NON-COMBATIVE PROBLEM-SOLVING

GPP does not inhibit any nation's military defence choices; in fact it makes national security more achievable and affordable by actively spreading a non-combative problem-solving culture world-wide. Problems that are prevented or resolved co-operatively never reach the stage of requiring armed threat and bloodshed. Nations would see other nations switching investments from the old security paradigm to the new. Those who have felt abandoned and alienated would gain hope from tangible opportunities for collaborative engagement. Agitated young men in every country, prone to carrying knives or worse, would see governments practising what they preach about non-violence and collaborative values. Terrorist recruiters and street gangs would progressively lose their recruiting power. Peace would be given a chance.

6. Fifth Policy Switch: Including Guardianship within Ownership of the Earth

6.1. THIS ONE'S FINISHED, CAN WE HAVE A NEW PLANET PLEASE?

A study involving more than 1,360 experts worldwide over 4 years warned of an "increasing likelihood of nonlinear changes in ecosystems including accelerating, abrupt, and potentially irreversible changes" (Millennium Eco-system Assessment 2005). The latest international scientific synthesis report about climate (International Alliance of Research Universities 2009) warns of "an increasing risk of abrupt or irreversible climatic shifts". This means that humanity is undefended against the day when critical ecosystem services are no longer available and not replaceable at any cost. If so, people will have won every individual battle against nature and then suddenly, tragically lost the war. Nature would endure but civilisation would not.

6.2. VALUING NATURE?

Pavan Sukhdev, author of the EU-commissioned study The Economics of Ecosystems and Biodiversity (TEEB 2008) reports that the world is losing more wealth from the disappearance of forests than from the credit crunch, "...at today's rate we are losing natural capital at least between US\$2–5 trillion every year." The relentless conversion of nature to cash provides only a facade of wealth-creation that masks the reality of collective impoverishment. Historically much effort has been devoted to nature conservation but action is now needed with unprecedented speed and effectiveness. Schemes to value nature by paying for it (including the above precycling insurance) can help but they risk reinforcing the commodification of the Earth. All such schemes are up against continuing large-scale exploitation and destruction that excuses itself very simply by saying, "It's mine".

6.3. BELONGING

"A person lives on the land for a brief time and is gone, but the land endures. So people must be careful to preserve it - to live by the old Native saying that, 'The real owners of the land have not been born yet." Among Native people, the land and all that grows upon it is treated with the greatest respect. It, and everything in it, is sacred, and it's up to the people who use it to protect it as well." (Gale 2002). This native Canadian quote is typical of indigenous cultures' views on 'belonging'. Any serious attempt at prospering in partnership with nature requires a rapid switch of emphasis from assumed ownership of the Earth to a sense of belonging to the Earth. A culture of belonging and guardianship is equally suited to private, state and commons areas of the Earth. Such a culture is a precondition for reversing the loss and degradation of the ecosystems on which everyone's life and livelihood depends.

6.4. OWNERSHIP CAN EVOLVE FROM MASTERY TO GUARDIANSHIP

Existing practices of ownership of the Earth's surface haven't worked since they rely on every individual owner respecting a rarely observed line between natural capital and the sustainable 'interest' of renewable harvests. This line and a sense of belonging to the Earth can be restored with a policy switch within the cultural and legal meaning of ownership. Ownership of a piece of the Earth can be reinterpreted by international treaty as a duty of care to future generations. All land, sea and non-renewable resource ownership title can be interpreted as a title of guardianship of ecological capital. All rights for access and use of natural resources can be interpreted as applying only to the renewable harvest, to diminish neither biological diversity nor ecosystem services. Use of non-renewable resources can encompass a compensating expansion in ecosystems and a guarantee (such as precycling insurance) of protecting the resources within circular flows.

6.5. LET'S TRY FORWARD GEAR

Reversing the loss of nature is not a bad deal for owners, as can be explained by farmers of barren lands and fishermen of barren seas. Making this switch is like finding a car rolling back towards a cliff edge and helping the sleepy driver to locate forward gear. Although the driver may be startled by the intrusion, they would be pleased to be able to move on safely. Society would discover that abundance and prosperity accord with an expansion of nature, rather than its subjugation. The battle with nature can be ended quickly and permanently. One class of owner will remain unhappy; the minority with no intention other than to convert their corner of the world into private profit. The political choice is between catering for this exploitive minority or expanding nature's abundance for the benefit of all.

6.6. COMPENSATION

This policy switch effectively gifts the world to the unborn future. In compensation, the present gets to have a future. Nations would gain new reasons to co-operate more and fight less. Populations characterised by separateness would learn to create and share abundance. Depleted soils and waters would be restocked with diverse life. For those with less interest in such tangible compensations there are more direct options. Those who have degraded ecosystems may be relieved of the privilege of ownership. Those without an interest in guardianship could bid for funds to compensate them for the transfer of title to a community-based trust of landless people. Funds could also be provided for bids to permanently leave undisturbed high-risk non-renewable resources, such as fossil fuels and heavy metals.

7. Sixth Policy Switch: Recruiting the Mega-Rich to Inspire a Mega-Transformation

7.1. ACKNOWLEDGE THE WINNERS

If economics is a board game where players set out to own more than others then the mega-rich are the world's winners. The richest 2% of adults own more than half of all global wealth (United Nations University 2008b) and the winnings of the wealthiest continue to rise because wealth enables more wealth. The success of the winners should be acknowledged and then a new 'game' compatible with global security should be started, with benefits to all including today's winners. Central to this new game is how to bring surplus wealth (wealth beyond the needs of its owners) back into play. Politics, being dependent on contributions from those with the means to pay them, has in general failed to implement effective progressive policies for either incomes or assets, so a deeper systemic change is required.

7.2. IT'S CRUNCH TIME EVEN FOR THE MEGA-RICH

The planet crunch cannot be eluded by any fortress mentality, such as skepticism, materialist escapism or security fencing. Pandemics, civil unrest, political instability, armed struggle, ecosystem collapses and financial turmoil can harm anyone anywhere and those with the most have the most to lose. The steady worldwide diversion of wealth towards the wealthiest was previously financed by borrowing from the future. A bubble of credit and ecological debt made it possible for the majority of the world's population to meet their basic needs, for governments to obtain economic growth and for wealth to concentrate. Further financial or ecological borrowing from the future is a strategy for collapse so any future society that is stable enough to run at all can run only from currently generated wealth. The world must quickly change to pay-as-you-go and past accumulations of assets are the only store of wealth sufficient to support this change.

7.3. SWITCHING TO A NEW WINNING STRATEGY

The winning strategy for the mega-rich is a policy switch from accumulation to a new game of creating a world where wealth can retain meaning; a world where the planet crunch is replaced by global security. By peer-pressure and peer-dialogue, the mega-rich can coordinate their collective abilities and assets in expanding the world's ambitions from 'less bad' to positive development. Without this leadership, with its mindset of opportunity and abundance, it will be too easy for the public and institutions to remain absorbed in small plans that don't matter and big plans that don't work. Today's markets await an underlying vision of a successful future with real lasting wealth to replace volatile speculative bubbles and accounting tricks. Today's money awaits the backing of activities that work to generate believable value. The massive stockpile of problems awaits being matched with the massive stockpile of wealth.

7.4. TRANSFORMATIONAL PHILANTHROPY

"The problem of our age is the proper administration of wealth, so that the ties of brotherhood may still bind together the rich and poor in harmonious relationship." Andrew Carnegie's (1889) call for philanthropy may be heard today as a last call. Carnegie foresaw the accumulation and hoarding of "intense individualism" being systematically corrected by the most wealthy self-organising to administer their surplus wealth "to produce the most beneficial results for the community". Modern philanthropists such as Doug Tompkins echo the call, "What if the 10,000 richest people in the world would do what I did? Bill Gates, Warren Buffet, the Sultan of Brunei, the Saudi princes – they could change the entire world." (Zeller 2005). Philanthropy on a scale sufficient to ensure global security requires no self-sacrifice, just enlightened self-interest.

7.5. PRESERVING WEALTH BY SHARING IT

Money and all other stores of value rely upon functioning stable economic, ecological and social systems. If the planet crunch proceeds then at an unpredictable moment these systems will not function and the meaning of money and assets will vanish. The status of the wealthy would vanish. Native cultural traditions such as potlatch, that accorded most status to those who were able to share the most, provide a model to support modern 'recycling' of surplus wealth (Trosper 1998). Paradoxically, accumulated wealth can now be preserved only by being shared. Governments can act to support and invoke a peer-led global philanthropic transformation in many ways. For example, the international trend toward opening up 'tax havens' provides a strong incentive, if financial authorities agree to provide favorable treatment to such funds used philanthropically.

8. Seventh Policy Switch: Local and Central Creation of Money

8.1. WHERE DOES MONEY COME FROM?

Almost all new money is issued into national economies as credit via the 500 year old fractional reserve system (UK Parliament 1931) that was set up for metal money. In this digital era governments issue only small portions of the money supply as notes and coins. The Federal Reserve Bank of Dallas (2009) explains the recursive lending in fractional reserve banking, "Banks actually create money when they lend it. Most of a bank's loans are made to its own customers and are deposited in their checking accounts. Because the loan becomes a new deposit, just like a paycheck does, the bank ... holds a small percentage of that new amount in reserve and again lends the remainder to someone else, repeating the money-creation process many times."

8.2. HARVESTING THE MONEY TREES

This basic information about where money comes from is generally unknown to the public and rarely mentioned by financial professionals as it makes money seem unreal, like something growing on trees. (The 'money trees' of money issuance are known to economists technically as 'seigniorage'.) Banks don't keep the new money they create by fractional reserve banking but they do profit from the interest charged when loaning it out. The creation of almost all money as credit means that as money supply grows so does the debt of the economy. The pressure of keeping up with repayments of loans and interest squeezes economic activity towards speculation and unsustainability. During global recession, debts by vulnerable individuals, organisations and nations will be unpayable on a scale that invites global shocks. Governments and communities can choose between waiting to experience these shocks and rethinking money.

8.3. OVER-HARVESTING

Allowing financial institutions to harvest the money trees creates incentives for over-issuance of credit by competing banks (Gersbachd 1998). Other players in the economy have the incentive of easy credit and governments have the incentive of tax revenues from permitting ever more credit used for ever more spending. Unreal salary and bonus incentives for financial bosses and whizz-kids ensured that the money trees were over-harvested until the financiers themselves lost faith in the process and the credit crunch began. The credit crunch and recession might not be escapable without sufficient attention to the underlying systemic errors. The rethinking of finance should start with asking whether the money trees (money creation) would be best harvested for private or public benefit?

8.4. MONEY TREES CAN PROVIDE PUBLIC BENEFIT

"The privilege of creating and issuing money is not only the supreme prerogative of government, but it is the government's greatest creative opportunity." wrote Abraham Lincoln (1865), who won the American Civil War by printing and spending into circulation $450 million of green printed ('greenbacks') full legal tender treasury notes. Hitler built his ill-fated inter-war popularity in Germany by creating money used for employment in work-creation projects (Liu 2005). The UK's recent practice of 'quantitative easing' (making up for the scarcity of money supply from bank credit) is money creation by the Bank of England. The public benefit in this case depends upon the extent that the new money begins to support 'bottom-up' employment and productive activity rather than 'top-down' bail-outs and institutional asset purchases.

8.5. THE MONEY TREES CAN BE SWITCHED FROM PRIVATE TO PUBLIC BENEFIT

The financial benefit to banks from being keepers of the money trees is modest, estimated at only £21 billion annually for the UK (Huber and Robertson 2000). It is also evident that a private casino ethos is unsuited to the responsibilities of preserving economic stability for the benefit of bankers and non-bankers alike. Money creation can be permanently switched from private to public benefit by creating money in future centrally (by a public owned and accountable body) and locally (by local public-interest bodies or self-organised communities). The resulting public benefit would be the sum of all the new money that is created by spending it into the economy, plus the savings on interest by reducing debts. It is also possible for central and local bodies to create money as interest-free credit. Local currencies can avoid the trap of being 'play money' (vouchers swapped for national currencies) by spending, discounting or loaning new money into productive use.

8.6. FROM MORE DEBT TO MORE POSITIVE DEVELOPMENT AND GLOBAL SECURITY

Any economy dependent on creating money as interest-bearing debt cannot escape a future of mounting debt. Even when the entire planet is exploited for short-term profit the debts still grow. Keeping the economy going has become a confidence trick of preserving faith despite the spread of distrust, systemic instability and predatory profiteering. Given this situation, harvesting the money tree for public benefit may seem too good to be true. Yet it can be done without fear of inflation or disruption to banking business (Huber and Robertson 2000). It can provide the necessary supply and circulation of money, solidly underpinned by the lasting value and meaning of activities that contribute towards global security.

9. Conclusion

The flood of problems, locally, regionally and globally, already exceeds the capacity of researchers, policy-makers and the wider society to cope. The strategy of handling problems one at a time, a bit at a time, has not worked and cannot work. The solutions are not to be found where they are being sought, within the narrow territory of each symptom. Time is short but a new strategy is available: it is the 'system' that must be fixed, not all the separate symptoms. The boundaries of what is realistic need not be set by the struggles of the past nor current economic conditions. What is possible must match what is necessary. Habits of thought, world-views and economic rules-of-the-game can be switched from accelerating the planet crunch to reversing it. The goal of global security, implemented by the proposed policy switches, offers the possibility of future stability, security and prosperity. These switches redirect attention from the daunting list of changes that would be seen if things were working, to the manageable list of changes to make things work. This is not academic nor complicated and if acted upon people would soon wonder why it wasn't done long ago.

References

Armstrong, F. 2009. The Age of Stupid. Movie documentary. Spanner Films/Passion Pictures. www.ageofstupid.net
Birkeland, J. 2008. Positive Development: From Vicious Circles to Virtuous Cycles through Built Environment Design. Earthscan, London.
Boulding, K. 1966. The Economics of the Coming Spaceship Earth, in H. Jarrett (ed.), Environmental Quality in a Growing Economy, pp 3–14. Resources for the Future/Johns Hopkins University Press., Baltimore, MD www.eoearth.org/article/The_Economics_ of_the_Coming_Spaceship_Earth_(historical).
Carnegie, A. 1889. Wealth. The North American Review, 148 June, 1889. http: //www.visionsofgiving.org/document.php?loc=3&cat=4&sub=12. Accessed 7 June 2009.

Churchman, C. W. 1979. The Systems Approach and Its Enemies. Basic Books, New York.

European Union 2002. Waste Electrical and Electronic Equipment. European Union, Directive 2002/96/EC. http://europa.eu/legislation_summaries/environment/waste_management/l21210_en.htm. Accessed 7 June 2009.

European Union 2008. Waste: revision of the Framework Directive. Directive 2008/98/EC. www.europarl.europa.eu/oeil/file.jsp?id=5303132

Federal Reserve Bank of Dallas 2009. Everyday Economics. http://www.dallasfed.org/educate/everyday/ev9.html. Accessed 7 June 2009.

Gale, T. 2002. Guardians of the land. 2002 Canada and the World. Thomson Corporation. May 2002, pp 3.

Gersbachd, H. 1998. Liquidity Creation, Efficiency, and Free Banking. Journal of Financial Intermediation. Vol 7, Issue 1, January 1998, pp 91–118.

Greyson, J. 2007. An Economic Instrument for Zero Waste, Economic Growth and Sustainability. Journal of Cleaner Production. Vol 15, pp 1382–1390.

Greyson, J. 2008. Systemic Economic Instruments for Energy, Climate, and Global Security, in F. Barbir and S. Ulgiati (eds.), Sustainable Energy Production and Consumption. NATO Science for Peace and Security Programme, Springer, The Netherlands, pp 139–158.

Greyson, J. 2009a. Creative Enquiry Project at Somerhill Junior School, Hove, England. Creative Partnerships. www.creative-partnerships.com

Greyson, J. 2009b. From credit crunch to planet crunch – or revival? Middle East Waste Summit, Turret ME, May 2009. http://www.wiserearth.org/resource/view/6cde9add775de8a2ead56e6234d9ec7a. Accessed 7 June 2009.

Hansen, J., Sato, M., Kharecha, P., Beerling, D., Berner, R., Masson-Delmotte, V. et al. 2008. Target atmospheric CO2: Where should humanity aim? The Open Atmospheric Science Journal. Vol. 2, pp 217–231. http://arxiv.org/abs/0804.1126 Accessed 7 June 2009.

Heap, B. 2009. Don't Forget the Science Bit. NATO Review, March 2009. http://www.nato.int/docu/review/2009/0902/SCIENCE/EN/. Accessed 7 June 2009.

De Hoop Scheffer, J.07. Managing Global Security and Risk. Speech at The International Institute for Strategic Studies Annual Conference, 7 Sept 2007. http://www.nato.int/cps/en/natolive/opinions_8489.htm. Accessed 7 June 2009.

Huber, J., Robertson, J. 2000. Creating New Money. New Economics Foundation 2000, pp 29–32. http://www.neweconomics.org/gen/z_sys_PublicationDetail.aspx?PID=81. Accessed 7 June 2009.

International Alliance of Research Universities, 2009. Richardson, K., Chair of the Scientific Steering Committee and Synthesis Report; Climate Change: Global Risks, Challenges and Decisions. http://en.cop15.dk/files/pdf/iaru_synthesis_report_2009_press_release.pdf. Accessed 20 June 2009.

Jackson, T. 2009. Prosperity without Growth? UK Sustainable Development Commission. www.sd-commission.org.uk/publications.php?id=914. Accessed 7 June 2009.

Kashdan, T. 2009. Curious? HarperCollins, New York, pp 7.

Kettles, N. 2009. In a Class of Their Own. Ecologist, UK, March 2009, pp 45–47. www.newschoolthinking.com. Accessed 7 June 2009.

Keynes, J M. 1936. The General Theory of Employment, Interest and Money. Macmillan, London, pp vii (reprinted 2007).

Kuznets, S. 1934, National Income, 1929–1932. 73rd US Congress, 2nd Session, Senate Document no. 124, pp. 7.

Leonard, A. 2007. The Story of Stuff. Short animated film. Free Range Studios 2007. www.storyofstuff.com. Accessed 7 June 2009.

Lincoln, A. 1865. Personal letter published in Ludwig E, Lincoln. Little Brown & Company 1930. http://www.xat.org/xat/usury.html. Accessed 7 June 2009.

Liu, H. 2005. Nazism and the German Economic Miracle. Asia Times, May 24, 2005. http://www.atimes.com/atimes/Global_Economy/GE24Dj01.html. Accessed 7 June 2009.

Meadows, D. 1999. Leverage Points: Places to Intervene in a System. Sustainability Institute, Vermont, p. 1. www.sustainabilityinstitute.org/tools_resources/papers.html

Millennium Ecosystem Assessment, 2005. Ecosystems and Human Well-being: Synthesis. Island Press, Washington, DC.

O'Rorke, M. 1988. Public information campaign on precycling. Prepared for City of Berkeley, California.

Pring, R., Hayward, G., Hodgson, A., Johnson, J., Keep, E., Oancea, A. et al 2009. Education for All (Nuffield Review). Routledge, London, pp 86.

Reuters, 2008. "New Deal" needed for climate change. Wed Oct 22, 2008. http://uk.reuters.com/article/domesticNews/idUKTRE49L7B520081022. Accessed 7 June 2009.

Schmidt-Bleek, F. 2008. Future, Beyond Climatic Change. Factor 10 Institute, France pp 7. http://www.factor10-institute.org/files/FUTURE_2008.pdf. Accessed 7 June 2009.

Stålenheim, P., Kelly, N., Perdomo, C., Perlo-Freeman, S. and Sköns E. SIPRI Yearbook 2009: Armaments, Disarmament and International Security. Oxford University Press. http://www.sipri.org/yearbook/2009/05. Accessed 7 June 2009.

Sukhdev, P. 2008. http://www.unep.ch/etb/publications/TEEB/TEEB_interim_report.pdf. Accessed 7 June 2009.

The Economics of Ecosystems and Biodiversity (TEEB) 2008. Sukhdev, P. http://ec.europa.eu/environment/nature/biodiversity/economics/, reported at http://news.bbc.co.uk/1/hi/sci/tech/7662565.stm. Accessed 7 June 2009.

Trosper, R L. 1998. Incentive Systems that Support Sustainability: A First Nations Example. Conservation Ecology, Vol 2, Issue 2, pp 31–44.

UK Parliament 1931. Report of the Committee of Finance and Industry (Macmillan report) Parliament Paper no 3897 H. M. Stationery Office, London, pp 34.

United Nations Charter, Chapter 5. San Francisco, 1945. http://www.un.org/en/documents/charter/chapter5.shtml. Accessed 7 June 2009.

United Nations University, 2008a. Naudé W, Santos-Paulino A, McGillivray M. Fragile States. Research Brief. University World Institute for Development Economics Research. March 2008. http://www.wider.unu.edu/publications/policy-briefs/en_GB/policy-briefs/. Accessed 7 June 2009.

United Nations University 2008b. Davies B, Sandström S, Shorrocks A, Wolff E. The World Distribution of Household Wealth. World Institute for Development Economics Research. Feb 2008, pp 7. http://www.wider.unu.edu/publications/working-papers/discussion-papers/2008/en_GB/dp2008-03/. Accessed 7 June 2009.

US Joint Forces Command 2008. The Joint Operating Environment 2008. Center for Joint Futures. pp 50. http://transnet.act.nato.int/WISE/JointOpera/. Accessed 7 June 2009.

Zeller, F. 2005. Buy Now and Save! World Watch. Vol 18, No 5, July 2005, pp 24–29. http://www.worldwatch.org/system/files/EP184C.pdf. Accessed 7 June 2009.

Zhou, H. 2006. Circular economy in China and recommendations, Development Research Center of the State Council. Ecological Economy. Vol 2, pp 102–114.

ENERGY SECURITY AND THE SOCIAL USE OF ENERGY

IGOR MATUTINOVIĆ*
GfK – Center for market research, Zagreb, Croatia

Abstract   When energy and natural resources are becoming scarce on the global scale, attention should be focused on how energy is being used to meet a variety of social needs and priorities – not just the political priority of growth. New energy accounting is needed to distinguish the energy used for operation, maintenance, and development of capital goods and infrastructure from that used to pursue economic growth. This would serve both developed and developing economies to make informed decisions on their own socioeconomic priorities under the constraints of growing resource scarcity and divergent demographic trends. Without such informed policy making, Western societies will be hardly able to respond to national security requirements, and to contribute to the international quest for peace and security.

Keywords: Energy use, energy accounting, net capital formation, economic development

1. Introduction

The fundamental link between energy and society has been known in social sciences at least from the writings of cultural anthropologists Leslie White who related the evolution of cultural complexity with energy per capita consumption and productive efficiency in a simple equation: *Culture = Energy × Technology* (White 1949; Trigger 1989, p. 127). His message was clear: societies may evolve towards higher levels of complexity to the extent that they are able to harness more energy by means of ever more advanced technologies. The corollary to this conjecture is that the more complex a society is the more energy it will need to sustain itself. And not just any

---

* To whom correspondence should be addressed: Igor Matutinović, GfK – Center for market research, Draškovićeva 54, 10000 Zagreb, Croatia; E-mail: igor.matutinovic@gfk.com

type of energy would do the job, but only those primary sources which yield *high energy return on investment* (EROI) (Giampietro 2008).

Rapid technological advances and socio-economic complexification of Western societies and Japan after the World War II, was largely based on the primary energy source with the high EROI – the oil. Along their development path, Western economies increased materially several times by attracting into their dissipative loop disproportionate amounts of fossil fuels and natural resources. The social use of energy was initially directed more towards building new infrastructure – urban, industrial, electro-energetic, transport, and communication – and later towards increasing per capita personal consumption. In the post-war period per capita material consumption in industrialized world was growing faster than at any time in human history. By the mid 1960s the new consumer society started to emerge – one that went beyond the final stage of economic growth envisaged by W. Rostow (1960) – that of mass consumption. With the aid of sophisticated marketing techniques, mass-media, and a cascade of technological innovations the industry created an autocatalytic loop where economic growth was fueled by fast product diversification and an equally fast obsolescence, thus creating a new form of highly dissipative behavior known as the *"throw away society"*.

While the *"throw away society"* was emerging in the most advanced economies, the earth was still considered an "empty world" – the one in which man-made capital was the crucial limiting factor in economic development (Daly 1996). Rapid population growth on the world scale, which started after the WW II, in concert with aid and development policies that Breton Woods institutions applied to Third-World economies, transformed rapidly the once empty world to the full world – overpopulated and heavily weighting on its supportive ecosystems.

The fact that the notion of the full world has been with us for several decades did not affect the overall dynamics of our economies: it is almost certain that the global capitalist economic system[1] has an intrinsic tendency to continue on its long-term growth trajectory until it reaches energetic or biophysical limits of the Earth, whichever happens to occur first. At the societal level, the causal mechanism behind this tendency of growth can be recognized in the systemic interplay of the Western worldview, technologies and capitalist institutions coupled with autocatalytic dynamics at the

---

[1] Market economy is not necessarily capitalist (Matutinović 2010) in the same way as capitalist economy is not necessarily democratic (Berger 1986). Presently the only the only type of market economy running in the world is capitalist, China and India included.

microeconomic level (Matutinović 2005, 2007). From another, complementary, perspective, growth may represent an expression of the second law
of thermodynamics whereby the economy organizes its activities in such a
way as to degrade free energy along the steepest gradients (Annila and
Salthe 2008).[2] Whichever theoretic or philosophical perspective we may
take regarding the final causes of economic growth – socioeconomic or
thermodynamic – there is no doubt that global ecosystems cannot sustain it
indefinitely (Schaffer et al. 2001, Monney et al. 2005) nor can it draw on
inexhaustible natural resources and carbon-based primary energy. Therefore, the conclusion that "we failed to adapt our current socio-ecological
regime from an empty world to a full world" (Beddoe et al. 2009) appears
uncontroversial.

This conclusion has far reaching consequences on national security and
international peace and stability. It is clear that competition for finite
resources in a full world is replete with political tensions which may
explode into violent conflicts (Stoett 1994, Homer-Dixon 1999, Abbot et al.
2006, Sachs and Santarius 2007, p. 84). Competition for fossil fuels, fresh
water, arable land, fisheries is coupled with a growing rate of degradation of
global ecosystems, which undermine the ecological basis of support for
human societies (Schaffer et al. 2001, Monney et al. 2005) and where some
safe plantetaryboundaries, including those that refer to climate change, have
been most probably transgressed (Rockström et al. 2009).

Uneven access to natural resources and energy, combined with ecosystem
degradation and climate change effects, contribute to the marginalization of
those populations who already live near physical subsistence levels, and
who represent a large proportion of the global population. The marginalization of entire social and demographic strata, and indeed of entire nations,
provides a fertile ground for terrorism and thus reinforces risks of international conflicts in an unpredictable way. These interconnected and complex
global processes were identified by the Oxford Research Group study as the
five major global treats: competition over resources, climate change, and
marginalization of the world's majority populations, international terrorism,
and global militarization (Abbot et al. 2006).

This is the background for discussion of the social use of energy under
the full world constraint, which accounts also for the fact that depletion of
finite fossil energy sources is increasing while renewable energy is far from

---

[2] In that sense the capitalist system has been naturally selected among alternative
socioeconomic formations because of its capacity to degrade comparatively much faster
available energy gradients.

being able to compensate it in the foreseeable future. Present work is the first step in the direction of addressing conceptual and methodological issue related to societal energy use – one that may prepare the ground for subsequent empirical investigations.

2. Social Use of Energy

Socioeconomic systems are dissipative structures that require constant inflow of free energy to ensure their smooth day-to-day functioning. Georgescu-Roegen (1971, p. 277) brought this awareness to economics by pointing out that *"our whole economic life feeds on low entropy"* (italics are in the original). Socioeconomic systems are *autopoietic* in the sense that they succeed in constantly renewing their functional components and their (infra)structure, for which they need to dissipate additional exergy. Societies are evolving systems where each substantial step towards higher complexity is based upon tapping a new energy source or on improving the efficiency rate of the previously used ones. In short, socioeconomic systems are open thermodynamic systems which keep themselves far from equilibrium thanks to uninterrupted inputs of low entropy and whose evolution depends on being able to secure certain minimal free energy inputs indefinitely.

There are four basic dimensions of energy use in a socioeconomic system: (1) operations, (2) maintenance, (3) growth, and (4) development (Fig. 1). Growth and development have been closely interrelated since the time of industrialization (Rostow 1960, Herrick and Kindelberger 1983). However, there is a need to analyze them separately in order to understand how to conceive the future in which there is still development, but without growth – the key issue of sustainability (Daly 1996).

2.1. OPERATIONS

We distinguish two basic dimensions of using energy for supporting socio-economic operations. The first relates to energy used for satisfying *basic human needs* like supply of daily food, heating, water, and public transport. That a society must secure continuous inflow of high quality energy to cover basic needs of prevalently urban population does not need further explication. A society that fails to that task for a prolonged period of time is the collapsed one.

The second aspect concerns the *standard operating energy* – the one that is required to run daily economic activities, public services, and all household activities beyond the basic human needs. We are accustomed to use energy for needs that largely transcend food, daily transfer to work and

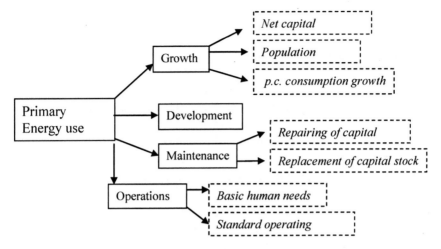

Figure 1. Four basic dimensions of social energy use

heating, and refer to the fulfillment of these needs as to "normality". The recurrent blackouts in California in the period 2000–2001 are an example of displacement from such normality, which affected, among other things, traffic flow regulation, elevators, cooling of residential and commercial buildings and Internet access. The 2009 natural gas crisis in Romania presents a case of more serious consequences. It is an example of a double failure: on the one hand the government failed in providing for basic human needs when the household heating system was out of gas for weeks; on the other hand the energy was not available to the gas-using industry, which had to close down production. In all these cases it was not just *any* energy that was missing: in one case it was electricity – the highest quality energy carrier, and in the other, natural gas – a primary energy with high EROI. The residents in Romanian cities reverted temporarily to wood as an alternative fuel for heating, while Californian residents had no other way to respond to many problems that appeared during the blackouts but to patiently wait.

2.2. MAINTENANCE

We distinguish two basic aspects of energy use maintenance of capital stock. The first relates to energy needed for *repairing* the extant infrastructure[3] – roads, railroads, airports, ports, public institutions, housing, manufacturing

---

[3] The focus is on public and industrial infrastructure because it is the most important from the societal point of view and it needs relatively much more energy than maintenance of machinery.

plants, power plants, electric grid, transport equipment, and telecom-munications – during its operating lifetime. This aspect of energy use is simple and it needs relatively low inputs of exergy.

The second refers to *depreciation*: to the fact that infrastructure and other capital goods wear off with time and use and must be eventually replaced. This is what goes under the voice of depreciation in national accounts. Georgescu-Roegen referred to it as "decumulation" and pointed that what is being "decumulated" is the *fund of services* that a given capital stock provides to a society. The scale and importance of infrastructure depreciation is perhaps not so visible in rich industrialized countries so we can take the housing replacement in a large and prevalently poor city like Cairo, with its 20 million inhabitants, as an example. Driving around Cairo one will immediately note that a great proportion of residential buildings shows a high degree of wear and tear – a breathtaking sight of the entropy law at work – where entire city blocks appear to be on their way to natural decomposition. However, if one then takes a look from the higher city points, he will note that large residential quarters are being replaced with the new housing (see the block of white houses in the Fig. 2). The inability to counter the entropy process with capital investments that require both new material and exergy would lead, eventually, to the complete wear out and physical collapse of the city, and eventually, as we have learned from history, of the civilization itself.

Figure 2. Infrastructure depreciation: residential block replacement in Cairo (Photo taken by the author, Cairo 2009)

By putting together these two aspects of infrastructure maintenance – *repairing* and *depreciation* – we see that societies must secure a stable and uninterrupted influx of energy and material just to stay at their present level of infrastructural complexity. While this may have not been an issue in the empty world, it becomes of crucial importance in the full world, burdened with a still growing population.

2.3. GROWTH

The discussion starts first with the growth – the dimension of energy use that distinguishes most the capitalist economy from any other historic socioeconomic formation. There are three basic aspect of energy use for growth: (a) net capital formation, (b) population growth, and (c) growth in per capita consumption.

2.3.1. *Net capital formation*

Economic growth is often equated with net capital formation – addition of new plant, machinery, buildings, railways, vehicles etc. – less their disposal and depreciation. Net capital formation is the precondition for capital accumulation and creation of material wealth of a society. There are two basic dimensions of net capital formation: one refers to social overheads or infrastructure and the other to manufacturing. Both are essential for economic growth to take off and to continue in a sustained way (Rostow 1960). While the state has an important role in providing initial investments into the basic infrastructure like roads, railways, energy utilities and transmission networks, the private sector of the capitalist economy makes use of these overheads to generate new capital for further investments, especially in the manufacturing sector. Once engendered, this process continues in an auto-catalytic mode, where the end result is sustained economic growth (Matutinović 2002, 2005). Net capital formation in the manufacturing industry, which includes also food processing, serves to satisfy the material needs of the population. Since the time of industrial revolution it has been the basis of the improvement of the material standard of living. In the background of the process of net capital formation in infrastructure and in manufacturing lies the capital formation and productivity growth in agriculture and extractive industries, without which the former process cannot take-off or sustain itself. When net capital formation involves introduction of new technologies or substantial improvement of the existing ones then it is equivalent to the process of economic development.

Net capital formation and capital accumulation enable a society to tackle more intensively diverse energy sources, which creates a loop of increasing returns between capital formation and energy appropriation. On the other hand, this positive feedback leads to increasing socioeconomic complexity which needs ever more energy to sustain itself.

2.3.2. Population growth

Even in the absence of autocatalytic growth, which is characteristic only for capitalist economies, the sole presence of demographic growth requires minimal capital addition in order to increase the output fulfilling the basic needs of the population. The inability of the ruling political structure to provide for this socially requisite economic growth leads inevitably to increasing poverty, famine and social unrest. As most of the Western societies appear to have either stabilized their populations or have depopulation trends, we can suppose that this driver of growth in rich societies has been exhausted.[4] However, population growth in the South will continue and probabilistic projections of total world population in 2100 point at the 8.4 billion[5] (Lutz et al. 2001). To accommodate this population growth, which will happen in an environment of dwindling natural resources, Southern economies will face a serious challenge to secure enough capital invest-ments in social overheads and manufacturing. This still does not address the issue of alleviating the existing poverty which is again a function of new capital investments. Even a partial fulfillment of Millennium development goals entails considerable degree of economic growth and a non-trivial increase in the per capita energy use.

2.3.3. Growth in per capita consumption

Conventional economic wisdom says that when the rate of capital formation increases output faster than the rate of population growth, the society will achieve an increase in per capita personal consumption (Rostow 1960, p.20). Productivity growth and technological change, which are an intrinsic property of the autocatalytic dynamics of the capitalist economy, have been enabling, so far, the sustained growth in per capita consumption and, con-sequently, in per capita energy use. Historic growth in material standard of living in Western capitalist democracies created equally strong anticipation that this trend will continue also in the future – the so called "revolution of rising expectations" (Zinam 1989). This, originally, Western belief that all

---

[4] Recent research showed, however, that at least some developed nations with high human development index will increase their fertility rates to the replacement level (Tuljapurkar 2009).

[5] This is the median value of simulated projections with the 80% prediction interval bounded by 5.6 and 12.1 billion (Lutz et al. 2001).

human beings have the basic right to expect the attainment of ever higher material well-being has been successfully transplanted globally, and especially to China and India, who are investing heavily in their energy sectors to support anticipated economic growth. The trickle-down of wealth creation trough economic growth to low-income social strata secured political stability to capitalist societies and served as a tacit precondition for historic agreement between capital and labor. In addition to alleviating extreme poverty, economic growth serves the same political purposes for the new elites in the South. Currently, one can not see the approaching of the saturation level in per capita consumption in the industrialized world. Meanwhile, developing countries will continue to put extraordinary efforts to reduce the large gap in material standard of living that divides them from the rich West.

2.4. DEVELOPMENT

In the ancient world net capital formation meant adding more of the same – more pyramids, more stone temples, more aqueducts. In modern economies, development and growth go hand in hand as technological change makes new capital additions often different from those that are replaced or depreciated. Development refers to change in the economic structure and is often equated with technical progress and productivity growth. Unlike economic growth, which basically refers to adding more of the same, development is more about adding entirely new things or processes, and in general has positive connotations: e.g. better hospital buildings with advanced equipment, further the national health care system, and prolong life expectancy at birth.

Having spare energy at disposal to develop new technologies or infrastructure is essential to meet unknown, future challenges from the environment (like, for example, climate change) or from within the societies, like demographic changes (aging of the population). *National security may be jeopardized if all of the available energy ends up serving only consumption and maintenance purposes.*

3. Discussion

The Western civilization finds itself in a paradoxical situation. On one hand, science points unequivocally that anthropogenic activities has been stretching to the edge the resilience of world ecosystems (Schaffer et al. 2001) and are rapidly reducing their basic regulative functions, concerning among other things climate, soil erosion, air and water quality, pest control and natural catastrophes (Monney et al. 2005). On the other hand, our societies persist in pursuing the "business as usual" scenario – by sticking to the patterns of production and consumption that are at the very origin of current problems and for which we have almost reached public consensus that are unsustainable.

The most important reason of persisting with the business as usual mode lies in the fact that the Western or any other capitalist society cannot tolerate for long the absence of economic growth. During the current global recession hundreds of billions of US dollars have been spent in an attempt to get the world economy back on the growth track, including the huge program for aiding the automotive industry and consumer spending on new cars. The scope of investments and the degree of political cooperation in addressing environmental deterioration (e.g. deforestation of the Amazon basin or in averting the climate change threat) has been nowhere close to that employed in mitigating the last economic crisis. This shows clearly where the priorities lie.

The problem does not exhaust itself in this huge gap of political commitment and action when it comes to deal with short term economic problems and the long-term sustainability issues. The reasons for concern come from the societal inability to conceive and plan for a different future – the one that would be more in accordance with the constraints of the full world. The projections of international and regional agencies regarding economic dynamics and energy consumption in the developed Western economies consistently point in the same direction – that of growth. Take, for example, the European Union which projects an increase of its GDP in the period 2005–2030 by 71% and of the primary energy use by 11%. Much of the growth of energy demand is expected in the transport sector (28%), services (26%), industry (20%), and households (12%). The projections for the period 2007–2030 in the United States are similar: an increase in GDP of 57.5% which is supported by a growth in primary energy use of 11.5% (EIA 2009). What is common to these projections is that (1) the figures address only the energy use by sector but do not make transparent the different aspects of the social use of energy, and (2), nowhere is made clear what is the social purpose of this large growth of GDP. Let me briefly touch these two issues.

3.1. SOCIAL ENERGY USE ACCOUNTING

Publicly available information concerns the aggregate energy use by sectors – transport, residential and commercial, industry, and electric power – to which different energy supply sources are associated (petroleum, coal, natural gas, renewable, and nuclear electric power) (see, for example, EIA 2009a, b). Present energy use accounting presumably takes into account only the energy required to support socioeconomic operations (the first dimension of the social use of energy). Nothing is known about the quantities and qualities of energy that were dissipated in the recent past or will be used in the foreseeable future for social uses like repairing and depreciation, or for net capital formation in infrastructure and manufacturing.

With only sector-aggregated data one is not led to think about the marginal utility of alternative energy uses at the societal level. For example, compare marginal utilities of energy to be spent for increase in the per capita consumption, e.g. for private transport, with the energy to be spent in net capital formation for the infrastructure that serves wider social needs. In the full world, where the world energy consumption is expected to rise between 45% and 50% in 2030 (EIA 2008, IEA 2008) and the world net electricity generation to almost double (EIA 2008), the societies will certainly face such choices, and the present day reporting of energy use by sectors is no longer enough to support decision making. However, this kind of know-ledge is essential both for strategic planning at the national level and for development of alternative future, security-related scenarios at the global level.

National and international energy agencies should complement their reports by splitting the projected energy use according to the rough scheme shown in Fig. 1: it would make transparent the future social use of energy and the qualities of energy that are needed to support its different dimensions. Such transparent reporting, once publicly available, may become a vehicle of raising the awareness among the wider public, and thus the subject of open debate – an indispensable prerequisite for political action in the right direction. The new energy use accounting is complementary to yet another type of accounting – one that addresses the EROI and the different qualities of energy carriers needed to meet different energy uses (Gardner and Robinson 1993, Giampietro 2008).

3.2. MEASURING SOCIETAL WELL-BEING

Concerning the societal rationale for further growth of GDP in the developed economies, the projected rates of growth in the EU and in the US are not justified by demographic trends.[6] The projected growth of their respective economies in the 2030 is simply the cumulative result of annual

---

[6] The projected GDP growth is not substantiated by population growth rates: EU27 population is projected to increase by 5.2% – from 495 million on 1 January 2008 to 521 million in 2035 (Eurostat News Release, STAT/08/119, 26 August 2008), and the US population by 22% from 302.7 million in 2005 to 369.9 million in 2030 (UN World Population Prospects, 2009. http://esa.un.org/unpp/).

growth rates that are perceived by the planners as socially and politically indispensable to keep the society functioning. This is the reflection of the earlier mentioned qualification that capitalist economies have the intrinsic propensity to grow and cannot find spontaneously the steady-state mode of functioning or systemic maturity. This understanding, however, should not prevent questioning the adequacy of GDP as a sole measure of economic success and societal prosperity. In fact, it has been common knowledge for several decades that GDP is far from being an adequate indicator of human and societal well-being, and that it should be replaced with alternative indicators. These should address more precisely and transparently the direction in which societies are moving, and the environmental damage they leave behind (van den Bergh 2009).[7] By the same token the social benefit of further increase in per capita energy consumption in developed societies is not substantiated by empirical research which points at saturation levels after a threshold has been crossed (Martinez and Ebenhack 2008).

4. Conclusions

The Western civilization has been increasing its socioeconomic complexity over the past three centuries by degrading the available free energy and by depleting natural resources. It has also succeeded to export its patterns of production and consumption on the world scale. In the process, the world has undergone the transition from the empty to the full state with the number of consequences for the environment and for natural resources that we have been able to measure and understand. This understanding of changing environmental and energy constraints calls for developing scenarios and visions of the future, substantially different from those presently offered in national and international energy reports. To be able to accomplish this task for any practical purposes, new accounting methods are necessary that would address the use of energy[8] (in its different qualities) for different social purposes – operations, maintenance, growth, and development, and account for the effects of economic processes, fueled with this energy, on the societal well-being and prosperity (new indices like ISEW that would replace GDP).

---

[7] This issue, however, shows signs of possible improvements in the near future as the French government decided in September 2009 to engage a group of scientists led by the Nobel laureate Joseph Stigltz to develop an indicator which would replace GDP as the key socioeconomic indicator of "progress and well-being".

[8] Whatever is said about energy is valid also for other, critical natural resources that enter economic process.

When energy and natural resources are becoming scarce on the global scale, attention should be focused on how these are being used to meet a variety of social needs and priorities – not just the political priority of growth. By distinguishing between the energy used for growth and for the other three, essential, societal purposes, Western societies may be able to make informed decisions on their own socioeconomic priorities in the context of yet another global constraint: the political and moral imperative to allow Southern societies to use economic growth as one of the vehicles for reducing poverty of their growing populations. Without such informed policy making, Western societies would be hardly able to respond to national security requirements and to contribute to the international quest for peace and security. Southern economies may use this accounting method for setting development goals and for establishing investment priorities that can reasonably be supported by quantities and qualities of energy available in the foreseeable future.

Acknowledgments

I thank Velimir Pravdić for his valuable comments and help, which improved substantially the text; the GfK Group for support.

References

Abbot, C., Roger, P., and Sloboda J., 2006. Global Responses to Global Treats: Sustainable Security for the 21st Century. *ORG Briefing Paper*, June. Oxford Research Group.

Annila, A. and Salthe, S. 2009. Economies evolve by energy dispersal. *Entropy*, 11, 606–633: doi:10.3390/e11040606.

Beddoe, R., Costanza, R., Farley, J., Garza, E., Kent, J., Kubiszewski, I., Martinez, L., McCowen, T., Murphy, K., Myers, N., Ogden, Z., Stapleton, K., and Woodward, J. 2009. Overcoming systemic roadblocks to sustainability: The evolutionary redesign of worldviews, institutions, and technologies. *PNAS*, 106 (8): 2483–2489.

Berger, P.L. 1986. *The Capitalist Revolution: Fifty Propositions about Prosperity, Equality, and Liberty*. New York: Basic Books.

Daly, H. 1996. *Beyond Growth*. Boston, MA: Beakon Press.

Energy Information Administration (EIA) 2008. *International Energy Outlook 2008 Highlights*. http://www.eia.doe.gov

Energy Information Administration (EIA) 2009a. *Annual Energy Outlook 2009 with Projections to 2030*. http://www.eia.doe.gov/oiaf/aeo/economic.html

Energy Information Administration (EIA) 2009b. *Annual Energy Review 2008*. Report No. DOE/EIA-0384(2008). Release Date: June 26, 2009. http://www.eia.doe.gov/emeu/aer/pecss_diagram.html

European Commission, 2007. *European Energy and Transport: Trends to 2030 - Update 2007*. Directorate – General for Energy and Transport. http://ec.europa.eu/dgs/energy_transport/figures/trends_2030_update_2007/energy_transport_trends_2030_update_2007_en.pdf

Gardner, D.T. and Robinson, J.B. 1993. To what end? A conceptual framework for the Analysis of energy use. *Energy Studies Review*, 5 (1): 1–12.

Georgescu-Roegen, N. 1971. *The Entropy Law and the Economic Process*. Cambridge: Harvard University Press.

Giampietro, M. 2008. Studying the addiction to oil of developed societies using the multi-scale integrated analysis of societal metabolism (MSIAM). In Eds. F. Barbir and S. Ulgiati, *"Sustainable Energy Production and Consumption: Benefits, Strategies and Environmental Costing"*. New York: Springer.

Herrick, B. and Kindelberger, C. 1983. *Economic Development*. London: McGraw-Hill

Homer-Dixon, T. 1999. *Environment, Scarcity, and Violence*. Princeton, NJ: Princeton University Press.

International Energy Agency (IEA) 2008. *World Energy Outlook Fact Sheet: Global Energy Trends*. www.iea.org/weo/docs/weo2008/fact_sheets_08.pdf

Lutz, W., Sanderson W., and Scherbov S. 2001. The end of population growth. *Nature*, 412: 543–545.

Martinez, D.M. and Ebenhasck, B.W. 2008. Understanding the role of energy consumption in human development trough the use of saturation phenomena. *Energy Policy*, 36: 1430–1435.

Matutinović, I. 2005. The microeconomic foundations of business cycles: from institutions to autocatalytic networks. *Journal of Economic Issues*, 39 (4): 867–898.

Matutinović, I. 2007. An institutional approach to sustainability: a historical interplay of worldviews, institutions and technology. *Journal of Economic Issues*, XLI (4): 1109–1137.

Matutinović, I. 2010. Economic complexity and markets. *Journal of Economic Issues*, 44, 1: 31–52.

Mooney, H., Cropper, A. and Reid, W. 2005. Confronting the human dilemma. *Nature*, 434: 561–562.

Rockström, Johan, Steffen, W., Noone, K., Persson, Å., Stuart Chapin, F., Lambin, E.F., Lenton, T.M., Scheffer, M., Folke, C., Schellnhuber, H.J., Nykvist, B., de Wit, C.A., Hughes, T., van der Leeuw, S., Rodhe, H., Sörlin, S., Snyder, P.K., Costanza, R., Svedin, U., Falkenmark, M., Karlberg, L., Corell, R.W., Fabry, J., Hansen, J., Walker, B., Liverman, D., Richardson, K., Crutzen, P. and Foley, J.A. 2009. A safe operating space for humanity. *Nature*, 461: 472–475.

Rostow, W. W. 1960. *The Stages of Economic Growth: A Non-Communist Manifesto (third edition)*. Cambridge: Cambridge University Press.

Sachs, W. and Santarius, T. (eds) 2007. *Fair Future: Resource Conflicts, Security, and Global Justice*. London: Zed Books.

Schaffer, M., Carpenter, S., Foley, J., Folke, C. and Walker, B. 2001 Catastrophic Shifts in Ecosystems. *Nature*, 413: 591–596.

Stoett, P.J. 1994. Global environmental security, energy resources and planning. *Futures*, 26 (7): 741–758.

Triger, B.G. 1998. *Sociocultural Evolution: Calculation and Contingency*. Oxford: Blackwell Publishers.

Tuljapurkar, S. 2009. Babies make a comeback. *Nature*, 460: 93–694.

Van den Bergh, J. 2009. The GDP paradox. *Journal of Economic Psychology*, 30: 117–135.

White, L. 1949. *The Science of Culture: A Study of Man and Civilization*. Farrar: Straus & Giroux

Zinam, O. 1989. Quality of life, of the individual, technology and economic development. *The American Journal of Economics and Sociology*, 48 (1): 55–68.

TECHNOLOGY WEDGES FOR LOW CARBON-STRATEGIES IN INDUSTRY

HANS SCHNITZER[*] AND MICHAELA TITZ
Graz University of Technology, Institute for Process and Particle Engineering, Graz, Austria

Abstract   The paper deals with the application of technology wedges in industry for the reduction of greenhouse gases. According to the needs defined by energy services in industry, four wedges are defined: cogeneration, process intensification and heat integration, renewable energy and passive house standard for production halls. With these technologies, industry could cover its part in meeting international emission goals while making economical investments. Such a reduction of the dependency of industry from external energy sources is a substantial contribution to safety issues in nations and regions.

Keywords: Energy efficiency, renewable energy, industry, technology wedge, solar thermal energy

1. Introduction

Energy has become an important issue of discussion in every field of life. The world's energy demand will increase significantly due to a growing population, fast economic growth in large countries (IEA 2008, IEO 2009) and because globalization increases the amount of goods and people transported, although there was a first decrease in world wide oil consumption in 2009 due to the financial crisis.

At present, world energy supply is mostly fossil-based and will remain so for decades. Most of the primary energy is coming from countries with potentially politically instable governments or weak democracies. The

---

* To whom correspondence should be addressed: Hans Schnitzer, Institute for Process and Particle Engineering, Inffeldgasse 21B, A8010 Graz, Austria; E-mail: hans.schnitzer@tugraz.at

energy-related worldwide environmental impacts will continue to grow: climate change and extreme weather situations. Energy will continuously increase in price, and the access to affordable energy will not stay uniform. Problems with resource availability (energy, water, food, materials, …) aggravated by climate change will most probably lead to regional and international conflicts.

Even a change-over to renewable resources is no safe way out. The increased utilization of plant oil, corn and sugar for bio-fuels already led to higher food prices – or is at least is blamed for them.

1.1. EUROPEAN GOALS AND STRATEGIES FOR ENERGY AND CLIMATE

Energy supply problems cannot be encountered on the local or regional scale only, although such initiatives should not be underestimated. To render changes successful and recognized by the people, international agreements, legislation and economic incentives are necessary. On the one hand they have to define the goals, but more importantly they have to set the boundary conditions regarding prices, subsidies and regulations. Many states have agreed to reduce the emission of global warming gases under the Kyoto Protocol (United Nations 1992).

The European Union has set up a number of targets for 2012, 2020 and 2050 in order to reduce its emissions (Commission 2008). These political efforts include:

- Reducing greenhouse gas emissions by at least 20% by 2020
- Improving energy efficiency by 20% by 2020
- Raising the share of renewable energy to 20% by 2020

EU and several other organizations are aware of the fact that a further reduction of energy consumption and GHGs has to be aimed at. The "European Council recognized the vital importance of limit worldwide greenhouse gas emissions to an amount that would restrict global temperature increase to 2°C compared to pre-industrial levels. The Council affirmed that developed countries should take the lead by committing to collectively reducing their emissions of greenhouse gases in the order of 30% by 2020 compared to 1990, with intent to reduce emissions by 60% to 80% by 2050" (European Commission 2007).

The Tyndall decarbonization scenarios project has outlined alternative pathways whereby a 60% reduction in CO_2 emissions from 1990 levels by 2050, a goal adopted by the UK Government, can be achieved (Mander et al. 2008).

1.2. ENERGY AND INDUSTRY

Industry will play a major role in determining how energy targets will be met. The future strength of European countries on the world market will largely depend on the extent to which they will be able to provide products and services with a substantially reduced resource throughput.

TABLE 1. Distribution of final energy demand in the EU (Commission 2005)

2002	Buildings (residential and tertiary)		Industry		Transport		All final demand sectors	
	Mtoe	% of final demand	Mtoe	% of final demand	Mtoe	% of final demand	Mtoe	% of final demand
Solid fuels	12.2	1.1	38.7	3.6	0.0	0.0	50.9	4.7
Oil	96.8	8.9	46.9	4.3	331.5	30.6	475.2	43.9
Gas	155.6	14.4	105.4	9.7	0.4	0.0	261.5	24.2
Electricity (incl. 14 % from RES)	121.3	11.2	91.2	8.4	6.0	0.6	218.5	20.2
Derived heat	22.8	2.1	7.5	0.7	0.0	0.0	30.3	2.8
Renewables	29.0	2.7	16.2	1.5	1.0	0.1	46.2	4.3
Total	**437.8**	**40.4**	**306.0**	**28.3**	**338.9**	**31.3**	**1,082.6**	**100**

As shown in Table 1, in EU25, industry consumes 28% of total energy (Eurostat 2009), of which about 70% are for heat production, the rest goes to electricity consumption, yet 57% of the heat demand is at low and medium temperature (Werner 2006).

Speaking of energy efficiency and the use of renewable forms of energy in industry, it is important to know about the energy services and the temperature level required. The fractions of heat at the various temperature levels differ from sector to sector. Obviously, the "Basic Metals" and the "Non-Metallic Materials" (iron, glass, cement, fire proof bricks, ...) amount for an important share of high temperature heat demand, as can be seen from Fig. 1. Most of the other sectors mainly need energy at levels below 250°C.

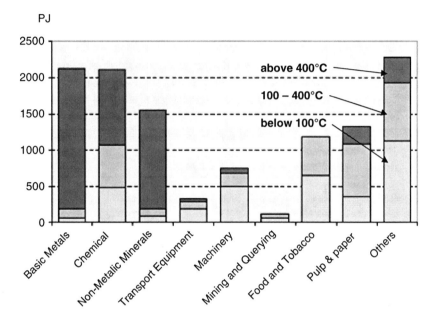

Figure 1. Estimated industrial heat demands by quality for EU25 + ACC4 + EFTA3 during 2003 (Werner 2006)

The use of energy is no end in itself. Energy is always needed to fulfill a service. Such a service might be a separation of substances (distillation, evaporation, membrane processes, filters, cleaning, ...), mixing of substances, transport, or a general heat treatment (pasteurization, sterilization). Typical low temperature processes are:

- Drying and dehydration processes
- Evaporation and distillation
- Pasteurization, sterilization
- Washing and cleaning
- Chemical reactions

These processes occur in a number of industrial sectors in different environments, but follow the same principles.

So the concept of "Unit Operations" can be applied to the energy use in industry. Several processes can be found in the various sectors with similar principles. Like with equipment design, experience with energy efficiency can therefore be transferred from one sector to an other.

Within the EINSTEIN-project (Brunner 2008), this principle has been programmed into an expert system, based on WIKIWEB. Since many unit operations with high energetic relevance can be found in the sectors analyzed, the information can be structured in an operation/sector matrix (Fig. 2).

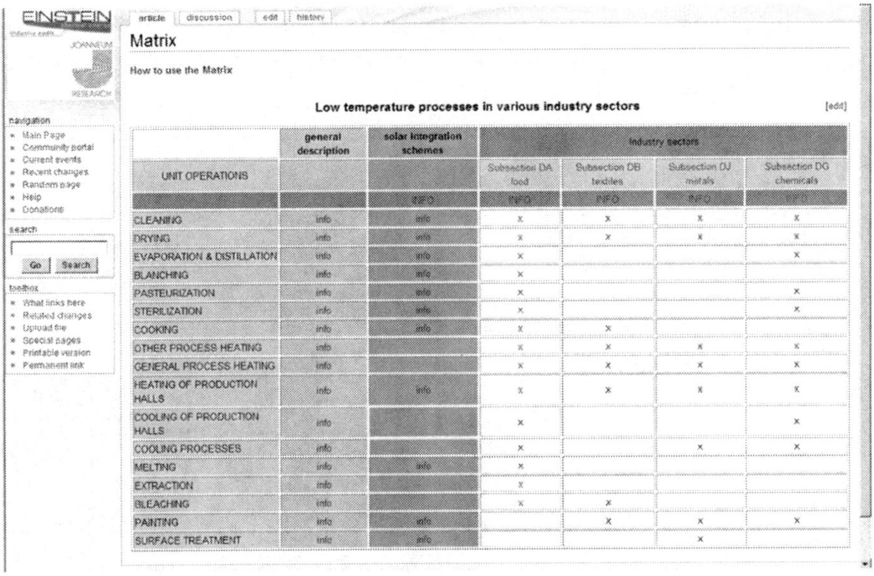

UNIT OPERATIONS	general description	solar integration schemes	Industry sectors: Subsection DA food	Subsection DB textiles	Subsection DJ metals	Subsection DG chemicals
		INFO	INFO	INFO	INFO	INFO
CLEANING	info	info	x	x	x	x
DRYING	info	info	x	x	x	x
EVAPORATION & DISTILLATION	info	info	x			x
BLANCHING	info	info	x			
PASTEURIZATION	info	info	x			x
STERILIZATION	info	info	x			x
COOKING	info	info	x	x		
OTHER PROCESS HEATING	info		x	x	x	x
GENERAL PROCESS HEATING	info		x	x	x	x
HEATING OF PRODUCTION HALLS	info	info	x	x	x	x
COOLING OF PRODUCTION HALLS	info		x			x
COOLING PROCESSES	info		x		x	x
MELTING	info	info	x			
EXTRACTION	info		x			
BLEACHING	info		x	x		
PAINTING	info	info		x	x	x
SURFACE TREATMENT	info	info			x	

Figure 2. Matrix of energy intense operations with industrial sectors (Energy in Industry, 2009)

Using this matrix, energy conserving measures can be transferred between different sectors.

2. Technology Wedges

There exist hundreds of possible measures to reduce the energy demand and GHG emissions in industry. According to the political and environmental goals, industry has to contribute according to its share in consumption. Martin et al. (2000) list 54 energy-efficient emerging technologies.

2.1. THE CONCEPT OF TECHNOLOGY WEDGES

To assess the potential of various strategies to reduce the energy demand, the concept of "technology wedges" can be used. The difference between the inclining path of energy demand currently predicted (BAU – business as usual) and the flat path from the present to the future urges a triangle of necessary reductions. This "reduction triangle" can be divided into several triangles – or "wedges" – of equal area as shown in Fig. 3. Globally seen, each wedge should result in a reduction of 1 billion ton of carbon per year by 2054, or 25 billion tons over 50 years. Fifteen carbon-reduction strategies have been examined by Socolow et al. (2004), each of which is based on a known technology, is being implemented somewhere at an industrial scale, and has the potential to contribute a full wedge to carbon mitigation (Pacala and Socolow 2004).

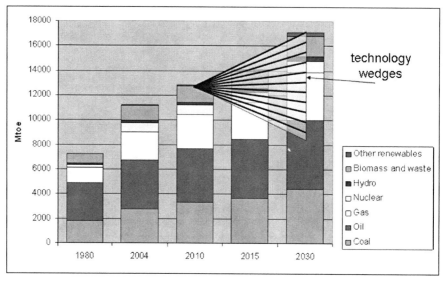

Figure 3. Technology wedges would stabilize and reduce the energy consumption (Data from IEA 2006)

Changes are possible in energy services, application technologies and transformation technologies at a given life style. Only a few technologies will fulfil the requirements for wedges.

In Austria, a group of researchers under the lead of WIFO is working on the definition of technology wedges that will be able to meet Austria's national GHG targets. For Austria a wedge should have the size of about 1 GgC (Energy Transition 2009). They consider the following sectors:

- Housing and living
- Mobility
- Industry
- Materials
- Transformation technologies

In each of the sectors, a set of wedges will be defined and described according to its technological characteristics and economic potential.

2.2. TECHNOLOGY WEDGES FOR THE PRODUCTION SECTOR

The production sector will have to deliver at least three wedges of the relevant size. At the moment, the following possible technology wedges are under discussion:

- Energy efficient technologies/increased efficiency in fuel and electricity use
- Cogeneration of heat/electricity/cold/compressed air
- Heat integration and energy recovery, process intensification
- Renewable energy: solar heat for industrial processes, fuel switch to biogenic resources and organic waste, plant based materials, green electricity
- HVAC – passive house technologies for offices and factory buildings, ...

2.2.1. *Energy effective technologies*

The effective transformation of energy into energy services is one of the key wedges. Depending on the service inquired and the energy source, the most effective[1] technology can therefore save a lot of energy and emissions.

Most case studies show a high potential of energy saving in the field of motors, light and heating. Not all approaches have to be purely technical ones. Many projects dealing with "Cleaner Production" prove a substantial potential for "Good Housekeeping" measures, like training of employees, advanced control strategies and awareness rising. In many production halls and offices electric light is on while the sun shines outside. Day-lighting systems are able to offer good working conditions while saving electricity.

In many applications it is possible to replace thermal processes by mechanical ones (drying in centrifuges or presses replace hot-air dryers, distillation is replaced by membrane processes, washing by brushing ...). High efficient burners (condensing boilers) can reduce the consumption of fuels considerably.

---

[1] In the case of conversion of energy carriers (oil, electricity, gas, wood,..) into energy services (information, mobility, ...) we speak of effectiveness and not of efficiency.

2.2.2. *Cogeneration*

Practically in any company, there is a need for several forms of energy like heat, cold, electricity and compressed air. Cogeneration and trigeneration systems usually will offer higher efficiencies than simple transformation equipment.

Cogeneration of heat and electricity can always be applied, if there is a need for both forms of final energy. While there is always a grid for electricity, the demand for heat can vary considerably. In order to make a cogeneration widely applicable, the guiding principle should be called "No Heat without Electricity" rather than using the waste heat of electricity generation. Cogeneration units work with all kinds of fuels using a set of proven technologies like steam power plants, gas turbines, internal combustion engines, organic Rankine cycles or Stirling motors.

Heat recovery from air compressors can be seen as a cogeneration of compressed air and heat. This technology will only provide low temperature heat, but as has been demonstrated, most sectors of industry require great amounts of low grade heat. Chillers and refrigerating plants can be used "on both ends", thus operating as a heat pump simultaneously (Moser 1985). While the evaporator provides "cold", the condenser will deliver "heat". If both sides can be integrated into a production system, the system will turn out to be very economic. In case a power station provides electricity, cooling and heat, this is called trigeneration. Technologically it is a question of connecting an absorption chiller to a cogeneration unit. Waste heat from production processes or solar thermal energy can also be used for the production of cold as has been investigated by Moser 1985 in detail.

2.2.3. *Heat integration and energy recovery, process intensification*

In the chemical industry heat recovery from hot streams for meeting the demand of cold streams is quite traditionally done. Many other sectors are not so familiar with this technology. One of the reasons might be that heat recovery is much easier and more economical in continuous processes than in batch processes. According to the economic requirements and the operating reliability there is a hierarchy of integration schemes:

- Heat recovery from hot streams within the production process investigated
- Heat exchange with another process in the same company, but in another production line
- Heat pumps (compression and absorption)
- Waste heat driven ORCs
- Heat delivery to customers outside company (other company, fish farm, district heating, …)

The term 'Process Intensification' describes the strategy of making dramatic (100–1,000-fold) reductions in plant volume in order to meet a given production objective. When the concept of process intensification was developed within ICI in the late 1970s, the main intention was to make big reductions in the cost of processing systems, without impairing their production rate. While cost reduction was the original target for process intensification, it quickly became apparent that there were other important benefits, particularly in respect of improved intrinsic safety, reduced environmental impact and energy consumption. The high heat and mass transfer coefficients which can be generated in intensified equipment can be exploited to reduce the concentration/temperature driving forces needed to operate energy transformers such as heat pumps, furnaces, electrochemical cells etc. This enhances the equipment's thermodynamic reversibility and hence its energy efficiency (Ramshaw 1999).

Process intensification addresses the need for energy savings, CO_2 emission reduction and enhanced cost competitiveness throughout the process industry.

The potential benefits of PI that have been identified are significant (Senter Novem 2007):

- Petro and bulk chemicals (PETCHEM): Higher overall energy efficiency – 5% (10–20 years), 20% (30–40 years)
- Specialty chemicals, pharmaceuticals (FINEPHARM): Overall cost reduction (and related energy savings due to higher raw material yield) – 20% (5–10 years), 50% (10–15 years)
- Food ingredients (INFOOD):
  o Higher energy efficiency in water removal – 25% (5–10 years), 75% (10–15 years)
  o Lower costs through intensified processes throughout the value chain – 30% (10 years), 60% (30–40 years)
- Consumer foods (CONFOOD):
  o Higher energy efficiency in preservation process – 10–15% (10 years), 30–40% (40 years)
  o Through capacity increase – 60% (40 years)
  o Through move from batch to continuous processes – 30% (40 years)

2.2.4. *Renewables: Solar heat for industrial processes, fuel switch to biogenic resources and organic waste, green electricity*

The application of renewable forms of energy provides a chance for the reduction of Green House Gases, but not for energy savings in general.

Since also renewables will be scarce as soon as used in a higher amount, there have to be rules on how to use them:

- No fuels for temperatures below 100°C (only flameless technologies like solar energy, waste heat, heat pumps, ...)
- No heat without cogeneration of electricity and vice versa
- No food for energy
- No processing of agro-products (food, feed, materials, fuels, ...) without utilization of the whole plant (biorefinery – approach)
- EFFICIENCY FIRST

The best track to energy services depends on the availability of the resources, but Fig. 4 gives an indication of the technology paths that are available.

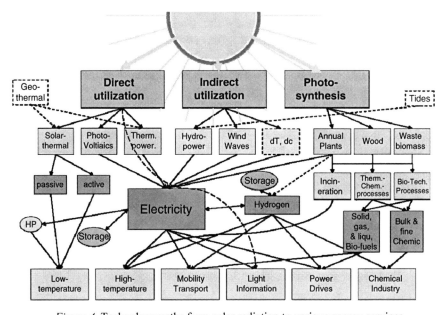

Figure 4. Technology paths from solar radiation to various energy services

It can be seen that electricity will play an increasing role in an energy system based on renewables, since it will penetrate the heating market through heat pumps and the mobility market through electric vehicles. Fuels are banned from the low temperature sector and hydrogen can function as a fuel in future but will never be an energy source.

Biomass heating systems for high temperature applications (including cogeneration systems) will be based on annual plants, wood and waste bio-mass. Biotechnological processes like fermentation in a 100% solar system will provide biogas and ethanol, as long as the do not compete with the demand for food. Liquid and solid fuels (charcoal) can be generated from wood and organic waste, as soon as the necessary technology is available on a commercial scale.

For low temperature processes, solar thermal energy can be used. The collectors can be integrated into the existing cycle of the heating medium (steam, hot water) or they can be linked to the production process directly as shown in Fig. 5. The first version has the advantage of less investment costs and a safer operation, but the temperatures required are high compared to those needed in the process. The second version is more expensive in terms of installation costs but thermodynamically favourable. A more detailed analysis of various integration schemes of thermal collectors into the production systems can be obtained from Müller (2004).

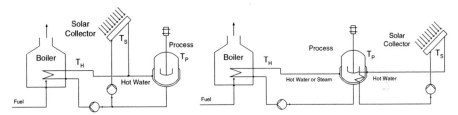

Figure 5. Solar thermal collectors can be integrated into the existing heating cycle or into the process itself

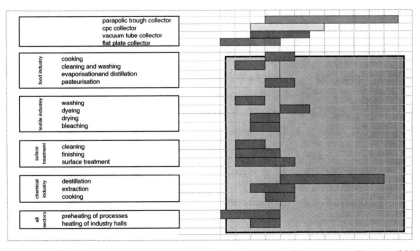

Figure 6. Typical low temperature processes and the suitable collector types (Brunner 2007)

Figure 6 contains a number of frequently occurring processes in various industrial sectors with the corresponding temperatures. This graph also shows that most of the process temperatures required can be reached with standard solar thermal collectors.

A study carried out for the International Energy Agency (IEA-SHC-Program, task 33) showed that there is a potential of 160 GW/year for solar process heating in the European Union (Table 2, Vannoni 2008).

TABLE 2. Potential for solar thermal process heat in Europe (Vannoni 2008)

Country	Industrial energy demand in 2002 (Source EUROSTAT) (PJ/year)	Solar process heat potential (including medium temp. applications) (PJ/year)	Solar process heat potential (GW)	Solar process heat potential (Mio.m²)
Austria	220	5.4	4,3	3
Spain	841	17	10	7
Portugal	184	4	2,5	1.7
Italy	1,136	32	15.4	10.8
Netherlands	425	1.9	1	0.7
Germany	1,575	50	35	24.5
EU 15	**7,880**	**199**	**138**	**97**
EU 25	**9,145**	**230**	**160**	**112**

In biomass refineries not only fuels, but also chemicals may be obtained from the integral utilisation of whole green plants, cellulosic plants or waste biomass. The products obtained will include materials (fibers, bio-polymers), basic chemicals like lactic acid and ethanol, fine chemicals and finally fertilizers. These technologies should utilize the whole plant and represent ZERO emissions approaches, including the recycling of water and inorganic matter for agriculture.

2.2.5. *HVAC – passive house technologies for offices and production halls*

Like in the domestic sector, office buildings and production halls should not require primary energy in the sense of fuels. They can be heated by waste heat or solar sources. Several case studies show that due to the low temperatures required for highly efficient low-energy buildings solar heating systems can entirely eliminate conventional heating systems.

3. Conclusions

Energy efficiency and the use of renewables are essential elements of a national and regional security program. The strong dependency on foreign resources is a permanent threat for industrialized and developing countries. A transition to a high fraction of renewables in the supply mix is only possible if the total amount of energy is reduced. This can be done without a reduction of energy services, if proper technologies are applied.

Out of practical reasons it can be favorable to use the concept of technology wedges for political decisions. Wedges are technological or organizational measures that reduce the GHG emissions significantly. For industry such wedges are cogeneration units, renewables for energy and materials, passive house technologies for production halls and process intensification. They are able to provide a win-win situation for the companies and the environment.

Acknowledgements

The research work about technology wedges in industry is carried out within two research project. "Energy transition" is a project financed by the Austrian Climate and Energy Fund managed by WIFO (http://energytransition. wifo.ac.at/). The second project, coordinated by Graz University, deals with the design of a Regional Styrian Climate Protection Plan. Special thanks to the Wegener Centre at Graz University and the Regional Government of Styria.

References

Brunner, C., Schnitzer, H., Slawitsch, B., Schweiger, H., Vannoni, C, 2008, Einstein – Expert-system for an intelligent supply of thermal energy in industry, in: *Proceedings "Advanced Energy Studies"*, Graz (A), June 2008.

Brunner, C., Schnitzer, H., Slawitsch, B., Weiss, W. 2007, Solar Heat for Industrial Processes, in: *Proceedings "Asian Pacific Roundtable Sustainable Consumption and Production"*, Hanoi, Vietnam, May 2007.

Commission of the European Communities, 200, 20 20 by 2020 – Europe's climate change opportunity, Communication from the commission to the European Parliament, the council, the European Economic and Social Committee and the Committee of the Regions COM(2008) 30.

Commission of the European Communities, 2005, Green Paper on Energy Efficiency or Doing More With Less, COM(2005) 265.

Energy in Industry, 2009, Graz (June 9, 2009) http://energy-in-industry.joanneum.at/.

Energy Transition, 2009, Vienna, (June 2, 2009) http://energytransition.wifo.ac.at/.

European Commission, 2007, Towards a Post-Carbon Society, in: *Proceedings "European research on economic incentives and social behaviour"* Brussels (B), October 2007.

Eurostat, 2009, Brussels, (June 8, 2009), http://epp.eurostat.ec.europa.eu.

IEA – International Energy Agency, 2006, *World Energy Outlook 2006*, International Energy Agency (IEA), Head of Publications Service, Paris France.

IEA – International Energy Agency, 2008, *World Energy Outlook 2008*, International Energy Agency (IEA), Head of Publications Service, Paris France.

IEO2009 is available on the EIA Home Page (http://www.eia.doe.gov/oiaf/ieo/index.html) by May 2009, including text, forecast tables, and graphics. To download the entire publication in Portable Document Format (PDF), go to http://www.eia.doe.gov/oiaf/ieo/pdf/0484(2009).pdf.

Mander, S.L., Bows, A., Kevin. L. Anderson, L., Shackley, S., Agnolucci, P., Ekins, P., 2008, The Tyndall decarbonisation scenarios—Part I: Development of a backcasting methodology with stakeholder participation. *Energy Policy* 36, 3754–3763.

Martin, N., Worrell, E., Ruth, M., Price, L., Elliott, R.N., Shipley, A.M., Thorne, J., 2000, Emerging Energy-efficient Industrial Technologies, *Ernest Orlando Lawrence Berkeley National Laboratory*.

Moser, F., Schnitzer, H. 1985, *Heat Pumps in Industry*, Elsevier, Amsterdam.

Müller, T., Weiß W., Schnitzer H., Brunner C., Begander U., Themel O., 2004, Produzieren mit Sonnenenergie – Potenzialstudie zur thermischen Solarenergienutzung in österreichischen Gewerbe- und Industriebetrieben, BMVIT – Berichte aus Energie- und Umweltforschung 1/2004, Wien.

Pacala, S., Socolow, R., 2004, Stabilization wedges: Solving the climate problem for the next 50 years with current technologies, in: *Science*, American Association for the Advancement of Science, 305 (5686) Washington, 968–972.

Ramshaw, C., 1999, Process intensification and green chemistry. *Green Chemistry* 1 (1), 15–17.

SenterNovem, 2007, The Hague (June 9, 2009) European Roadmap for Process Intensification, http://www.senternovem.nl.

Socolow, R., 2005, Stabilization Wedges: Mitigation Tools for the Next Half-Century, in: *Proceedings "Avoiding Dangerous Climate Change"*, Exeter (UK), February 2005.

Socolow, R., Hotinski, R., Jeffery, B. Greenblatt, J.B., Pacala, S., 2004 Solving the climate problem -Technologies available to curb CO_2 emissions, *Environment* 46 (10), 8–19.

United Nations, 1992, United Nations Framework Convention on Climate Change, Document FCCC/INFORMAL/84 GE.05-62220 (E) 200705.

Vannoni, C., Battisti, R., Drigo, S., 2008, Potential for Solar Heat in Industrial Processes, *Ciemat*, Madrid.

Werner, S., 2006, The European Heat Market: Workpackage 1, *Final Report, EcoHeat & Power*, Brüssel.

SUSTAINABLE ENERGY FOR WORLD ECONOMIES

TARIQ MUNEER[*]
School of Engineering and Built Environment,
Edinburgh Napier University, Edinburgh, UK

Abstract Energy is inevitable for human life and a secure and accessible supply of energy is crucial for the sustainability of modern societies. Yet mankind now faces challenges of depletion of fossil fuel reserves, global warming and other environmental concerns, geopolitical and military conflicts, and of late, continued and significant fuel price rise. Renewable energy (RE) is a key solution to this challenge. This article provides an overview of the current and projected energy scene. This article also proposes a palette of RE technologies that are available to meet current and future energy demands.

Keywords: Renewable energy technologies, super-insulated windows, lightpipes, photovoltaics, wind energy, energy cooperatives

1. Introduction

Energy drives human life and is extremely crucial for continued human development. Throughout the course of history, with the evolution of civilizations, the human demand for energy has continuously risen. The world heavily relies on fossil fuels to meet its energy requirements – fossil fuels such as oil, gas and coal are providing almost 80% of the global energy demands. On the other hand presently, renewable energy and nuclear power are, respectively, only contributing 13.5% and 6.5% of the total energy needs.

Presently employed energy systems will be unable to cope with future energy requirements – fossil fuel reserves are depleting, and climate change

---

[*] To whom correspondence should be addressed: Tariq Muneer, School of Engineering and Built Environment, Edinburgh Napier University, Edinburgh, UK; E-mail: T.Muneer @napier.ac.uk

F. Barbir and S. Ulgiati (eds.), *Energy Options Impact on Regional Security,*
DOI 10.1007/978-90-481-9565-7_6, © Springer Science + Business Media B.V. 2010

has become a serious issue. Fossil fuel and nuclear energy production and consumption are closely linked to environmental degradation that threatens human health and quality of life, and affects ecological balance and biological diversity.

Renewable energy is abundant, inexhaustible and widely available. These resources have the capacity to meet the present and future energy demands of the world as indicated in Fig. 1. The cost of energy generated from these renewable resources is significantly coming down while the cost of fossil fuel produced energy is in an increasing mode. Over the last 2 decades solar and wind energy systems have experienced rapid growth. This is being supported by several factors such as declining capital cost; declining cost of electricity generated and continued improvement in performance characteristics of these systems.

This article aims to present the challenges facing secure and sustainable supply of energy.

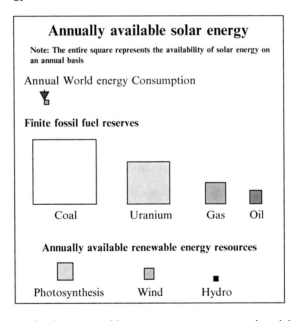

Figure 1. Potential of various renewable energy sources as compared to global energy needs

2. Human Civilization and Energy Use

Energy is one of the most basic of human needs. The accomplishments of civilization have largely been achieved through the increasingly efficient and extensive harnessing of various forms of energy to extend human capabilities.

Providing adequate and affordable energy is essential for eradicating poverty, improving human welfare, and raising living standards worldwide. With the exception of humans every organism's total energy demand is its supply of energy in the form of food derived directly or indirectly from sun's energy. For humans the energy requirements are not just for heating, cooling, transport and manufacture of goods but also those related to agriculture. However, the ingenuity of humans is such that throughout history there has been an exponential increase in the carrying capacity of land, particularly during the past few decades. Table 1 presents the remarkable progression of the agricultural land carrying capacity. What is now needed is for mankind to achieve a 'sustainable energy revolution' that matches the dynamism of its agricultural counterpart.

TABLE 1. Progression of agricultural land carrying capacity

Mode of agriculture	Number of humans supported per km^2
Hunter-gatherer	1
Domesticated animals	2
Shifting cultivation, S.E. Asia	10
Medieval agriculture (1200 CE, UK)	20
Intense tropical peasant agriculture (New Guinea)	50
Modern agriculture, UK	150
Modern agriculture, USA	500
Modern agriculture, Japan	2,500
Phytofarming	10,00,000

It has been estimated that the global population in 1800 was approximately one billion, an uncertain estimate given that the first population census had just been introduced around that time in Sweden and England. Estimates of past energy use based on historic statistics and current energy use in rural areas of developing countries suggest that energy use per capita typically did not exceed 20 GJ as a global average. Over 200 years later, the global population has risen by a factor of 6 while the per capita energy consumption is estimated to have risen by a factor of 20 (Grubler 2004). Figures 2 and 3 show an inherent link between prosperity and energy use. Given the past record of developed countries in their profligate use of energy, developing nations tend to mimic their energy consumption pattern to match those of the developed nations. What Fig. 3 does is it helps us increase our understanding of the expected global energy consumption increase as nations across the world target their quest for increased prosperity. If increased life expectancy is accepted as an important index of well-being, then Fig. 2 shows that the well-being is achievable at much lower levels of energy use index, any further energy use may then be regarded as either sheer luxury or energy wastage. An example of this type of wastage could be single-occupancy, large automobiles being driven aimlessly. Note that Table 2 provides detailed information on energy use, GDP and life expectancy.

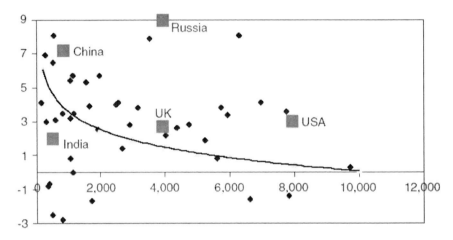

Figure 2. Per cent increase in energy consumption versus kg oil consumption/capita for 50 countries. Curve shown is based on Asif and Muneer (2007)

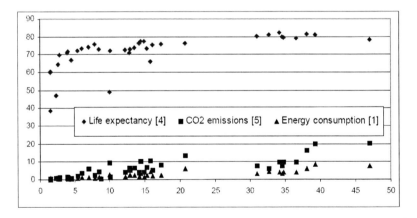

Figure 3. Life expectancy, carbon dioxide emissions and energy consumption link with per capita GDP. Note: units shown in Table 2

TABLE 2. Energy, economic and life-expectancy data for selected countries

Continent	Country	Population[1]	GDP[2]	Life expectancy[3]	CO_2 emissions[4]	Energy consumption[5]
Australia	Australia	21	38	81.6	16.3	5.9
Asia	Bangladesh	156	2	60.3	0.2	0.2
	China	1,339	6	73.5	3.8	1.3
	Indonesia	240	4	70.8	1.7	0.8
	Iran	66	13	71.1	6.3	2.4
	Japan	127	34	82.1	9.8	4.1
	Malaysia	26	15	73.3	7.0	2.4
	Pakistan	176	3	64.5	0.8	0.5
	S. Arabia	29	21	76.3	13.4	6.1
	Thailand	66	9	73.1	4.3	1.6
Europe	Bulgaria	7	13	73.1	5.5	2.6
	Croatia	4	16	75.4	5.2	2.0
	France	64	33	81.0	6.2	4.4
	Germany	82	35	79.3	9.8	4.2
	Italy	58	31	80.2	7.7	3.2
	Poland	38	17	75.6	8.0	2.4
	Romania	22	12	72.5	4.2	1.8
	Russia	140	16	66.0	10.5	4.5
	Spain	41	35	80.1	7.7	3.3
	UK	61	37	79.0	9.8	3.9
America	Argentina	41	14	76.6	3.7	1.6
	Bolivia	10	5	66.9	0.8	0.6
	Brazil	199	10	72.0	1.8	1.1
	Canada	33	39	81.2	20.0	8.5
	Chile	17	15	77.3	3.9	1.8
	Mexico	111	14	76.1	4.2	1.7
	USA	307	47	78.1	20.4	7.9
	Venezuela	27	14	73.6	6.6	2.3
Africa	Algeria	34	7	74.0	6.0	1.1
	Angola	13	9	38.2	0.5	0.6
	Egypt	83	5	72.1	2.2	0.8
	Ghana	24	2	59.9	0.3	0.4
	Libya	6	14	77.3	10.3	3.3
	Morocco	35	4	71.8	1.4	0.5
	Nigeria	149	2	46.9	0.8	0.8
	South Africa	49	10	49.0	9.2	2.7
	Tunisia	10	8	75.8	2.3	0.8
	Zambia	12	2	38.6	0.2	0.6

1. https://www.cia.gov/library/publications/the-world-factbook/rankorder/2119rank.html (population, millions)
2. https://www.cia.gov/library/publications/the-world-factbook/fields/2004.html (thousand $/cap, 2008)
3. https://www.cia.gov/library/publications/the-world-factbook/rankorder/2102rank.html (Life expectancy in years, 2005)
4. http://millenniumindicators.un.org/unsd/mdg/SeriesDetail.aspx?srid=751&crid= (tonnes/capita, 2004)
5. http://earthtrends.wri.org/text/energy-resources/variable-351.html (Energy consumption in Toe/capita, 2005)

3. Review of Nuclear-Fissile and Fossil Fuels

Tables 3 and 4 respectively compare the energy use and fossil fuel reserve data for India and UK. The latter choice of countries has been made to enable a comparison of a developed and developing economy. Note that Asif and Muneer (2007) provide a broader survey of energy research.

TABLE 3. Energy use data for developed and developing economies

Energy consumption within UK (2007) and India (2000), per cent breakdown		
	UK	India
Domestic	28.4	10
Transport	38.6	22
Industry	20.5	49
Other	12.5	19
Energy consumption within UK (transport sector), per cent breakdown		
Rail	2.4	
Water	2.7	
Air	23.4	
Road	71.5	
Energy consumption within UK (transport sector: road use), per cent breakdown		
Personal	63.3	
Freight	36.7	
Energy consumption within UK (domestic use), per cent breakdown		
Space heating	61	
Water heating	23	
Cooking	3	
Electricity – lighting	2.6	
Electricity – consumer electronics	2.4	
Electricity – other appliances	8.0	
Domestic sector: Space heating related data		
	Floor area, m^2	Heat-loss index, Wh/m^2/degree-day
Sweden	90	38
Netherlands	N/A	41
UK	85	50
Germany	78	70

Nuclear power is a largely carbon-free energy source that could in theory help phase out fossil fuels. More than 300 nuclear plants currently provide 15% of the world's electricity. But this energy source has been plagued by a range of problems, most fundamentally high cost and the lack of public acceptance, that have halted development for more than 20 years in most of Europe and North America.

TABLE 4. Energy use, fuel reserves and population growth data for developed and developing economies

Energy use	India (TWh)	UK (TWh)	World (TWh)
Population growth (%)	1.4	0.28	1.14
Total energy consumption	4,384	2,647	1,19,285
Electricity consumption	575	359	15,714
Coal consumption	2,389	445	32,412
Oil consumption	1,392	943	43,950
Gas consumption	337	1,029	28,238
Nuclear consumption	44	211	7,284
Renewable energy consumption	19	2	634
Total reserves	India (TWh)	UK (TWh)	
Coal	752,000	2,000	
Oil	9,000	7,000	
Gas	8,000	6,000	
Nuclear	6,000	n/a	
Years to exhaustion (based on a nil growth rate)	India	UK	
Coal	118	9	
Oil	19	6	
Gas	35	6	
Nuclear	140	n/a	

References: Asif and Muneer (2007) and BP Statistical Review of Energy (2008)

Over the past decade, global nuclear capacity has expanded at a rate of less than 1% a year. Major efforts are now under way to revive the nuclear industry – driven by a combination of high natural gas prices, concern about climate change, and a large dose of new government subsidies. Technology advances have led several companies to develop modestly revamped plant designs that are intended to make nuclear plants easier to control, less prone to accidents, and cheaper to build. A study by Keystone Center panel (Gardner and Prugh 2008) estimated the cost of new nuclear power at 8–11 US ¢/kWh, more expensive than natural gas and wind-powered generators. In view of large capital requirements and long lead times, nuclear plants face a risk premium that other generators do not. It has been estimated that

1,000–1,500 new reactors would be needed by 2050 for nuclear to play a meaningful role in reducing global emissions – a construction pace 20 times that of the past decade. Speed, however, is not one of nuclear power's virtues. Planning, licensing, and constructing any nuclear plant takes a decade or more, and plants frequently fail to meet completion deadlines. Due to the dearth of orders in recent decades, the world currently has very limited capacity to manufacture many of the critical components of nuclear plants. Rebuilding that capacity will take a decade or more.

Furthermore and in contrary to common perception nuclear power is no cleaner than technologies such as wind power. Depending upon the fuel enrichment process used, nuclear power has global warming potential (GWP) of 22–30 gCO_2/kWh. In comparison wind power has a GWP of 29 g CO_2/kWh, as reported by Hondo (2005). Furthermore, the social costs of nuclear power at (5.7–12 US ¢/kWh) have been estimated to be twice that of fossil fuels (2.2–5 US ¢/kWh) (Leggett 1990).

Figure 4 shows the result of a survey that was carried out in Britain with reference to people perception of risks from nuclear power. The result showed that an overwhelming majority of people has concerns in this respect (Tyndall Centre 2005).

Fossil fuels too have their fair share of problems and these may be summarized as follows: depleting reserves, high price fluctuations, climate change impact, security of supply, and potential for use as a political tool.

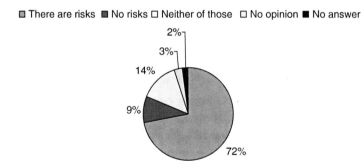

Figure 4. Result of a poll carried out to assess public's perception of risks from nuclear power plants

4.   Renewable Energy Technologies

The World Watch Institute has conservatively estimated that solar water heaters could provide half of the global hot water demand, the PV cells could contribute to 10% of the US electricity needs by the year 2030 and that just seven southern US states could generate 7 TW of solar thermal electricity (Gardner and Prugh 2008). The latter represents a capacity that is seven times the present total electricity requirements for the US. On the wind-powered generation front the above institute assesses that 20% of world's demands could be met whereas for the EU, with its much higher wind resource, the entire electrical energy needs could be satisfied.

Refer to Tables 3 and 4 that provide a breakdown of energy use and fast depleting fuel reserves in the two countries that have been chosen in this study as a means of demonstrating the applicability of RE technologies to displace fossil fuels. We note that transport and domestic sectors within India and UK respectively consume a third and two-thirds of the total energy use – a staggeringly high figure. Moreover, with the high rate of Indian population increase and added to that the increasing rate of energy use shown in Fig. 2 suggests an ever increasing total energy demand. With reference to Table 4 we see that UK is fast approaching a near complete fossil fuel exhaustion. In view of safety and environmental concerns pointed out in Section 3 and also due to a finite stockpile of fissile material nuclear energy is also not a viable option. Hence there is an express need to turn our attention to RE technologies. The following is a palette of present/futuristic RE technologies that are/will be available:

- Thermal energy generation
- Solar water heaters for domestic/commercial/industrial sectors
- Super-insulated windows
- Absorption space cooling
- Adsorption space cooling
- Daylighting
- Super-insulated windows with appropriate tint
- Light-guides
- Electrical energy generation (localised)
- BIPV
- Micro, or midi-wind turbines
- Electrical energy generation (centralised)
- Hyper-PV arrays
- Concentrating Solar Power
- Macro-wind turbines

In the rest of this section those technologies that are able to offset a substantial part of the load related to transport and domestic sectors within India and UK shall be discussed.

4.1. LIGHTING OF BUILDINGS

As shown in Table 3 lighting consumes a significant part of the national energy budget. One half of the national energy used within UK is related to buildings and a fifth of that energy is consumed by electric lights. A large part of that load can be reduced by the use of daylight – via enhanced use of super-insulated windows and light-pipes. Furthermore, help is also available in the form of energy efficient electric lights. The following section will cover this important cost centre.

4.1.1. *Super-insulated windows*

In the latter part of this section the luminous efficacy data for electric lighting appliances shall be presented. It will be shown that the luminous efficacy varies between 15 and 96 lm/W for most lamps. In contrast the luminous efficacy of daylight is much higher, i.e. for global (total) illuminance it has an average figure of 110 lm/W, while for sky-diffuse illuminance that latter figure rises to 120 lm/W. With the advent of super-insulated windows the inherent disadvantage associated with single glazing, i.e. daylight gain but at the cost of higher heat loss has been overcome. At the top end of the scale we now have silver-coated, xenon-filled triple-glazing that offers a U-value of just $0.4 W/m^2$-K. Thus daylight can now be introduced quite handsomely without any serious tax on the space-heating load. Figure 5 shows the evolutionary development of super-insulated windows. Note that Table 3 shows the disparate progress with respect to thermal insulation within the EU member states with Sweden registering a low heat-loss index for its building stock. There is clearly a case for other EU states to follow the latter example. A low heat-loss index ensures effective exploitation of passive solar solutions.

4.1.2. *Light-pipes*

The basic light-pipe design is a fairly simple concept. A highly transparent polycarbonate collector gathers light at rooftop level then channels light down a thin, highly reflective aluminium tube (with 95–99% reflectance). The pipe may also be made of plastic, but with a total internal reflective film bonded to it. At the bottom end of the pipe a diffuser disperses daylight throughout the desired area. Light-pipes have the advantage of capturing daylight from the entire sky hemisphere. Furthermore, due to the fact that most rooftops have clear, unobstructed views of the sky they are better suited to channel daylight in buildings. Windows on the other hand have the problem of shading, particularly for inner-city buildings.

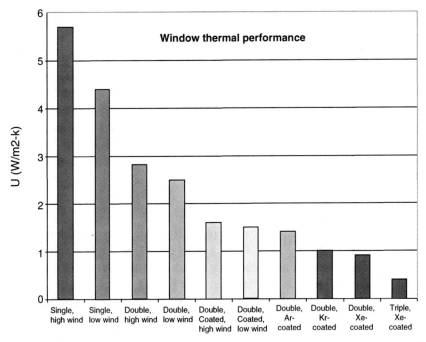

Figure 5. Evolution of super-insulated windows – by the introduction of inert gases and low-emissive glass coating the heat loss has been reduced by a factor of 14

4.1.3. *Electric lighting*

This section is largely based on material that was made available by Grinfeld and Grinfeld (2009). Lighting of buildings consumes 19% of global electricity generation, more than the combined output of hydroelectric and nuclear power stations. It produces CO_2 that is equivalent to 70% of emissions of global passenger vehicles. It has been estimated that 66% of all lamps installed in the EU at present are energy inefficient. The percentage for domestic use is even greater at 85%. If the whole of Europe switched to Compact Fluorescent Lamps (CFLs), there would be a saving of 28 million tonnes of CO_2, the equivalent to that produced by 12 power stations.

4.1.4. *Conventional lamps*

Incandescent (GLS) lamps use a tungsten wire that is heated to 2,700 K thus making it glow. The luminous efficacy is only 15 lm/W, and 95% of the energy is wasted as heat. Lifetime is only 750–2,000 h because of degradation of the filament, which also causes blackening of the bulb and reduces light output through its lifetime. Tungsten halogen lamps are similar in design but have a bromine compound added to the gas in the bulb which allows them to operate at a higher gas pressure and temperature. The luminous efficacy is increased to between 16 and 18 lm/W and the average lifetime to 1,000–2,000 h.

Linear fluorescents lamps are referred to by letter "T" followed by a number referring to the diameter of the lamp in eighths of an inch. For example, T8 has a 1-in. diameter. They operate by passing a current through a gas containing a mercury compound. The mercury atoms become excited and emit radiation, which is turned into visible light by a coating on the tube. As they operate at low temperatures they have a much higher luminous efficacy of between 60 and 96 lm/W, and a much longer lifetime of 12,000–20,000 h.

4.1.5. *Energy efficient lamps*

The EU Energy Performance in Buildings Directive aims to eliminate the sale of inefficient lamps by 2012. CFLs can be used in place of incandescent lamps. They have integrated ballasts, and this accounts for their relatively high initial cost. In terms of light output a 20 W CFL is equivalent to a 100 W tungsten lamp. However, CFLs contain small quantities of mercury and as such provision must be made for their disposal. The life of CFLs is 8,000–10,000 h. T5 fluorescent lamps are thinner than T8 lamps and also more efficient, offering a higher intensity of light output. Typical life for a T lamp is 20,000 h.

Light emitting diodes (LEDs) are electronic light sources. When current circulates through an LED, energy is released in the form of light, a phenomenon called electroluminescence. High energy LEDs produce 40–80 lm/W. LEDs do not burn out like GLS lamps, their light output declines slowly with time. In view of that behavior the concept of estimated useful life, i.e. the point at which lighting output has been reduced to 70% of the original, has been developed. For LEDs, L70 is 35,000–50,000 h.

4.1.6. *Energy efficient controls*

Effective lighting controls can reduce the artificial lighting load by 30–40%. Control strategies can be broken down as follows:

- *Manual controls.* Occupants will initially turn on lighting according to daylight levels, but will often turn on all the lighting and not turn it off again until everyone has left the room. A basic type of control can be used to create flexible and energy-saving lighting options, for example, half of the lamps in each fixture could be switched together, every other fixture could be switched together to uniformly reduce the light levels, or lighting near the windows could be turned off.

- *Timed systems* can turn off lights automatically at set times. Occupants can then switch them on again, but if natural lighting levels have improved they rarely do so. Clock switches can be used in conjunction with light-level sensors.
- *Photoelectric switching switch lights* on or off according to daylight levels at a control point. This may result in continual switching if natural lighting levels are oscillating around the cut-off value, due to partly cloudy conditions for example. Therefore, these systems are better suited for well-lit perimeter areas where the lights are normally off during the day.
- *Photoelectric dimming systems* also monitor natural lighting conditions, but use this information to alter artificial lighting levels to provide constant illuminance.
- *Occupancy sensors* serve three basic functions: to turn lights on automatically when a room becomes occupied; to keep the lights on without interruption while the controlled space is occupied; and to turn the lights off within a preset time after the space has been vacated. There are three types of occupancy sensors: passive infrared (PIR), which are triggered by the movement of a heat-emitting body through their field of view; ultrasonic sensors which emit an inaudible sound pattern and reread the reflection; and dual-technology occupancy sensors which use both passive infrared and ultrasonic technologies. PIR occupancy sensors are best suited to small, enclosed spaces such as private offices, where the sensor replaces the light switch on the wall and no extra wiring is required. PIR sensors cannot "see" through opaque walls or partitions, so occupants must be in direct line-of-sight view of the sensor. This is not a problem though with ultrasonic sensors.
- *Programmable logic controllers* (PLC5) are centralised lighting management systems that can reduce energy use significantly in buildings where the lights are often left on all night even when they are not needed. Sensors are installed to monitor daylight level, occupancy status, or both. A central controller that turns on, turns off, or dims lights in a building according to a preset program processes the information. For example, turning on the lights before employees arrive, dimming them during periods of high power demand or turning lights off at the end of the working day. By the exploitation of super-insulated windows, light-pipes and the above-surveyed high efficiency lamps and controls very significant reduction in lighting related electricity use can be achieved.

4.2. SOLAR WATER HEATING

The Energy Saving Trust in Scotland has estimated that solar water heaters are capable of displacing 50% of the hot water load within dwellings. Across the EU and the rest of the world that latter proportion is of course much higher owing to the fact that there is a healthier solar radiation budget in those locations.

The author has developed a highly efficient design of a solar water heater that is capable of achieving efficiencies exceeding 55% mark even when the water is to be heated to temperatures as high as 80°C. Such higher duty is expected in industrial sectors such as textile mills.

4.3. SOLAR SPACE HEATING: SUPER-INSULATED WINDOWS AS SOLAR ENERGY PROVIDERS

Within the United Kingdom, the European Union or the US and Canada, more energy is used to maintain comfortable internal environments in buildings than for any other single purpose. The energy bill associated with the energy consumption due to heating of buildings is over 25% for the UK and 35% for the US. Environmental conditioning in non-residential buildings accounts for about 15% of the UK energy consumption, nearly two thirds of which is used for space heating, one third for lighting, and about 5% for cooling.

Direct use of the sun's energy, through passive solar design in buildings, is one of the most economically attractive ways of utilising renewable energy in Europe and North America. Passive solar design uses a building's form and fabric to capture, store and distribute solar energy received, thereby reducing the demand for heat and artificial light. At its simplest it consists of siting buildings so that large glazed areas can face south, free from overshadowing, minimising glazing on north-facing walls and incorporating complementary energy efficiency features, such as adequate roof and wall insulation and automatic controls on heating systems. Without an added capital expenditure penalty, passive solar design enhances the internal environment of buildings. Passive solar concepts offer an ideal opportunity to reduce energy bills for industrial, commercial and domestic buildings. Moreover, this form of solar energy technology is both proven and economical.

The role of windows in the exploitation of passive solar energy, both for space heating and daylighting is unreservedly important. Most building regulations discriminate against windows in favour of ordinary opaque insulating materials. The result is that limiting window areas are recommended while prescribing minimum allowable U-values. The regulations make no allowance for the functions of the window, e.g. its transparency to solar heat and daylight.

4.4. SOLAR PHOTOVOLTAIC

In April 2005, a 160 m^2 photovoltaic (PV) array was added to the façade of one of the buildings that belongs to Edinburgh Napier University. AC power from this PV facility is fed into the University grid. Careful and continuous monitoring of the latter facility has been conducted whereby the DC and AC PV energy output and solar irradiation on the plane of the tilted PV panels have been measured at a frequency of 15 min. This activity was completed over a period four years. The study by Muneer et al. (2007) provides a detailed assessment of the latter facility. Some of the most important conclusions of that study, adjusted for the current grid electricity price are as follows:

(a) The PV array delivers 11 MWh per annum of AC electricity.
(b) Based on a weighted-averaged figure of 460 gCO_2/kWh for UK grid, the CO2 saved is 5 tonnes per annum.
(c) The embodied energy payback time was found to be 8 years.
(d) On the basis of an LCA audit, it was found that the facility under discussion has a carbon footprint of 44 gCO_2/kWhe. This value is higher than hydroelectric power plants at 17 gCO_2/kWhe and wind power at 29 gCO_2/kWhe, but it is much lower than that of coal-, oil-, and liquefied natural gas fired power plants with respective values of 920, 760, and 560 gCO_2/kWhe.
(e) The proportional cost breakdown for the PV facility was PV panels (70%), design and project management (6%), reinforcement of older wall structure (9%), and installation costs (15%).
(f) The pay back period is around 50 years. Note that since the latter installation in 2005 the price of PV panels has dropped very sharply (see Fig. 6) and also the grid electricity price is steadily on an incline. Hence it is difficult to 'pin down' a concrete value for the pay back period. The current price of PV panels, quoted by Chinese manufacturers is around 1.7€/W for orders worth 1 MW or more.

In Section 5 of this article the concept of 'Energy cooperatives' shall be introduced. With a third of UK electricity consumption being due to the housing sector alone, it might be worthwhile for installation of PV systems for complete housing colonies that would be able to buy bulk and save costs.

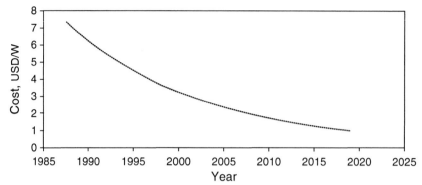

Figure 6. PV price profile. By mid-2009 Chinese manufacturers were quoting a price of 1.7€/Wp for orders over 1 MW

The downward trend for PV price would be most welcome for developing countries with very many locations simply not grid connected. Table 5 shows the potential for PV and other technologies under discussion for India. Note that in the warmer countries of the developing economies such as India and most of Africa the PV cell temperature can reach much higher than Normal Operating Cell Temperature (NOCT). For example, a Libyan research team (Eljrushi and Zubia 1995) has reported a measured operational cell temperature of 127°C! All manufacturers' data is related to the latter NOCT and hence significant correction in the cell power output has to be made. Generally speaking, in the 80–90°C temperature range a silicon PV module loses 0.5% efficiency per Celsius of cell temperature rise.

Table 6 that shows the amount of area required for building sustainable energy farms. In the preparation of these tables it has been assumed that only 50% of the energy needs will be met by the proposed energy farms, with the rest of the demand being met by either decentralized renewable energy production and/or other sustainable means such as hydroelectric or bio-mass based projects. Furthermore, bearing in mind the availability of a given resource, i.e. wind or solar for any given country and also keeping the relevant future economics of the two resources a solar/wind generation ratio has been proposed within Table 6. Note that the proposed area for collection of solar and wind energy by means of ultra-large scale farms in fact will occupy a mere fraction of the available land and near-offshore area for the respective countries, e.g. a solar PV electricity farm of 26 km^2 required for India represents 0.01% land area for Rajasthan desert. The above areas required for the farms may be further split to form a cluster of smaller energy farms.

TABLE 5. Status of RE technology in India

Energy sources	Potential	Achievement as on 31.12.2002	India's rank in the World
Biogas plants	12 Million	3.37 Million	Second
Improved Chulhas	120 Million	33.9 Million	Second
Wind	45,000 MW	1,702 MW	Fifth
Small hydro	15,000 MW	1,463 MW	Tenth
Biomass power/cogeneration	19,000 MW	468 MW	Fourth
Biomass gasifiers		53 MW	First
Solar PV	20 MW/km^2	107 MW$_p$*	Third
Waste to energy	2,500 MW	25 MW$_e$	
Solar water heating	140 million square metre collector area	0.68 Million square metre collector area	

*Of this, 46 MW$_p$ solar PV products have been exported.

TABLE 6. Solar and wind farm sizing for a developed and a developing economy

	UK	India
Thermal energy requirements for 2020, TWh	6,021	2,903
Electrical energy requirements for 2020, TWh	909	456
Available solar energy, kWh/m^2-annum	2,599	1,155
Annual-average wind speed, m/s	5	12
Efficiency of wind turbine	0.2	0.4
Available wind energy per m^2 of intercepted area, kWh/m^2-year	39	1,090
Solar/wind generation ratio	3/2	1/4
Solar PV farm size for generating required electricity, km^2	26	16
Solar thermal collection farm size for generating required thermal energy, km^2	54	56
Total area of interception required by wind turbines, km^2	68	13
Number of nominally rated wind turbines of 5 MW capacity	5,430	1,034
Wind farm length for generating required electrical energy, km	42	18
Wind farm width for generating required electrical energy, km	23	10

4.5. WIND TURBINES

Figure 7 shows the latitude-dependent availability pattern for solar and wind energy for locations across the globe. In general terms locations within the tropics would benefit greatly from solar energy – water heating for thermal demand and PV for their electricity needs. Most of the developed economies lie in the belt where wind power has now become

economically competitive. However, there still exists a significant inertia in
the installation of wind farms on a sizable scale. Table 7 shows the present
status of the wind turbine installed generation capacity in the EU. A great
disparity is seen in this table, even for countries with like GDP/capita (refer
to Table 2). Furthermore what is ironical is that UK with the highest
potential for wind power lags far behind countries such as Denmark, Ireland,
Germany and Spain. A study carried out by Asif and Muneer (2007) has
shown that large-scale wind turbines can deliver 50% of the UK electricity
demand with only a fraction of offshore/onshore area exploited.

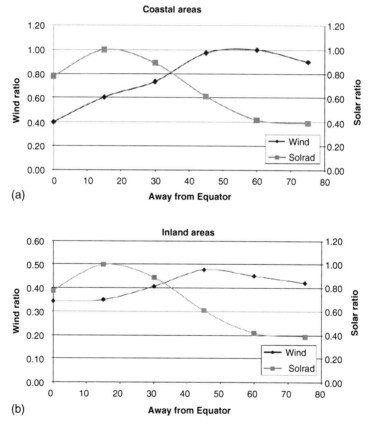

Figure 7. Latitudinal effect on proportion of available solar and wind energy (a) coastal, and
(b) inland locations

A point that is worth mentioning here is that there is a hype that micro-
wind turbines installed on rooftops, particularly in urban or suburban areas,
may deliver significant amount of energy. However, a study, named Warwick
trials, undertook the monitoring of rooftop mounted micro-wind turbines
across the UK and showed that the annual energy output ranged only

between 18 and 657 kWh, in contrast to the average electricity demand of 3,000 kWh for a UK home. The corresponding wind turbine capacity factors varied between less than 1% to a maximum of 7%.

TABLE 7. End of year 2007 EU wind turbine installed generation capacity, MW

Country	Capacity	Population, million	W/capita
Austria	982	8	120
Belgium	287	10	28
Bulgaria	70	7	10
Czech	116	10	11
Denmark	3,125	5	571
Estonia	58	1	44
Finland	110	5	21
France	2,454	64	39
Germany	22,247	82	270
Greece	871	11	81
Hungary	65	10	7
Ireland	805	4	196
Italy	2,726	58	47
Latvia	27	2	12
Lithuania	50	4	14
Luxembourg	27	0	56
Netherlands	1,746	17	105
Poland	276	39	7
Portugal	2,150	11	202
Romania	8	22	0
Slovakia	5	5	1
Spain	15,145	40	374
Sweden	788	9	87
UK	2,389	61	39
Totals for EU27	56,535	490	115

4.6. SOLAR POWERED PERSONAL TRANSPORT

It is becoming increasingly clear that unrestricted and unplanned growth of automobile ownership and use is unsustainable. Currently, there are more than 600 million vehicles on roads across the world, with a further 60 million new vehicles being added each year. About two-thirds of the world's oil is consumed by the transport sector, with road vehicles accounting for 40% of the total output. In terms of greenhouse gas emissions the transport sector in general and road vehicles in particular, respectively contribute 25% and 22% of the total.

It was shown in Table 3 that nearly two-fifths of UK's national energy budget is due to transport. For India the latter figure is one-fifth. In UK personal road transport alone is responsible for 16% of the total energy use. Among the developing economies India is one of the fastest-growing markets. As India's GDP continues to rise by some 8% per year over the next decade, the number of drivers in India is set to soar from seven in 1,000 today to 11 in 1,000 by 2010 – a doubling since 2000 when 0.5% of Indians had a car (Bouachera and Mazraati 2007). Furthermore, India is ranked second in the world after the USA, in terms of transport network, encompassing 3.3 million kilometres of roads. The road transport is contributing about 80% of land transport (passenger and freight) compared to all other modes. The transport infrastructure in the road sector has expanded considerably over the last decades in order to maintain the required service demand and economic growth. In 2004 there were 17.4 million vehicles and 47.5 million motorcycles (two wheelers), which consumed about 34% of total oil demand of India (IEA, 2006). In the past few years the growth rate for two-wheelers was 11%, which constituted more than 70% of total vehicles.

In view of the unchecked rise of vehicle emissions and as a potential solution, experiments have been undertaken at Edinburgh Napier University with the view to explore the use of more sustainable forms of urban transport (Muneer et al. 2009). Hence an electric scooter that is charged with the University's own solar photovoltaic facility has been employed to investigate its emissions. An instrumented electric scooter was driven along selected routes within the City of Edinburgh with a view to measure and record key parameters such as clock time, scooter speed, electric motor power demand and the distance covered. The road gradient information was obtained from detailed maps. An MS-Excel/VBA simulation program was developed to simulate the four modes of driving that may be encountered on any given route, i.e. cruise on a level road, gradient climbing, deceleration and acceleration, taking account of the road friction and aerodynamic drag. The simulation program returns the cumulative energy value for the driving cycle with an accuracy of 96% or more. It was also found that, like with like, the scooter requires only a sixth of the energy used by an average-sized automobile. However, further reductions in the energy demand of scooters may be achieved by technological improvements such as the use of electrical capacitors that may be charged during downhill sections of a route and which also conserve energy during braking modes. The capacitors may then be used to prevent rapid discharge from the battery during acceleration and hill-climbing modes. It was also noted that for the Edinburgh driving cycle, with an idling period that occupies a time fraction of 27% of the total journey time, the present fuel consumption of automobiles may be reduced by 9.6%, if engines were switched off during idle mode. The energy consumed for the scooter is 0.056 kWh/km (5 km/MJ) with the respective monetary and environmental costs of 0.8 US ¢ and 33.3 gCO_2/km.

Estimates of well-to-wheel efficiency have been presented in literature (Randall, 2009a, b) that quote figures of 0.9- and 0.3 km/MJ for electric and diesel cars respectively. In the former case the estimate was based on the assumption that electricity was sourced from gas-fired generation. The environmental impact is also in favour of electric cars with an emission of 65 gCO_2/km as opposed to that of the most efficient diesel vehicle at 100 g/km. If however, PV-electricity was used instead for charging of electric cars then the latter performance figure would drop from 65 to 7 gCO_2/km. Note that the corresponding figure for the PV-electricity charged scooter is 1.3 gCO_2/km.

Solar powered scooters (and automobiles) would therefore be a particularly attractive solution to developing economies such as India and China with most of their major conurbations experiencing a mild climate almost throughout the year.

5. Energy Cooperatives

In the above section a case was made for large-scale introduction of RE technologies within the developed and developing economies. Developed economies with their established infrastructures are well placed to fully exploit the benefits of RE technologies. Developing economies on the other hand stand on cross roads, i.e. their many locations do not have access to grid electricity and hence they can go down the way of RE with a de-centralized approach. Most developing economies lie within the temperate belt and hence with their rich harvest of solar energy they would benefit most significantly from the RE revolution.

Some 5,000–7,000 years ago the Aryans from the northern countries moved southwards to the lush and fertile Ganges plains in search of fodder for their cattle. It is ironical that we are now seeing a return to those times in the form of EU energy producers aiming to move to deserts of Middle East and North Africa (MENA) for very large scale generation of electricity using hyper-PV and Concentrating Solar Power (CSP). The White Paper produced by TREC (2008) describes this scheme in greater detail. The author's research team (Aldali et al. 2009) is indeed engaged in one such project for the Libyan Desert at Kufra oasis. The latter location with large water aquifers and abundant irradiation is ideal for CSP application. The main findings of the latter design exercise were as follows: with an absorber tube with an inner and outer diameter respectively equal to 5 and 7 cm, and glass cover of 9 cm diameter it is possible to produce superheated steam at 500°C at a pressure of 100 bar. The length of tubing required would be 636 m (53 modules, each of 12 m length). It was found that just under half of the time during its 7AM-5PM operation, the power plant would produce superheated steam with only a small temperature lift required from the supplementary gas boiler to take the steam to its design operating condition of 500°C. The overall result is that a solar fraction of 76% is obtained, thus producing a steady power of 897 kW. The required desert area would be 4,500 m². Similar projects, whether they are desert- or elsewhere based, may be developed to exploit large-scale PV or wind turbine technology. The funding and management of such decentralized projects may be achieved through creation of 'Energy cooperatives'. This would be of particular significant economic benefit to communities within the developing economies. Clean and sustainable energy would then truly become a tradable commodity. The concept of Energy cooperatives could be further extended so that the communities start producing the fabrication plants to produce solar hot water modules, wind turbine components and even PV panels. The cost of those technologies would then drop sharply as all activity would be under-taken on a non-profit basis with the sole beneficiary being the global environment!

6. Conclusion

A review of the globally used nuclear-fissile and fossil fuels and the sectorial demands for developed and developing economies has been presently undertaken. It has been shown that the present rate of consumption and exhaustion of the latter fuels is unsustainable. If the latter fuels are to be replaced by renewable energy sources there will have to be a palette of technologies that will enable the switch over. Presently, the above tech-nologies have been presented and their appropriateness discussed. The latitude dependence of the availability of solar and wind energy has also been discussed. It was shown that while wind energy is able to satisfy nearly all of the energy demand for developed economies such as the EU, solar offers an attractive option for most of the developing economies. It was also demonstrated that large-scale wind generators, as opposed to micro wind turbines is the way forward.

Passive solar for space heating, solar water heating and energy efficiency within the lighting sector have also been shown to have come to maturity.

It has been shown that solar-PV charged electric vehicles, particularly two-wheelers for the developing economies, offer an order of magnitude reduction of carbon emissions.

References

Aldali, Y., Henderson, D. and Muneer, T., 2009, Proceedings of ES2009, ASME 3rd International Conference on Energy Sustainability, July 19–23, 2009, San Francisco, California, USA (accepted for publication).

Asif, M. and Muneer, T., 2007, Energy supply, its demand and security issues for developed and emerging economies. Renewable and Sustainable Energy Reviews, 11, 1388–1413.

Bouachera, T. and Mazraati, M., 2007, Fuel demand and car ownership modelling in India. OPEC Review. Organization of the Petroleum Exporting Countries, Vienna.

Eljrushi, G. S. and Zubia, J. N., 1995 Photovoltaic power plant for the southern region of Libya. Applied Energy, 52, 219–227.

Gardner, G. and Prugh, T., 2008, State of the world 2008, Innovations for a sustainable economy, World-watch Institute, W.W. Norton, New York.

Grinfeld, M and Grinfeld A., 2009, Energy efficient lighting, 43–45, CIBSE: BSD Journal, May 2009, London.

Grubler, A., 2004, Transitions in energy use. Encyclopedia of energy, vol. 6, Elsevier, London.

Hondo, H., 2005, Life cycle GHG emission analysis of power generation systems: Japanese case study. Energy, 30, 2042–2056.

http://www.berr.gov.uk/energy/statistics/source/total/page18424.html Energy consumption by final user (energy supplied basis) 1970 to 2007 (DUKES 1.1.5). Accessed on May 1, 2009.

IEA, 2006, International Energy Agency, online IEA data, www.data.iea.org. Accessed on May 1, 2009.

Leggett, J., 1990, Global warming – the Greenpeace report. Oxford University Press, Oxford, UK.

Muneer, T, Younes, S, Lambert, N and Kubie, J., 2006, Life cycle assessment of a medium-sized photovoltaic facility at a high latitude location. Journal for Power and Energy, 220, A6, 517–524.

Muneer, T., 2004 Solar radiation and daylight models, Elsevier, Oxford.

Muneer, T., Abodahab, N., Weir, G. and Kubie, J., 2000, Window in Buildings, Elsevier, Oxford.

Muneer, T., Clarke, P. and Cullinane, K., 2009, The electric scooter as a means of green transport. Proc. IMechE Conf. on Low-Carbon vehicles 2009, 27–36. Automobile Division, Institution of Mechanical Engineers, London.

Randall, B., 2009a, Professional Engineering, 20 May 2009, p. 17. Institution of Mechanical Engineers, London.

Randall, B., 2009b, Electric cars – are they really green? Proc. IMechE Conf. on Low-Carbon vehicles 2009, 17–26. Automobile Division, Institution of Mechanical Engineers, London.

Solar Energy International, 2004, Photovoltaic design and installation manual, Carbondale, Colorado, USA.

TREC, 2008 Clean power from deserts: The DESERTEC concept for energy, water and climate security (White Paper). Trans-Mediterranean Renewable Energy Cooperation, Hamburg, Germany.

Tyndall Centre, 2005 Public Perceptions of Nuclear Power, Climate Change and Energy Options in Britain http://www.tyndall.ac.uk/publications/energy_futures_full_report.pdf Accessed on May 1, 2009.

CADMIUM FLOWS AND EMISSIONS: CDTE PV FRIEND OR FOE?

MARCO RAUGEI[*]

*Environmental Management Research Group (GiGa),
Escola Superior de Comerç Internacional,
Universitat Pompeu Fabra, Pg. Pujades 1, 08003 Barcelona,
Spain*

Abstract CdTe PV has been growing exponentially since its recent introduction to the market, and already accounts for almost 5% of the global PV sector. In order to estimate the potential impact that this technology may end up having on the global cadmium flows, the author has drafted three scenarios for this technology up to 2050, considering potential market growth and technological improvements. On the Cd demand side, a future large-scale deployment of CdTe PV was found to be potentially beneficial, since it would entail sequestering a non-negligible fraction of the Cd that will be mined as a by-product of Zn. On the emission side, potential global Cd emissions to air from CdTe PV in 2050 are expected to remain over four orders of magnitude lower than current documented Cd emissions to air in the EU-27.

Keywords: Renewable energy, photovoltaics, CdTe, cadmium, LCA, prospective analysis

1. Past Growth Record and Future Projections

Photovoltaics (PV) stands among the most lively and promising technologies in the whole renewable energy sector, and has been growing exponentially ever since the early 1990s (Fig. 1).

---

[*] To whom correspondence should be addressed: Marco Raugei, Escola Superior de Comerç Internacional, Universitat Pompeu Fabra, Pg. Pujades 1, 08003 Barcelona, Spain; E-mail: marco.raugei@admi.esci.es

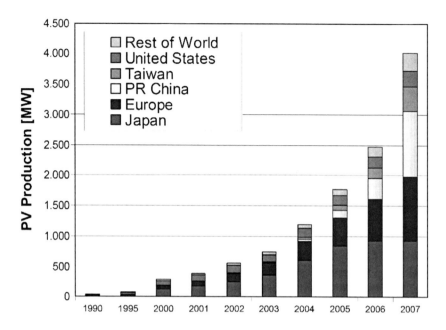

Figure 1. Global PV sector growth record (1990–2007) (Jaeger-Waldau 2008)

Within the PV sector, non-Si-based thin film technologies have been gaining in importance due to their comparatively lower production costs and lower energy and materials demands during production.

In particular, comparing Figs. 2a b shows that the share of cadmium telluride (CdTe) PV has grown from 1.6% to almost 5% of the global PV market in as little as 2 years (EPIA 2006, 2008). When combined with the expansion of the PV market as a whole, this translates into a nearly 400% increase in installed power per year (Fig. 3).

In order to estimate the potential impact that this technology may end up having on the global cadmium flows, far-reaching projections are needed in terms of market development. Research Stream 1a of the EU project NEEDS – New Energy Externalities Developments for Sustainability (Frankl et al. 2008) produced three such development scenarios for the PV sector as a whole, as well as for the three main 'families' of PV technologies (Si-based, thin film and third generation). The three scenarios presented here for CdTe PV are largely based on those, but have been updated and revised by taking into account the very latest market and technological developments. The main underlying assumptions are listed below:

1. 'Pessimistic' scenario: economic incentives to PV are assumed to be curtailed before it achieves grid parity, which results in stunted growth of the whole sector, and slow technological improvement and low market penetration for CdTe PV.
2. 'Reference' scenario: PV market growth is assumed to be in line with the 'moderate' scenario drawn by the European Photovoltaic Industry Association until 2025 (EPIA 2008), gradually slowing down afterwards. The market share of CdTe PV is assumed to keep growing rapidly as it has done in the last 5 years, with concurrently relevant technological improvements.

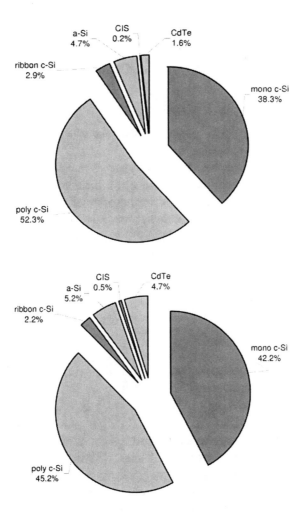

Figure 2. Global market shares of PV technologies in 2006 (a) and in 2008 (b) (After EPIA/ Greenpeace 2006, 2008)

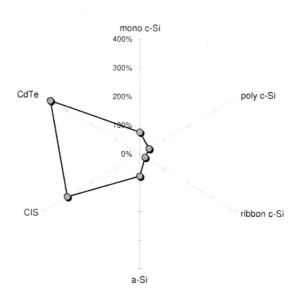

Figure 3. Respective relative growth rates of PV technologies from 2006 to 2008

3. 'Optimistic' scenario: bolder annual growth rates are assumed throughout (this scenario remains however still a bit conservative compared to
EPIA's 'advanced' scenario); in terms of relative market share, CdTe
PV is assumed to keep growing as in the reference scenario, before
reaching an upper limit of 1 TWp of installed power, which is assumed
to be consistent with likely limits in tellurium supply.

The outcomes of these three scenarios in terms of CdTe PV installed
power are illustrated in Fig. 4.

2. Demand-Side Projections

Cadmium sulphide is virtually the only chemical form in which Cd appears
in nature in concentrated form, and is not generally present in significant
quantities in isolated deposits on its own, but it is nearly always associated
with zinc sulphide (sphalerite).

The average Cd/Zn ratio in commercially exploited mineral ores is
0.003 (Liewellyn 1994, Ayres 2002), and a similar proportion is carried
over to Zn concentrates. It is extremely relevant to note that since the two
elements are always found together, Zn producers do not have the option of
not producing Cd, at least to the stage of the impure Cd metal sponge.

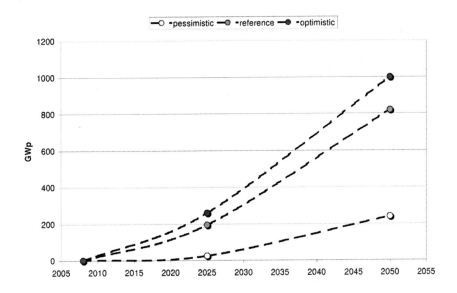

Figure 4. Global scenarios of CdTe PV installed power (2008–2050)

Following a steady increase in demand, worldwide primary Zn production has been growing exponentially in the last 3.5 decades, reaching almost 11 million tonnes/year in 2007 (Fig. 4).

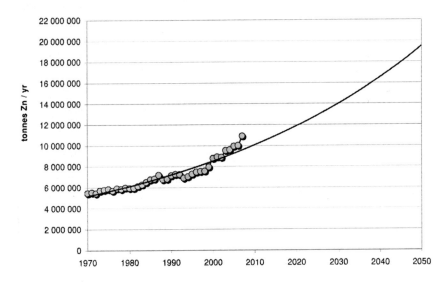

Figure 5. World production of primary zinc (1970–2005), plus exponential trend-line (After USGS, 2009a, b)

Global cadmium production, instead, has remained fairly constant over the same period, at approximately 20,000 tonnes/year (Fig. 6). Subtracting 17.5% thereof for secondary (recycled) Cd (UNEP, 2006), one obtains 16,500 tonnes of primary Cd.

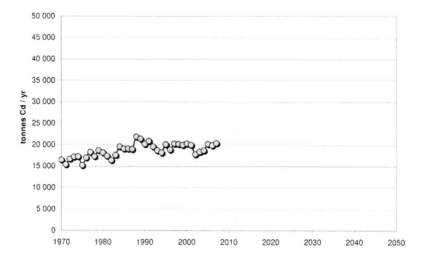

Figure 6. World production of cadmium (primary plus secondary) (After USGS, 2009a, c)

This figure is however well below the theoretical recoverable amount that can be calculated on the basis of primary Zn production and the average Cd content of Zn ores. In fact, the latter amounts to $11,000,000 \times 0.003 = 33,000$ tonnes/year, i.e. almost double the actual marketed amount. Since no market exists for this large surplus of Cd, it ends up not being officially accounted for, and a large part thereof probably remains stockpiled in unrefined form at the mines. In the opinion of the author, the lack of strict control over such a huge flow of Cd should however remain a serious reason for concern.

In the last few decades, there have been four principal industrial uses for cadmium: rechargeable (NiCd) batteries, pigments, plating, and plastic stabilizers. All other applications collectively accounted for less than 0.5% of total Cd consumption in 2005.

Since the late 1980s, there has been a clear exponential reduction in cadmium use for all applications, except for the battery sector, where Cd demand has levelled off to an almost steady state at around 14,000 tonnes/ year.

It is then interesting to extrapolate these trends to the future, and combine them with the CdTe PV scenarios discussed in Section 1, assuming global Cd recycling to grow from 17.5% to 30% in 2050, and assuming improvements in CdTe PV manufacture to attain a gradual reduction in Cd demand per Wp. The resulting projected global shares of primary Cd demand in 2050, according to the 'reference' and 'optimistic' scenarios are illustrated in Fig. 7a and b, respectively.

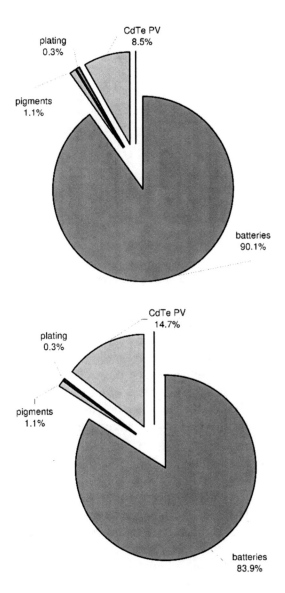

Figure 7. Prospective analysis: global shares of primary Cd demand by application in year 2050, according to 'reference' (a) and 'optimistic' (b) scenarios

Figure 8 then illustrates the resulting estimates in terms of projections of global primary Cd demand, compared to the total amount that would be theoretically recoverable as a by-product of Zn, assuming a constant Cd/Zn ratio in the ores and an exponentially growing production of primary Zn (Fig. 5).

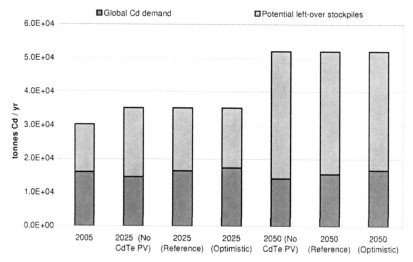

Figure 8. Prospective analysis: projections of global primary Cd demand versus total amount theoretically recoverable as by-product of Zn production, showing potential left-over stock-piles of Cd as difference

Of course, this kind of analysis is inevitably affected by a number of large uncertainty factors; however, one central point stands out clearly: Zn demand will go on producing larger and larger amounts of left-over cadmium for which there will likely be no ready commercial use. In this sense, the future large-scale deployment of CdTe PV may be regarded as having a beneficial effect on the global environment, by effectively sequester-ing a non-negligible fraction thereof and putting it to comparatively safe use.

3. Emission-Side Projections

In spite of commonly voiced concern about the presence of cadmium in CdTe PV, the use phase of the modules presents virtually no risks, and extremely low experimental levels of cadmium emissions to air have even been reported in simulated fires (Fthenakis et al. 2004). This is largely due to the fact that the modules actually only contain few grams of cadmium per square metre, and all that is there is present in the form of two thermally and chemically stable binary compounds (CdTe and CdS), which in case of fire would remain safely enclosed in a matrix of molten glass.

From a life cycle point of view, three more stages need to be taken into consideration, namely: Cd extraction and refining, CdTe production and module manufacturing, and module decommissioning.

Since Cd production is entirely driven by Zn demand, it is reasonable to allocate any Cd emissions that may occur during the mining phases to Zn.

Cd emissions during the following phases of CdTe production and module assembly were analyzed by Fthenakis (2004). Extrapolating his results to the future, under the same assumptions that were made for the purposes of making the demand-side projections illustrated in Section 2, it has been possible to evaluate the potential global Cd emissions to air associated to the large-scale deployment of this technology.

Figure 9 compares such results for the 'optimistic' scenario (i.e. largest potential emission flow, associated to 1TWp of installed power in 2050) to the current routine Cd emissions to air in the European Union in the year 2000, on the basis of the outcome of the European project ESPREME (2006). The results are striking, in that the former turn out to be over four orders of magnitude lower than the latter.

A further integration of this analysis is under way by Raugei and Fthenakis, whereby these results will also be extended to include the module decommissioning phase (including CdTe recycling), and the associated emissions to water.

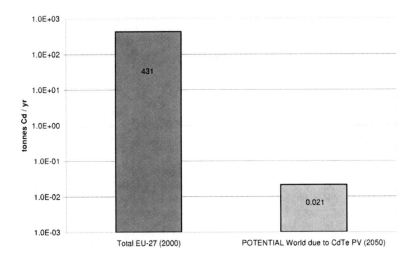

Figure 9. Routine Cd emissions to air in EU-27 in 2000 (After ESPREME 2006), compared to potential global Cd emissions to air due to CdTe PV in 2050 ('optimistic' scenario = 1 TWp installed power)

4. Conclusions

As is common for all long-term prospective analyses, the results of this study are inevitably affected by large degrees of uncertainty, and quantitative comparisons between scenarios that stretch several decades into the future should always be taken with a grain of salt.

Nonetheless, in the opinion of the author, two important qualitative results are worthy of careful consideration.

On the one side, it has been shown that the potential large-scale deployment of CdTe PV could actually end up having a positive effect in terms of sequestration of left-over unrefined Cd.

On the other side, even the rosiest scenario in terms of commercial success of CdTe PV is not likely to make a significant contribution to the global Cd emission flows to the atmosphere, which are dominated by far more offensive sectors such as fossil fuel-fired power plants and boilers and cement production.

Acknowledgements

The author wishes to acknowledge the financial support provided by Brookhaven National Laboratory (USA), and the valuable collaboration of Dr. Vasilis Fthenakis in carrying out parts of this study.

References

Alsema E., 2000, Energy pay-back time and CO_2 emissions of PV systems. *Prog. Photovoltaics Res. Appl.* **8**(1):17–25.

Ayres R. U., Ayres L. W. and Råde I., 2002, The life cycle of copper, its co-products and byproducts. MMSD – Mining, Minerals and Sustainable Development; http://www.iied.org/mmsd/mmsd_pdfs/ayres_lca_main.pdf

Brown T.J., Hetherington L.E., Hannis S.D., Bide T., Benham A. J., Idoine N.E., Lusrty P.A., 2007, World mineral production 2003 – 2007. British Geological Survey; http://www.bgs.ac.uk/mineralsuk/downloads/wmp_2003_2007.pdf.

EPIA/Greenpeace, 2006, Solar generation III – solar electricity for over one billion people and two million jobs by 2020. Greenpeace and European Photovoltaic Industry Association, The Netherlands/Belgium.

EPIA/Greenpeace, 2008, Solar Generation V – Solar electricity for over one billion people and two million jobs by 2020. Greenpeace and European Photovoltaic Industry Association, The Netherlands/Belgium.

ESPREME, 2006, EU project 'Estimation of willingness-to-pay to reduce risks of exposure to heavy metals and cost-benefit analysis for reducing heavy metals occurrence in Europe'; http://espreme.ier.unistuttgart.de.

Frankl P., Menichetti E., Raugei M., 2008, Final report on technical data, costs and life cycle inventories of PV applications, NEEDS deliverable 11.2 – RS Ia; http://www.needs-project.org.

Fthenakis V.M., 2004, Life cycle impact analysis of cadmium in CdTe PV production, *Renewable Sustainable Energy Rev.* **8**:303–334.

Fthenakis V.M., Alsema E., 2006, Photovoltaics energy payback times, greenhouse gas emissions and external costs: 2004–early 2005 status. *Prog. Photovoltaics Res. Appl.* **14**:275–280.

Fthenakis V.M., Heiser, J., Fuhrmann, M. and Wang, W., 2004, Experimental Investigation of Emissions and Redistribution of Elements in CdTe PV Modules During Fires, 19th European Photovoltaic Energy Conference, Paris.

Fthenakis V.M., Kim H.C., Alsema E., 2008, Emissions from photovoltaic life cycles. *Environ. Sci. Technol.* **42**:2168–2174.

Jaeger-Waldau J., 2008, PV Status Report 2008 – Research, solar cell production and market implementation of photovoltaics. European Commission Joint Research Centre Technical Notes. EUR_23604EN_2008; http://sunbird.jrc.it/refsys/pdf/PV%20Report%202008.pdf.

Kato K., Hibino T., Komoto K., Ihara S., Yamamoto S., Fujihara H., 2001, A life-cycle analysis on thin film CdS/CdTe PV modules. *Sol Energ. Mater. Sol. Cell.* 67:279–287.

Liewellyn T.O., 1994, Cadmium (Material Flows). United States Department of the Interior, Bureau of Mines Information Circular 9380.

Raugei M., Bargigli S., Ulgiati S., 2007. Life cycle assessment and energy pay-back time of advanced photovoltaic modules: CdTe and CIS compared to poly-Si. *Energy* **32**(8):1310–1318.

UNEP, 2006, Interim review of scientific information on cadmium. United Nations Environmental Programme; http://www.chem.unep.ch/Pb_and_CD/SR/Files/Interim_reviews/UNEP_Cadmium_review_Interim_Oct2006.pdf

USGS, 2009a, Kelly, T.D., and Matos, G.R., comps., Historical statistics for mineral and material commodities in the United States; http://minerals.usgs.gov/ds/2005/140.

USGS, 2009b, Minerals Yearbook – Zinc. United States Geological Survey; http://minerals.usgs.gov/minerals/pubs/commodity/zinc.

USGS, 2009c, Minerals Yearbook – Cadmium. United States Geological Survey; http://minerals.usgs.gov/minerals/pubs/commodity/cadmium.

INCREASING RES PENETRATION AND SECURITY OF ENERGY SUPPLY BY USE OF ENERGY STORAGES AND HEAT PUMPS IN CROATIAN ENERGY SYSTEM

GORAN KRAJAČIĆ[1]*, BRIAN VAD MATHIESEN[1,2],
NEVEN DUIĆ[1,3], AND MARIA DA GRAÇA CARVALHO[3]
[1]*Department of Energy, Power Engineering and Environment;*
University of Zagreb, Zagreb, Croatia
[2]*Department of Development and Planning;*
Aalborg University, Aalborg, Denmark
[3]*Department of Mechanical Engineering, Instituto Superior*
Técnico, Lisbon, Portugal

Abstract In this paper integration of wind power generation into the Croatian electricity supply is analysed using available technologies. The starting point is a model of the energy system in Croatia in 2007. Comprehensive hour-by-hour energy system analyses are conducted of a complete system meeting electricity, heat and transport demands, and including renewable energy, power plants, and combined heat and power production (CHP) for district heating. Using the 2007 energy system the wind power share is increased by two energy storage options: Pumped hydro and heat pumps in combination with heat storages. The results show that such options can enable an increased penetration of wind power. Using pumped hydro storage (PHS) may increase wind power penetration from 0.5 TWh, for existing PHS installations and up to 6 TWh for very large installations. Using large heat pumps and heat storages in combination with specific regulation of power system could additionally increase wind penetration for 0.37 TWh. Hence, with the current technologies installed in the Croatian energy system the installed pumped hydro-plant may facilitate more than 10% wind power in the electricity system. Large-scale integration of wind power in the Croatian energy systems requires new technologies in other parts of the energy system.

---
* To whom correspondence should be addressed: Goran Krajacic, Department of Power Engineering and Environment, Faculty of Mechanical Engineering and Naval Architecture, Luciceva 5, 21000 Zagreb, Croatia; E-mail: goran.hrajacic@fsb.hr

F. Barbir and S. Ulgiati (eds.), *Energy Options Impact on Regional Security*,
DOI 10.1007/978-90-481-9565-7_8, © Springer Science + Business Media B.V. 2010

Keywords: Renewable energy sources, penetration, wind energy, pumped hydro, heat pumps

1. Introduction

The primary energy imports dependence of European Union is currently around 53%, and it is expected that in the next 20–30 years it will reach or surpass 70% if no measures to prevent this are made. The situation in Croatia is similar. In 2007 the import dependence was 53.1% while for 2030 it is predicted to reach 72%. Such import dependence leads to decreased security of energy supply, due to current geopolitical situation in which main sources of fossil fuels are in unstable regions and in which the competition for those resources from developing countries is growing. The EU energy strategy, and a comparable Croatian strategy, is focused on policies and measures that will bring increase of share of renewable and distributed energy sources, increase in energy efficiency and energy savings and decrease in green house gas emissions.

The results of previous research has shown that in order to increase efficiency and viability, there is need for energy storage, in the primary or secondary form, in order to transfer energy surplus form period of excess to the period when there is a lack. The problem of storage systems is that they increase the cost of already expensive distributed and renewable energy sources, making them, in market circumstances, even less economically viable. Although there are a number of storage technologies, as chemical, potential or heat energy, not all those technologies are optimal for each energy system. Several authors have shown that by integration of energy and resource flows it is possible to decrease costs and that by rational energy managing and financial support that takes into account externalities, it is possible to devise such a system to be environmentally, economically and socially acceptable (Duic et al. 2003, 2004, 2008; Krajacic et al. 2008; Lund and Mathiesen 2009; Aalborg University 2009; Lund et al. 2007, 2009; Lund 2005, 2006; Lund and Munster 2003a, b; Lund and Kempton 2008; Mathiesen 2008; Mathiesen and Lund 2009; Blarke and Lund 2008; Lund and Salgi 2009; Lund and Clark 2002).

As the import of fuel to Croatia is increasing one of the measures to increase the security of supply is to increase the share of renewable energy in the electricity sector. Wind power generation are especially promising, as there are good wind sites in Croatia and as this technology is one of the best developed renewable energy technologies.

This paper presents analyses of different scenarios for development of Croatian energy system with integration of wind using energy storages. The current amount of wind is very small in Croatia, hence this paper focuses on solutions that could be implemented for increasing the amount of wind power in the short term. Emphasis is put on fuel efficiency for the integration technologies as the aim is to enhanced Croatia's security of supply. The storage technologies used in the analyses are pumped hydro and heat pumps.

2. Energy System Analyses Tool

Detailed energy system analysis are performed by use of the freeware model EnergyPLAN (Aalborg University 2009). The model is an input/output model that performs annual analyses in steps of 1 h. Inputs are demands and capacities of the technologies included as well as demand distributions, and fluctuating renewable energy distributions. A number of technologies can be included enabling the reconstruction of all elements of an energy system and allowing the analyses of integration technologies. The model is specialised in making scenarios with a large amount of fluctuating renewable energy and analysing CHP systems with large interaction between the heat and electricity supply. EnergyPLAN was used to simulate a 100% renewable energy-system for the island of Mljet in Croatia (Lund et al. 2007) and the entire country of Denmark (Lund and Mathiesen 2009). It was also used in various studies to investigate the large-scale integration of wind energy (Lund 2005), optimal combinations of renewable energy sources (Lund 2006), management of surplus electricity (Lund and Munster 2003a), the integration of wind power using electric vehicles (EVs) (Lund and Kempton 2008), the potential of fuel cells and electrolysers in future energy-systems (Mathiesen 2008; Mathiesen and Lund 2009), and the effect of energy storage (Blarke and Lund 2008), compressed-air energy storage (Lund and Salgi 2009; Lund et al. 2009) and thermal energy storage (Lund 2005; Lund H, Clark 2002; Lund and Munster 2003b).

In the model it is possible to use different regulation strategies putting emphasis on heat and power supply, import/export, and excess electricity production and using the different components included in the energy system analysed. Outputs are energy balances, resulting annual productions, fuel consumption, and import/exports.

It provides the possibility of including restrictions caused by the delivery of ancillary services to secure grid stability. Hence, it is possible to have a minimum capacity running during all hours and/or a percentage running from a certain type of plants required to secure voltage and frequency in the electricity supply grid.

3. Reference Energy System

The Croatian energy system for 2007 has been reconstructed in the EnergyPLAN model. Energy consumption and supply data have been taken from (Energy in Croatia 2008; Vuk and Simurina 2009) while hourly load data for Croatian power system have been provided by UCTE (UCTE (ENTSO-E). Basic data about power producing units have been obtained from Croatian utility company (http://www.hep.hr/proizvodnja) and from Energy in Croatia (2008). Hourly production data for hydro power plants have been reconstructed from monthly values provided in UCTE (ENTSO-E) while hydro storage capacities have been taken from Geres (2007). Load curve for hourly district heating demand was calculated according yearly consumption in Croatia (Vuk and Simurina 2009) and according patterns of hourly heat demand in Denmark that are provided by EnergyPLAN model.

Hourly wind power production was calculated by use of hourly wind speeds provided by program METEONORM (http://www.meteonorm.com/pages/en/meteonorm. php) for year 1995. Croatia was divided into eight regions (Dubrovnik, Istria, Knin, Lika, Senj, Sibenik, Split, Zadar) and for each region METEONORM provides wind speeds that are based on real measurement data or in case that measured data do not exists for exact location, METEONORM provides interpolation of wind speeds from the three nearest meteorological stations. To obtain hourly wind power production, hourly wind speeds are processed in H_2RES program. Firstly the average wind speed for each region is adjusted to match speeds that are stated in Schneider et al. (2007) and then, by use of power curve for 2 MW Vestas V90 wind turbine, total installed power of wind generators is increased in each region to match installed power presented by authors in Bajs and Majstrovic (2008). Resulting curve for hourly distribution of wind power production is used in EnergyPLAN calculations.

4. Comparison of Reference Scenarios and Statistics

Reference scenarios that are calculated by EnergyPLAN model are compared to statistical data for Croatia in order to see if they could represent situation in 2007 (Table 1). The biggest difference is in hydropower total primary energy supply as EnergyPLAN do not multiplies energy delivered by hydropower plants. Difference in natural gas occurs as reference scenarios do not include non-energy consumption of natural gas in order to have better overview of results for energy sector. Difference in electricity production in closed system operation occurs mostly due to fact that in these calculations all electricity demand is satisfied primarily from local production units and not from the import.

TABLE 1. Comparison of statistics and calculations of reference scenarios

	Statistics (Energy in Croatia 2008; Vuk and Simurina 2009)	EnergyPLAN reference	Diff. from stat.	EnergyPLAN reference (closed system operation)	Diff. from stat.
Total primary energy supply (TWh)					
Coal and coke	8.04	7.98	0.7%	8.01	0.3%
Liquid fuels	54.03	52.6	2.6%	58.28	−7.9%
Natural gas[a]	31.73	26.64	16.0%	29.99	5.5%
Hydro power[b]	11.73	4.39	62.6%	4.39	62.6%
Renewables	3.89	4.15	−6.7%	4.15	−6.7%
Electricity-import	6.36	6.36	0.0%	2.71	57.4%
TOTAL	115.78	102.12	11.8%	107.53	7.1%
CO_2 – emissions (Mt)					
TOTAL	24.86	22.18	11%	24.39	1.9%
Electricity supply (TWh)					
Hydro power plants	4.40	4.39	0.2%	4.39	0.2%
Wind power plants	0.03	0.04	−14.6%	0.04	−14.6%
Thermal power plants	5.18	5.17	0.2%	8.57	−65.4%
Public CHP plants	2.12	2.11	0.3%	2.11	0.3%
Industrial CHP plants	0.51	0.51	0.6%	0.51	0.6%
Import–export	6.36	6.36	0.0%	2.97	53.3%
TOTAL	18.61	18.58	0.1%	18.59	0.1%
District heating supply (TWh)					
Public CHP plants	2.41	2.41	0.0%	2.41	0.0%
Public heating plants	0.83	0.83	0.0%	0.83	0.0%
TOTAL	3.24	3.24	0.0%	3.24	0.0%

[a]Non-energy consumption of natural gas not included in EnergyPLAN reference
[b]Due to different accounting. Croatian energy accounting uses factor to levelize hydro power production comparable with thermal power plants

5.   Energy Systems Analyses Results

The potential of introducing integration technologies into the reference energy system is analysed by varying the amount of renewable energy in the electricity system. In this study installed wind power generation is varied from 17 to 4818 MW that corresponds to electricity generation from 0.04 to 11.64 TWh. The total demand is 18.6 and 2.71 TWh is covered by import (nuclear).

The technical energy system analyses are conducted for a period of 1 year taking into consideration demands and renewable energy production during all hours. The ability of the reference energy system to integrate fluctuating wind power is analysed with (1) the capability of the system to avoid excess electricity production and (2) the ability of the system to reduce fuel consumption and thus improve fuel efficiency. The methodology applied to these analyses is presented here:

Hour-by-hour the electricity production from wind power is prioritised as well as the production of electricity at CHP plants, industrial CHP and import (nuclear). The remaining electricity demand is met by power plants and the remaining district heating demand is met by boilers. By utilising extra capacity at the CHP plants combined with heat storages, the production at the condensing power plants is minimised and replaced by CHP production. At times when the demand is lower than the production from CHP and wind power, the critical excess electricity production (CEEP) is minimised mainly by use of either heat pumps or pumped hydro. This constitutes an open energy system in which the technologies are utilised with the aim of supplying demands in the system. The measures introduced to secure the balance between the supply from CHP and wind power and the electricity demand described may not be sufficient to reduce electricity production, and thus forced electricity export will be the result. This type of technical energy system analyses enables the investigation of the flexibility of the seven integration technologies, focusing directly on the effect on excess electricity production, i.e. regulation strategy (1) of the two types of energy system analyses.

The results of modeling analyses are shown in Fig. 1. The x axis in Fig. 1 illustrates the installed wind power between 17 and 4,818 MW TWh, or production equal to a variation from 0% to 62.5% of the total demand (18.61 TWh) or 0% to 100% of the demand that is not covered by nuclear import or domestic hydropower production. The y-axis illustrates the excess electricity production in TWh. The less ascending curve illustrates a better integration of RES.

The second of the two types of energy system analyses, i.e. regulation strategy 3), builds on the first analyses. However, here any excess electricity production is converted or avoided; first, by replacing CHP production by boilers in the district heating systems and, secondly, by stopping wind turbines. The import/export is of course zero, as it is a closed system. All excess electricity production is converted or avoided and the entire primary energy supply (PES) excluding the RES (wind power) of the system is presented.

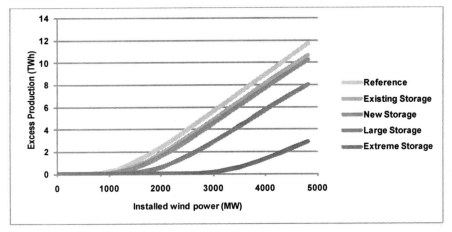

Figure 1. Installed wind power and excess electricity production in an open energy system analysis of the reference energy system without PHS and with different sizes of PHS.

The results of such analyses, shown in Fig. 2, represent a *closed* energy system. The x axis shows the installed wind capacity and the y axis illustrates the PES excluding the RES of the entire energy system. The less PES excl. the RES, the more flexible and fuel-efficient is the energy system. Again, a total of nine energy system analyses have been conducted hour-by-hour for a year for both types of CHP regulation. The advantage of presenting PES excl. the RES instead of PES incl. the RES is the fact that such results can reveal the ability of a technology to utilise RES, such as wind power, to efficiently replace other fuels.

The analyses are conducted with the following restrictions in order to secure the delivery of ancillary services and achieve grid stability (voltage and frequency). At least 40% of the power or as a minimum 490 MW (at any hour) must come from power production units capable of supplying ancillary services, such as central PP and CHP. The distributed generation from RES and small CHP units is not capable of supplying ancillary services in order to achieve grid stability. In the analyses here, the Croatian energy system is treated as a one point system, i.e. no internal bottlenecks are assumed.

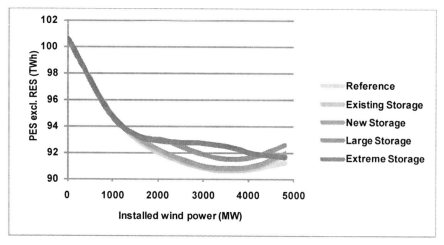

Figure 2. Primary energy supply (PES) in a closed energy system analysis of the reference energy system with and without PHS and CHP regulation

It is not the aim to recommend the precise optimal solutions for integration of RES in this paper. However, it is the aim to provide information on which technologies are fuel efficient and able to integrate RES and which approximate sizes are relevant.

Possibilities for reduction of CO_2 emission from energy sector is presented on Fig. 3. Diagram also shows that for different sizes of PHS there is optimum of installed wind due the fact that there is certain amount of stabilization load that needs to be satisfied from conventional units.

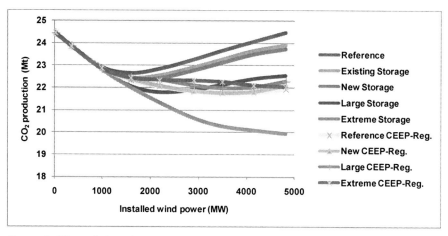

Figure 3. CO_2 emissions in the reference energy system with and without PHS and CHP regulation

6. Wind Power Generation Distribution Curves

Using the reference energy system for 2007 the amount of wind power is increased. Such analyses require considerations about the future distribution of the production from wind power in Croatia. The results of analysis will be more accurate if hourly wind production data will be available or at least hourly wind speeds that are measured on the sites where wind power plants will be constructed. In 2007, in Croatia there were only two wind farms with installed power of 17 MW which is not enough for good representation of wind power production for whole country, for example in 2007 Denmark had approximately 3,000 MW with long history of collecting data which enables good input for EnergyPLAN calculations. Constructed data for wind power production in Croatia in comparison with Danish are presented in Fig. 4.

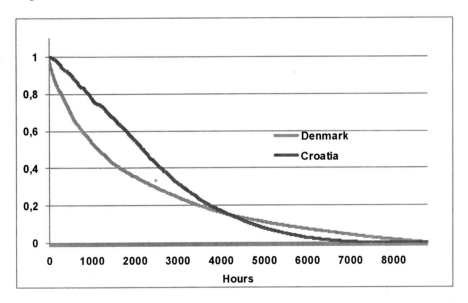

Figure 4. Comparison of constructed wind power production data for Croatia with real wind production in Denmark (EnergyPLAN wind distribution data)

7. Technologies Integration

In this section the technologies, capacities and efficiencies of the analysed technologies integration are presented.

In the reference scenario existing pumped hydro storage in the system is removed in order to have better overview of the results. Installed power the pumps and turbines, capacities of storages and efficiencies of technologies are given in Table 2.

In the group of calculations without pumped hydro storages, large scale-heat pumps and heat storages in district heat areas are used in order to increase wind power penetration and decrease CEEP. Heat storage capacity in all scenarios is set to 20,000 MWh, tested sizes of heat pumps are 25, 75 and 150 MWe and COP (coefficient of performance) is set to 3.

TABLE 2. Technology data in PHS scenarios

Scenarios Technology	Existing storage	New storage	Large storage	Extreme storage
Turbines (MW)	281.74	375.19	936.19	3,196.00
Pumps (MW)	246.00	344.90	1,018.90	3,169.00
Storages (GWh)	24.14	40.86	274.00	450.00
Turbines eff.	0.83	0.83	0.85	0.85
Pumps eff.	0.78	0.78	0.83	0.83

Finally, one calculation is made integrating both pumped hydro technologies with heat storages and large heat pumps. Sizes of PHS are the same as in New Storage scenario with 150 MWe in heat pumps and heat storage of 20 GWh. Used Regulation strategy is 3 with balancing heat and electricity demand with reduction of CHP and shutting down PP and RES units. Results for integrating scenario are presented in Table 3.

TABLE 3. Results of calculation in integrating scenario

Installed wind (MW)	CEEP (TWh)	PES without RES (TWh)	RES (TWh)	CO_2 Mt
17	0.00	100.61	4.44	24.46
360	0.00	98.41	5.27	23.85
1,000	0.08	94.81	6.80	22.89
1,600	0.36	92.99	7.96	22.42
2,200	0.53	91.79	8.70	22.12
2,900	0.65	90.87	9.26	21.89
3,500	0.75	90.40	9.56	21.77
4,150	0.92	90.42	9.72	21.78
4,818	1.43	91.34	9.83	22.03

8. Conclusions

Paper presents a new approach in planning of Croatian energy system with significant emphasis on integration of wind energy by use of different energy storage technologies and system regulation strategies. It shows that 10% of total electricity demand could be covered by wind energy without any significant change in current system and without exporting of electricity excess to neighbouring countries. With specific regulation of wind power plants, current large conventional power plants and CHP penetration could be doubled or at 20%. But this will lead to significant rejection of RES potential, around 30% and to utilize this excess, large pumped hydro storages should be build.

The better results for penetration could be achieved by decreasing stabilization load that is provided by conventional units or by replacing it with large hydro power stations. This option was not included in presented calculations so it will be worth of investigation. Moreover, use of existing CHP and installations of heat storages should be further investigated if the current district heating systems become based on lower temperature operation. Reason that the storage and heat pumps have so little effect (Fig. 2.4.1) is, that the district heating demand is small compared to the installed PP (of which the 500 MW CHP is part) and the electricity demand. Hence the results with storage does not change so much, because the alternative CHP could operate as condensing units.

Results of this study should be taken with certain reserve until constructed data for wind speeds and wind power production are verified by other calculations/measurements or until they are compared to similar results for other countries (Denmark, Ireland etc.). The calculated load factor for wind power plants in Croatia is at 27% which is to high as experts expect it to be somewhere between 20% and 25%. Another big issue is stabilization load which certainly could be provided by large hydropower so it should be somehow included in calculations as EnergyPLAN has few options to regulate that issue. It also shows importance that new power plants that will be built in Croatia in next 10 years should be made with bigger flexibility in operation. Moreover as there are 4,818 MW of wind power applications in current Croatian registry for wind power plants. Extreme pumped hydro energy storage might not be feasible do to economically and environmentally concerns but it could provide long term solution for 100% RES system and should be compared to investment into nuclear power plant.

Presented scenarios are just a part of current research efforts which investigate planning and development of Croatian energy system and it will cover other locally RES that could be utilized in Croatia (solar, geothermal, biomass, small hydro) and will include other energy consuming sectors (industry, transport, buildings). Further work will also include economic analysis, market optimization, analysis of environmental and social aspects of proposed development scenarios and improvement of reference energy system.

Acknowledgments

The authors would like to thank the National Foundation for Science, Higher Education and Technological Development of the Republic of Croatia for supporting the project "Role of the Smart Energy Storage in 100% Independent Energy Systems" that resulted in this work.

References

Aalborg University. EnergyPLAN: Advanced Energy System Analysis Computer Model. See also: http://energy.plan.aau.dk/, accessed on 10th October 2009.
Bajs D, Majstrovic G., 2008. The feasibility of the integration of wind power plants into the electric power system of the Republic of Croatia. Energija, 57(2):124–155.
Blarke MB, Lund H., 2008. The effectiveness of storage and relocation options in renewable energy systems. Renewable Energy, 33(7):1499–1507.
Data provided by UCTE (ENTSO-E), http://www.entsoe.eu/resources/data/packages/ .accessed February 5, 2010.
Duic N, Carvalho MG., 2004. Increasing renewable energy sources in island energy supply: case study Porto Santo. Renewable and Sustainable Energy Reviews, 8:383–399.
Duic N, Lerer M, Carvalho MG., 2003. Increasing the supply of renewable energy sources in island energy systems, International Journal of Sustainable Energy, 23(4):177–186.
Duic N, Krajacic G, Carvalho MG., 2008. RenewIslands methodology for sustainable energy and resource planning for islands, Renewable and Sustainable Energy Reviews Renewable and Sustainable Energy Reviews, 12(4):1032–1062.
Energy in Croatia – Annual Energy Report – 2007. Ministry of Economy Labour and Entrepreneurship, Zagreb (2008).
Geres D., 2007. Water resources and irrigation systems in coastal and karstic regions of Croatia. Priručnik za hidrotehničke melioracije: Vodnogospodarski aspekti razvoja navodnjavanja u priobalju i krškom zaleđu Hrvatske. Faculty of Civil Engineering, University of Rijeka, 23–68.
http://www.hep.hr/proizvodnja. accessed September 2009.
Krajacic G, Martins R, Busuttil A, Carvalho M.G., 2008. Hydrogen as an energy vector in the islands' energy supply. International Journal of Hydrogen Energy, 33(4):1091–1103.

Krajacic G, Duic N, Graça Carvalho Md. H$_2$RES, 2009. Energy planning tool for island energy systems – The case of the Island of Mljet. International Journal of Hydrogen Energy, 34(16):7015–7026.

Lund H., 2003. Excess electricity diagrams and the integration of renewable energy. International Journal of Sustainable Energy, 23(4):149–156.

Lund H., 2005. Large-scale integration of wind power into different energy systems. Energy, 30(13):2402–2412.

Lund H., 2006. Large-scale integration of optimal combinations of PV, wind and wave power into the electricity supply. Renewable Energy, 31(4):503–515.

Lund H, Clark WW., 2002. Management of fluctuations in wind power and CHP comparing two possible Danish strategies. Energy, 27(5):471–483.

Lund H, Munster E., 2003a. Management of surplus electricity-production from a fluctuating renewable-energy source. Applied Energy, 76(1-3):65–74.

Lund H, Munster E., 2003b. Modelling of energy systems with a high percentage of CHP and wind power. Renewable Energy, 28(14):2179–2193.

Lund H, Kempton W., 2008. Integration of renewable energy into the transport and electricity sectors through V2G. Energy Policy, 36(9):3578–3587.

Lund H, Mathiesen BV., 2009. Energy system analysis of 100% renewable energy systems – The case of Denmark in years 2030 and 2050. Energy, 34(5):524–531.

Lund H, Salgi G., 2009. The role of compressed air energy storage (CAES) in future sustainable energy systems. Energy Conversion and Management, 50(5):1172–1179.

Lund H, Duic N, Krajacic G, Graça Carvalho Md., 2007. Two energy system analysis models: A comparison of methodologies and results. Energy, 32(6):948–954.

Lund H, Salgi G, Elmegaard B, Andersen AN., 2009. Optimal operation strategies of compressed air energy storage (CAES) on electricity spot markets with fluctuating prices. Applied Thermal Engineering, 29(5–6):799–806.

Mathiesen BV., 2008. Fuel cells and electrolysers in future energy systems, PhD Thesis, Department of Development and Planning, Aalborg University, Aalborg, Denmark. See also: http://people.plan.aau.dk/~bvm/FinalWebVersion3.pdf.

Mathiesen BV, Lund H., 2009. Comparative analyses of seven technologies to facilitate the integration of fluctuating renewable energy sources. Renewable Power Generation, IET, 3(2):190–204.

METEOTEST's METEONORM – Global Meteorological Database for Engineers, Planners and Education, http://www.meteonorm.com/pages/en/meteonorm. php. accessed October 2009.

Schneider DR, Duić N, Bogdan Ž., 2007. Mapping the potential for decentralized energy generation based on renewable energy sources in the Republic of Croatia. Energy, 32(9):1731–1744.

Vuk B, Simurina I., 2009. Energy in Croatia 1945–2007. Energy Institute Hrvoje Pozar; Zagreb (May 2009) http://www.eihp.hr/hrvatski/projekti/euh45.html.

GOALS OF THE EUROPEAN UNION IN USING RENEWABLE ENERGY SOURCES AND ABILITIES OF THE REPUBLIC OF CROATIA IN REALIZATION OF THOSE GOALS

NIKO MALBAŠA[*]
Ekonerg – Energy Research and Environmental Protection Institute, Zagreb, Croatia

Abstract   Increased use of renewable energy sources is one of the key assumptions of EU sustainable development in energy sector and in fulfilling the request of decreasing greenhouse gases emissions. In its energy strategy, Croatia has accepted the basic EU guidelines in that area. It is also preparing to contribute to that European initiative. There are still some questions at the level of EU goals as well as in the Croatian future energy plan. Those questions relate to technical, economic and environmental parameters in connection with the future development. Feasibility of certain goals can be called into question, maybe not so much at the level of the entire EU but at the level of certain member states. What is the position of Croatia in that development? The article will briefly elaborate EU goals and the position of certain member states in implementation of those goals. The commitments of the Republic of Croatia and its possibilities and perspectives in realization of the set goals shall be analyzed and presented.

Keywords: Renewable energy, EU goals, energy statistics, wind power, biomass

1. Introduction

Energy is in all European countries traditionally treated as an activity of special national interest. That way EU energy policy has become a sum of energy policies of the member states. The sum of the national energy

---

[*] To whom correspondence should be addressed: Niko Malbasa, Ekonerg – Energy Research and Environmental Protection Institute, Koranska 5, 10000 Zagreb, Croatia, E-mail: niko.malbasa@ekonerg.hr

F. Barbir and S. Ulgiati (eds.), *Energy Options Impact on Regional Security*,
DOI 10.1007/978-90-481-9565-7_9, © Springer Science + Business Media B.V. 2010

policies is not the best policy for the EU as a whole. In addition, some new challenges cannot be solved without a comprehensive European energy strategy. Traditional attitude that energy is a matter of national energy security has led to diverging national energy policies that cannot solve some basic problems in energy by themselves, especially a chronic European dependence on import of practically all energy forms and the problem of reducing of greenhouse gas emissions.

The results of a number of initiatives that have been created in the last 15 years were very weak, almost negligible. The reason for that is a relative stability of world energy market and low prices of main energy sources as well as resistance of certain countries that had refused to give up their individual interests for mutual benefit.

Meanwhile a strong potential cohesive factor – a risk of climate changes caused by anthropogenic emissions of so-called greenhouse gases – had appeared. It is a global problem, the greenhouse gas emissions (mostly CO_2) equally affect regardless of its discharge into the atmosphere. Another characteristic of the greenhouse effect is that it is significantly a result of CO_2 emission in energy sector (about 35%).

Placing the problem of greenhouse gases emission in the first plan created the possibility of action in terms of producing the comprehensive European energy strategy as well as solving a number of challenges in the European energy sector, which such a strategy enables. Firstly, there are challenges associated with increased dependence on fossil fuels and their increasing import, increasing supply costs including issues related to the safety of supply.

Indeed, the EU's energy dependency on import of energy sources is very high and constantly growing. In 1990 it amounted to 44.6% and in 2006 it was 53.8%. The largest is import dependence on liquid fuel (83.7%) then gas (60.8%) and solid fuels (41.4%). The import of liquid fuels is dominated by imports from Russia (33.5%) and Norway (15.8%), followed by Libya (9.4%) and Saudi Arabia (9.0%). The largest import of gas is from Russia (42.0%), Norway (24.2) and Algeria (18.2%), and the import of coal is mainly from 6 countries (Russia 25.8%, South Africa 25.0%, Australia 12.8%, Colombia 12.3%, Indonesia 9.6% and USA 8.1%).

The key element towards sustainable development of European energy policy is identified renewable energy sources (RES). RES do not produce greenhouse gases, they are largely indigenous and do not depend on the future availability of fuels.

As the first step the Commission adopted a Green Paper on 20 November 1996. In that document the Commission sought views on setting an indicative objective of 12% for the contribution by renewable sources of energy to the

EU's gross national energy consumption by 2010. This was treated as an ambitious but realistic objective. The comparison with the obtained results shows that only 7.2% of renewable energy was realised by the 2006 with a small chance that the goal by 2010 will be achieved.

For the production of electricity from renewable sources, the Directive 2001/77/EC (Directive 2001) proposed 22.1% of electricity to be produced from renewable sources in 2010. Until 2006 an increase from 14.3% to 16.5% had been achieved.

The biofuel directive (Directive 2003) established a reference value of a 2% share for biofuels[1] in petrol and diesel consumptions in 2005 and 5.75% in 2010 compared to their share of 0.5% in 2003. The share achieved was very low, at 1%. Only three member states (Germany, France and Sweden) reached a share of more than 1%, Germany accounted for two thirds of total EU consumption.

In the meantime, the EU grew by 12 new members. Population has increased by 26.3%, from 392,262,000 to 495,578,000, and the surface for 33.6% (from 3,236,300 to 4,323,000 km^2). The total energy consumption (GIC) has increased by 18.4%, from 1,541.3 to 1,825.2 Mtoe (Statistical Pocketbook 2009), 2006.

Taking all these results into account European Council asked the Commission[2] to produce an analysis on how further to promote RES over the long term. The European Parliament has by an overwhelming majority called for a 25% target for renewable energies in the overall energy consumption by 2020 (resolution of 14 December 2006). Based on this, the Renewable Energy Road Map (2007) proposes that EU establishes a mandatory (legally binding) target of 20% for renewable energy share of energy consumption in the EU by 2020.

The main reasons for poor progress during previous years have been examined:

- RES were often not in the short term the least cost options. The failure to systematically include external costs in market prices gives an economically unjustified advantage to fossil fuels compared to renewables.
- The complexity, novelty and decentralized nature of most renewable energy applications result in numerous administrative problems.
- There are many examples of opaque and discriminating rules for grid access.

---

[1] Biofuels are liquid or gaseous fuels for transport produced from biomass
[2] Document 7775/1/06 Rev 10 of March 2006

The absence of legally binding targets, relatively weak EU regulatory framework for those of renewables in the transport sector and the complete absence of legal framework in the heating and cooling have caused the progress appeared only in a few Member States.

In addition to that the energy efficiency has not been as high as expected and the overall energy consumption therefore has been higher.

The year 2007 may be considered a turning point. On 14 February European Parliament adopted the resolution on climate change and in March the European Council (Heads of States and Governments of the 27 EU Member States) agreed to set precise, legally binding targets. The following key targets are the basis of a new policy:

- A reduction of at least 20% in greenhouse gases by 2020 – rising to 30% if there is an international agreement committing other developed countries to "comparable emission reduction and economically more advanced developing countries to contributing adequately according to their responsibilities and respective capabilities".
- A binding target of a 20% share of renewable energies in overall EU energy consumption by 2020.
- A 10% binding minimum target to be achieved by all Member States for the share of biofuels in overall EU transport petrol and diesel consumption by 2020, to be introduced in a cost-effective way. The binding character of this target is appropriate subject to production being sustainable, second-generation biofuels becoming commercially available and the Fuel Quality Directive being amended accordingly to allow for adequate levels of blending.

In addition to these binding targets the European Council "stresses the need to increase energy efficiency in the EU so as to achieve the objective of saving 20% of the EU's energy consumption compared to projections for 2020, as estimated by the Commission in its Green Paper on Energy Efficiency".

It is clear that the development of renewable energy sources will continue to be the central aim of EU energy policy. "A strong policy for RES deployment and energy efficiency will not only contribute to the dual objective of limiting EU's dependence on imported hydrocarbons and combating climate change, it can also contribute to competitiveness and Lisbon goals, in particular through the creation of high-quality jobs in Europe and in maintaining Europe's technological leadership in a rapidly growing global sector" (Renewable Energy Road Map 2007).

In order to answer the question whether the EU should adopt quantified targets for the share of renewable energy in 2020 and if so, for what amount and what form, several analyses and studies have been realized including the contribution of external experts.

For the modeling exercise various scenarios using the PRIMES and Green-X models have been carried out for the EU-27.

Finally, in April 2009, the European Commission adopted the Directive on the promotion of the use of energy from RES. It contains a series of elements to create the necessary legislative framework in order for 20% of renewable energy becomes a reality.

This Renewable Directive sets mandatory national targets for renewable energy shares of gross final consumption of energy (GFEC)[3] in 2020 which are calculated on the basis of the 2005 share of each country plus flat-rate increase of 5.5% and GDP-weighted additional increase. Interim targets are also set up for 2011/12, 2013/14 etc.

2. Status and Prospects of Renewable Energy Sources in EU

2.1. DETERMINATION OF GFEC IN 2005 AND AN ESTIMATE FOR 2020

Based on Eurostat data, Yearly Statistics 2006, GFEC[4] values for the member states of the European Union were determined for 2005. On average, GFEC is 4–5% higher than FEC. Total GFEC in 2005 for 27 EU member states was 1,227.5 Mtoe,[5] the highest in Germany, 228 Mtoe and the lowest in Malta, 0.6 Mtoe. On average, the consumption per capita is some 2.5 toe/year (from 1.24 toe/inh. in Romania to 9.3 toe/inh. in Luxemburg. For 1,000 € in GDP, on average it is spent 99 kgoe (from 67 in Ireland to 370 in Bulgaria). The five largest countries participating in the inhabitants with 62.5% and in GDP with 72.5% have approximately equal specific consumption of energy from 2,241 toe/inh. in Spain to 2,694 toe/inh. in France i.e. from 77 toe/1,000 € in Great Britain to 97 toe/1,000 € in Spain.

Various documents discuss energy efficiency and energy conservation but it is often forgotten that the energy conservation has its price that is sometimes much higher than the energy saved. In times of high prices of

---

[3] "Gross final consumption of energy" (GFEC) means the energy commodities delivered for energy purposes to industry, transport, households, services including public services, agriculture, forestry and fisheries, including the consumption of electricity and heat by the energy branch for electricity and heat production and including losses of electricity and heating distribution and transmission.

[4] GFEC was determined based on Eurostat data except for the losses in head production and distribution identified as a percentage of gross heat output (7% losses in production, 13% losses in distribution)

[5] Mtoe, million tons of oil equivalent, 41.868×10^6 GJ

energy, the energy conservation gains momentum but it is quite opposite when the energy prices are low.

An estimate of GFEC in 2020 as compared to 2005 has been made in this paper taking into account the following assumptions relating the increase in energy consumption:

Electricity consumption	1% p.a.
Fuel consumption in transport	0% p.a.
Consumption of conventional fuel in industry	−2% p.a.
Consumption of conventional fuel in services and households	0% p.a.
Heat consumption in services and households	0% p.a.
Consumption of renewable sources (excluding electricity)	5% p.a.
Consumption of other sources	0% p.a.

Decrease in higher consumption of electric energy cannot be avoided by 2020. Assumed increase of 1% per year is much lower than that achieved in the past. The same situation could be expected also in transport where the measures of energy conservation could have a significant impact but by 2020 it is hard to expect the change in the consumption trend. Stabilization of consumption at the level of 2005 seems to be very optimistic. On the other hand, much higher use of biomass in final consumption could be expected although a 5% annual increase is perhaps set too high. It is evident that the use of fossil fuels in industry will decrease and all the rest energy consumption should remain at the consumption lever from 2005.

Under those assumptions, GFEC would grow at an annual rate of 0.27% that would end up in 4% growth in 2020 as compared to 2005.

Table 1 shows the results of GFEC estimate in 2020.

TABLE 1. GFEC estimate in 2020

EU-27	GFEC (Mtoe)	
	2005	2020
Electricity consumption	237	275
Fuel in transport	347	347
Conventional fuel in industry	189	141
Conventional fuel in services and households	266	266
Heat consumption in services and households	30	30
Renewables in FEC (excluding electricity)	55	115
Heat and electricity looses	55	55
Other FEC	48	48
TOTAL	1,227	1,277

2.2. RES STATUS IN 2005

Renewable Energy Sources (RES) include energy from renewable non-fossil sources: wind, solar, aerothermal, geothermal, hydrothermal and ocean energy, hydropower, biomass, landfill gas, sewage treatment plant gas, and biogases.

Table 2 shows a status of RES in 2005 in the EU member states.

In 2005, totally realized participation of RES was 96 Mtoe, which was 7.8% of GFEC.[6] The participation of France is the highest (16.2 Mtoe), and then Germany (13.5 Mtoe), Sweden (11.2 Mtoe), Spain (8.5 Mtoe), Italy and Finland (6.5 Mtoe each), and Austria (6.1 Mtoe). Also important are (4.6 Mtoe), Poland (4.2 Mtoe), and Portugal (3.9 Mtoe).

Biomass had the highest share of RES in 2005, 60.9% altogether, of which the largest part (93%) accounts for wood and wood waste, which is followed by electric energy from hydroelectric power plants (28.5%), and from wind parks (6.0%). Biofuel accounts for 3.2% and the rest includes solar and geothermal energy (total 1.4%).

France is leading in using biomass with 10.8 Mtoe (2005), which is followed by Germany with 7.4 Mtoe, then Finland (5.6 Mtoe), and Sweden (5.5 Mtoe). Also significant are Poland (4.1), Spain (3.9), Romania (3,3), and then Portugal (2.8), Austria (2.7), and Italy (2.2).

In 2005, the participation of bio fuel in RES was only 3.2% i.e. 3.13 Mtoe. The highest production of bio fuel is in Germany (1.95 Mtoe), followed by France (0.41 Mtoe), Spain (0.26 Mtoe), Italy (0.18 Mtoe), and then Sweden).

The share of hydroelectric power plants in RES-u is 28.5% i.e. some 28 Mtoe. The highest share has Sweden (5.6 Mtoe), then France (5.2 Mtoe), Italy (3.7 Mtoe), Austria (3.2 Mtoe) in Spain (2.6 Mtoe). Significant potentials are in Germany (1.7 Mtoe), Romania (1.4 Mtoe), and in Finland (1.1 Mtoe) and Portugal (1.0 Mtoe).

Wind parks participate in RES with 6.0% i.e. 5.9 Mtoe. Approximately 77% of the output is in three countries: Germany (2.3 Mtoe), Spain (1.8 Mtoe) and Denmark (0.5 Mtoe). The remaining 1.3 Mtoe is distributed to another 15 EU member states.

---

[6] The results are based on Eurostat data. However, there is a certain difference with respect to the data published in [5] for some countries. The total amount obtained is 8% as compared to 8.6% in [5]. The highest difference is for Sweden (30.8% instead of 39.8%), Denmark (11% instead of 17%), and Finland (24.3% instead of 28.5%). It is not possible to identify the reason for those differences.

Solar energy participates in RES-u with some 0.8%, and geothermal energy with 0.6%, which is 1.44 Mtoe altogether. The participation of solar photovoltaic power plants is only 0.1% of RES i.e. 0.13 Mtoe.

TABLE 2. Renewable energy sources 2005, Mtoe

	Wood	MSW	Biogas	Biomass	Biofuel	Hydro	Solar PV	Solar Th	Geoth	Wind	All	GFEC
Belgium	0.70	0.07	0.05	0.86	0	0.028	0.000	0.003	0.001	0.020	0.91	39.3
Bulgaria	0.71	0.00	0.00	0.75	0	0.268	-	-	0.033	0.000	1.05	10.7
The Czech Republic	1.35	0.02	0.03	1.47	0.003	0.186	-	0.002	0	0.001	1.66	27.7
Denmark	1.01	0.23	0.04	1.34	0	0.002	0.000	0.010	0	0.523	1.87	16.5
Germany	6.17	0.52	0.40	7.42	1.948	1.713	0.110	0.243	0.127	2.266	13.82	228.0
Estonia	0.45	0.00	0.00	0.47	0	0.002	-	-	0	0.005	0.48	3.1
Ireland	0.18	0.00	0.02	0.20	0.001	0.057	-	-	0	0.088	0.35	12.7
Greece	0.96	0.00	0.02	1.02	0	0.365	0.000	0.102	0.001	0.115	1.60	21.9
Spain	3.58	0.08	0.09	3.91	0.259	2.627	0.004	0.061	0.008	1.767	8.64	101.5
France	9.51	0.69	0.12	10.78	0.408	5.218	0.001	0.021	0.130	0.093	16.65	166.7
Italy	1.82	0.23	0.10	2.24	0.179	3.739	0.003	0.027	0.213	0.221	6.62	136.5
Cyprus	0.01	0.00	0.00	0.01	0	0.000	0.000	0.041	0	0.000	0.05	1.8
Latvia	1.02	0.00	0.01	1.07	0.003	0.239	-	-	0	0.004	1.32	4.3
Lithuania	0.57	0.00	0.00	0.60	0.003	0.035	-	-	0	0.000	0.64	5.0
Luxembourg	0.02	0.00	0.00	0.02	0.001	0.008	0.002	0.000	0	0.005	0.04	4.5
Hungary	0.67	0.01	0.00	0.71	0.005	0.016	-	0.002	0.080	0.001	0.81	19.2
Malta	0.00	0.00	0.00	0.00	0	0.000	-	-	0	0.000	0.00	0.6
The Netherlands	0.52	0.24	0.09	0.89	0	0.008	0.003	0.019	0	0.174	1.09	53.6
Austria	2.59	0.03	0.01	2.75	0.042	3.180	0.001	0.091	0.006	0.117	6.19	28.2
Poland	3.90	0.02	0.02	4.11	0.046	0.188	-	-	0.009	0.017	4.37	62.8
Portugal	2.62	0.05	0.00	2.80	0	1.000	0.000	0.023	0.001	0.151	3.97	19.4
Romania	3.17	0.00	0.00	3.31	0	1.432	-	-	0.017	0.000	4.76	26.7
Slovenia	0.45	0.00	0.00	0.47	0	0.309	-	-	0	0.000	0.78	5.2
The Slovak Republic	0.31	0.00	0.00	0.33	0.011	0.385	-	-	0.001	0.001	0.73	11.3
Finland	5.29	0.06	0.02	5.60	0	1.074	0.000	0.001	0	0.013	6.69	26.6
Sweden	5.17	0.11	0.00	5.52	0.152	5.648	-	0.006	0	0.076	11.40	36.2
UK	0.59	0.25	0.46	1.36	0.070	0.382	0.001	0.029	0.001	0.247	2.09	157.5
Total	53.31	2.60	1.50	60.00	3.13	28.11	0.13	0.68	0.63	5.91	98.6	1227.5
%				60.9	3.2	28.5	0.1	0.7	0.6	6.0	100.0	

2.3. POSSIBLE DEVELOPMENT OF RES TO 2020

The use of biomass is expected to make the highest contribution to achieve 20% participation of RES by 2020 because biomass represents more than 60% of RES (2005). There are different attempts to estimate the participation of biomass in 2020. EU Biomass Action Plan (Biomass action Plan, COM 2005) estimates biomass potential as high as 215–239 Mtoe in 2020. This is an estimate from 2003 and it obviously represents an upper limit in using biomass that would not jeopardize the environment. A significant share in the increase includes agricultural energy crops that would require considerable changes in agricultural policy recently putting up with serious critics.

EREC in Energy Technology Roadmap (Renewable Energy Technology Roadmap, 2008) sets a target for 2020 of 175 Mtoe, which is also three times more than in 2005.

These documents assess the potentials but not real possibilities (technical, economic, and ecological) to realize the potential. The experience from previous 10 years shows much lower realization of the biomass potential than foreseen.

EFSOS Study[7] from 2005 gave an interesting analysis of future availability of biomass. The study assesses wood consumption in industry and wooden waste available for energy purposes. Additional consumption of wood produced only for energy needs has not been separately analyzed so one can say that the estimate represents the lower limit of wood availability in the energy industry. In Wood resources availability and demands (2007) there is an analysis (based on the available data) of the situation in some EU member states with respect to national goals in using renewable sources. A considerable lack of logic was noticed in the data and a large difference in the data quality between the member states. A "75% scenario" is recommended, according to which the participation of biomass in total RES consumption in 2000 would be three-quarter of the participation in 2005.

Here, the participation of wood in energy consumption is estimated as a mean value of EFSOS estimate and national estimates based on the wood potential in each country decreased by the wood used as a product in wood industry. Such an estimate (compare to Table 3) gives the wood use in energy industry of 96.9 Mtoe, which is by some 80% higher than in 2005 (annual growth of 4%).

Another two components of biomass are biodegradable part of waste and biogas. There are still many controversies in determination of those two components. On one hand, biodegradable part of municipal waste has not been precisely defined yet,[8] and industrial waste is statistically poorly processed in most countries. On the other hand, many wood processing

---

[7] European Forest Sector Outlook Study

[8] According to IPCC document "Good Practice Guidelines and Uncertainty Management in National Greenhouse Gases Inventory" biodegradable part of waste expressed as DOC (Degradable Organic Carbon) is defined by the following expressions $DOC = 0.4A + 0.17B + 0.15C + 0.3D$, where A – share in paper and textile waste, B – share in garden and similar waste, C – waste from food processing, and D – wood and agricultural waste.

plants generate biogas so there is a danger that the same energy is calculated twice, once as a municipal waste component and the other time as a biogas component. Still not fully clear is the position of the energy obtained by incineration of biodegradable component of the waste (that should be treated as the energy from renewable sources but it is still not a case).

The biodegradable part of waste in this situation is estimated based on an estimate of total municipal waste, participation of biodegradable component in the waste and use of biodegradable part for the energy purposes. The annual quantity of biodegradable part of waste is assumed to be from 150 to 230 kg/inh. (larger in developed and smaller in less developed countries) and the use for the energy purposes from 25% to 75%. The quantity of biogas is estimated at 40% of energy obtained from waste. Based on these assumptions energy from waste will amount to 16.5 Mtoe in 2020 and energy from biogas 6.6 Mtoe, total 23.1 Mtoe. Some 70% of that amount will involve the five largest countries.

By 2020, a significant increase in consumption should be achieved also in biofuel. Current situation does not inspire much optimism in terms of using the first generation raw material (mainly from sugar beet, sugar cane, feed-wheat, barley, maize and rapeseed) because of the problems involving the impact on food production, biodiversity, and alike. On the other hand, there are small chances that the second generation of biofuels produced from different forms of biomass will be competitive by 2020. Beside, higher consumption of biomass in biofuel production will reduce the availability of biomass for other energy purposes.

However, there is an optimistic assumption in this analysis that in 2020 it will be possible to achieve the participation of biofuel in transport of 10% giving 35.6 Mtoe, which is 10 times more than the quantity in 2005.

Hydroelectric energy, currently the main RES after biomass, will have an insignificant growth by 2020 because the hydro potentials have been generally used up. The estimate has been made by compilation of various sources and shows a total increase of some 5 Mtoe (from 28 Mtoe in 2005 to 33 Mtoe in 2020). A somewhat higher increase in the capacities is expected in pump storage power plants (due to increase in wind energy). However, this part does not belong to RES. The largest potentials are expected in Austria, Portugal, Spain, Sweden, and Romania.

Solar energy registers high rate of growth over the recent years mostly due to enormous incentives (300–500 €/MWh). The increase in construction is expected to result in considerable decrease in price and development of simpler and more efficient technical solutions. It is very likely that no radical change will happen by 2020. Even very optimistic forecasts by EREC (Renewable Energy Technology Roadmap) that anticipate an increase in electricity generation from solar power plants by 120 times, and the increase in energy from thermal solar systems by 18 times (in 2020 as compared to 2005) would not result in radical changes. This analysis assumes a growth of 22 and 17 times respectively (some 20%/year) that can still be considered a very optimistic estimate.

Another reason for skepticism with respect to the solar energy in EU results from climatic characteristics. Potential areas are in the Mediterranean region where there are good conditions for other RESs.

There are plans to construct large solar facilities for the needs of Europe in desert areas of North Africa but this is not definitely a project that could be realized until 2020.

A significant increase in using geothermal energy is not likely to happen because considerable natural potentials exist only in a few countries and the assumed energy of 9.6 Mtoe mainly involves the potentials in Italy.

Finally, there is wind energy that should contribute the most to realization of the required RES participation in GFEC. In many countries, leaded by Germany, Spain and Denmark, the construction of wind parks, supported by significant incentives, is speeded up. This is an energy source with a rapid technological progress. Unit power has significantly increased in the last 10–15 years, from some hundred kilowatts to some megawatts a unit that has considerably reduced specific costs.

The basic drawback of the wind potential is its unequal distribution giving the big advantage to the European Atlantic coast (from Portugal and Spain, via Ireland and U.K. to Nordic countries) as compared to the interior part of Europe.

Table 3 shows the required power of the wind parks (based on the average capacity factor of 0.23 which is some 2000 h/year). It is clear that some countries (Estonia, Italy, Cyprus, Lithuania, Austria, Slovenia, The Slovak Republic, Finland and Sweden) do not even require additional capacities because they can achieve their RES quotas from other sources, and some countries should build enormous capacities: Great Britain 42.5 GW, Germany 36.3 GW, the Netherlands 14.4 GW, Poland 13.4 GW, Belgium 10.9 GW, France 30.3 GW, Spain 14.6 GW etc. A lucky combination of circumstances (exceptions are Germany and Poland) is that it involves the countries that at least in some of their areas have good wind potential.

Totally required additional power of 244 GW in 2020 as compared to 2005 still seems to be very hard to implement because of a number of challenges to be solved in a very short time.

Table 4 gives a final layout of the achievements in 2005 and projections for the year 2020 according to the analysis made in this document.

TABLE 3. Assessment of RES, 2020

	GFEC	Goal 20%	Wood, Mtoe	MSW, Mtoe	Biogas, Mtoe	Biofuel, Mtoe	Hydro, Mtoe	Solar PV, Mtoe	Solar Th., Mtoe	Goeth., Mtoe	Wind, GW	Exist, GW	New Wind, GW
Belgium	36.5	5,1	0.60	0.44	0.18	0.98	0.03	0.00	0.00	0.00	11,07	0.01	10,87
Bulgaria	14.4	1,7	1.52	0.07	0.03	0.25	0.29	0.00	0.00	0.20	0,00	0.03	0,00
The Czech Republic	32.2	3,6	3.19	0.22	0.09	0.60	0.19	0.00	0.00	0.00	0,00	0.04	0,00
Denmark	17.8	5,0	1.27	0.22	0.09	0.52	0.00	0.00	0.00	0.00	10,88	3.13	7,61
Germany	228.0	41,0	12.41	3.38	1.35	6.08	1.71	0.20	0.40	0.50	57,55	20.62	36,29
Estonia	3.9	0,8	1.10	0.03	0.01	0.08	0.00	0.00	0.00	0.00	0,00	0.03	0,00
Ireland	14.7	2,0	0.48	0.18	0.07	0.50	0.06	0.00	0.00	0.00	2,83	0.49	2,29
Greece	25.4	3,9	0.22	0.24	0.10	0.81	0.39	0.50	1.00	0.20	1,88	0.49	1,36
Spain	109.4	20,3	3.05	0.97	0.39	3.91	3.43	1.00	1.00	0.10	24,72	9.92	14,55
France	166.7	38,3	15.44	2.54	1.02	4.89	5.22	0.30	0.30	0.40	31,55	0.72	30,30
Italy	136.5	23,2	3.76	2.45	0.98	4.30	3.93	0.30	0.50	7.00	0,00	1.64	0,00
Cyprus	2.1	0,2	0.00	0.02	0.01	0.10	0.00	0.20	1.00	0.00	0,00	0.00	0,00
Latvia	5.4	1,7	1.17	0.04	0.02	0.11	0.26	0.00	0.00	0.00	0,47	0.03	0,44
Lithuania	6.3	1,2	1.54	0.06	0.02	0.14	0.04	0.00	0.00	0.00	0,00	0.00	0,00
Luxembourg	4.5	0,5	-0.24	0.02	0.01	0.27	0.01	0.00	0.00	0.00	1,62	0.04	1,56
Hungary	22.3	2,5	1.22	0.22	0.09	0.41	0.21	0.00	0.00	0.30	0,21	0.02	0,19
Malta	0.6	0,1	0.00	0.01	0.00	0.03	0.00	0.00	0.00	0.00	0,06	0.00	0,06
The Netherlands	49.7	7,5	0.91	0.67	0.27	1.50	0.01	0.00	0.00	0.00	15,90	1.22	14,42
Austria	28.2	9,6	4.14	0.34	0.14	0.78	4.55	0.10	0.20	0.20	0,00	0.83	0,00
Poland	78.5	9,4	3.30	0.68	0.27	1.18	0.19	0.00	0.00	0.20	13,79	0.12	13,44
Portugal	20.9	6,0	1.53	0.23	0.09	0.70	1.93	0.20	0.30	0.30	2,81	1.06	1,72
Romania	33.4	6,4	3.03	0.19	0.08	0.41	1.88	0.00	0.00	0.20	2,37	0.00	2,33
Slovenia	6.0	1,3	0.49	0.04	0.02	0.15	0.65	0.00	0.00	0.00	0,00	0.00	0,00
The Slovak Republic	12.2	1,6	0.91	0.12	0.05	0.17	0.41	0.00	0.00	0.00	0,00	0.01	0,00
Finland	30.9	10,1	15.09	0.22	0.09	0.48	1.17	0.00	0.00	0.00	0,00	0.08	0,00
Sweden	33.6	17,7	18.44	0.38	0.15	0.83	6.20	0.00	0.00	0.00	0,00	0.49	0,00
UK	157.5	23,6	2.34	2.51	1.01	5.45	0.42	0.10	0.10	0.00	44,85	1.57	42,54
Total	1277.4	244,4	96.9	16.5	6.6	35.6	33.2	2.9	4.8	9.6	222,6	42.6	180,0

3. Review and Comments of Possible Obstacles and Inconsistencies

The analysis is based on available references, information and analyses, as well as on the author's own judgment of the whole situation. In such a short presentation many essential elements were left out. For example, an issue of environmental impact as the limiting factor in development of many RES was not discussed here. Emissions of some biomass plants can emit much more pollutants to the air than, for example, a coal fired power plant. Wind and solar power plants occupy a very large area.

TABLE 4. RES in GFES 2005 and projection for the year 2020

	Biomass	Bio fuel	Hydro[9]	Solar PV	Solar Th.	Geoth.	Wind	Total
	Million tones of oil equivalent (Mtoe)							
2005	60.0	3.13	28.11	0.13	0.68	0.63	5.91	98.6
2020	120.0	35.6	33.18	2.9	4.8	9.6	38.4	244.4
Δ	60.0	32.5	5.1	2.8	4.1	9.0	32.5	145.8

In addition, there is a review of possible main obstacles and issues referred to the RES concept within the EU energy sector.

1. Binding goals are strictly the result of political decision. This is definitely one of the most important decisions in the history of the EU that changes not only the direction in energy and economy of the EU, but has a significant influence worldwide. Although the set goals have been discussed in a series of previous elaborations, it feels that not everything needed was made in this preparation. It appears that the purpose of realized analyses was more to confirm an already brought political decision but to actually analyze it. The biggest drawback of these elaborations was in a very modest research in the Member States on technical, technological, environmental and economic characteristics of individual countries. In fact, a number of problems that may arise in licensing of individual objects are not always a consequence "of numerous administrative problems" or "opaque and discriminating rules for grid access". It is said in (Renewable Energy Road Map 2007) as one of the main conclusions that "greater use of variable energy is a question of economics and regulatory rules rather than of technical constraints". It is not clear what the basis of such conclusion was. The mentioned is also confirmed by the predicted procedure, which refers to the Member States:

(a) Based on mandatory national overall targets and measures for the use of energy from renewable sources each Member State shall adopt a national renewable energy action plan. By 30 June 2009 the Commission shall adopt a template for the national renewable energy action plans.

(b) Member States shall notify their national renewable energy action plans to the Commission by 30 June 2010.

---

[9] Excluding pumping.

It would be much better that the national plans – including any problems that may arise during licensing and construction of renewable energy sources – were the basis for determining the national capacities. They should have to be based on the agreed and adopted measures of cost-effectiveness and feasibility. This way the national plans will not be a part of the joint EU concept but will reflect the national preferences in the energy sector. That is what already existed and proved to be inadequate. In addition, development of national plans of high quality as required by the new EU policy is not possible in a short time of less than 1 year, because the member states have to change and harmonize regulations, the existing energy plans, physical plans, etc. It may happen to ignore the quality of the required infrastructure works due to short deadlines.

2. It is quite likely that the main RES would be wind power plants, at least in the period to 2020. According to very optimistic scenario referred to the share of biomass, biofuel and other RES, for fulfilling the expected goals it would be necessary to build wind power plants of new 180 GW till 2020. Naturally, it would be all under condition that the energy saving program would be realized and as such would bring an insignificant GFEC growth in 2020 in relation to 2005 (total of 4%)

3. A disadvantage of the concept is that proclaimed energy saving is related to GFEC (defined in Directive 2009/28/EC 2009]). Namely, losses in fuel and electricity production are not an integral part of GFEC, so an increase of efficiency in production of those final energy forms would be discouraged in thermal power plants, refineries, etc. Therefore, a large part of gross inland energy consumption (GIC) (around 35% or 650 Mtoe) had been left out the range of the whole concept. Available funds, incentives and similar resources will be directed pre-ferentially to decrease the final energy consumption and not to increase the plant efficiency for production of final energy forms and their modernization. For example, coal fired power plants in Europe at this moment operate with an average capacity of 38%. It is significantly better than in the USA or Japan, although the new power plants tech-nological solutions are able to achieve 45% and even 50%. Increasing the efficiency of coal fired power plants from 38% to 45% would bring to fuel saving of 21 Mtoe, which for example corresponds to total annual final energy consumption of Greece.

Gas is used in the EU in electricity generation with efficiency of 45%, although it is possible to achieve 60%. Total efficiency is reduced to 35%, where natural gas is used in more simple technological solutions (only with gas turbine etc.). There are also significant differences between countries. It is interesting to notify that the efficiency of gas usage in the energy sector in Germany,[10] as a country with large incentives towards the construction of RES objects, is only 34%, while in Great Britain, where the construction of RES objects is minor, the efficiency of gas power plants is approx. 51%. This example, which is surely conditioned by specific characteristics of two energy systems within two advanced and rational economies, could become a rule in case of construction of large capacities of wind and solar power plants, which availability is rather low. It could happen that the system is developed towards the increased energy consumption from RES, while on the other hand, more gas could be inefficiently consumed as a compensation in periods of unavailability of RES objects.

4. The target of 20% reduction in greenhouse gas emission refers specifically to renewable sources. Nuclear energy and carbon capture and storage (CCS) technologies are not included as they fall outside the definition of renewable energy. It could be put a question is it right to classify all first generation biofuels as RES and all fossil fuels as non-renewable even if used CCS facilities. This may discourage the development of those technologies in the near future.

Conditions for the construction of wind power plants are not the same everywhere. The largest wind potential is in northern and western areas of the EU. The development of wind power plants is especially planned on off-shore locations (at least one third of new capacities). Such a situation requires strong development of high-voltage electrical grid with a series of new technological solutions (smart grid), which will not be easily achieved in a relatively short period of time. There are also the problems of maintenance. Considerable challenges are set up in licensing of new facilities due to lack of adequate regulations or because of the necessity of its completion.

Construction of wind power plants, especially if it's about the large number of locations with many facilities, also causes a series of strictly environmental issues (space planning, restrictions relating to noise, visual problems, impact on birds, bets and protected areas, etc.) All this may affect the construction plans.

Large part of wind power in power generation system calls into question the stability of energy system in cases when there is no wind, or when its potential is significantly reduced. The construction of additional thermal power capacity (up to 20% of wind power plant

---

[10] Source: Eurostat data 2001–2006.

capacity) that will take the burden during these conditions is vital. The logic of the power system is that those facilities have low investment costs, so the majority of them will be gas turbines with or without combined cycle. Therefore, the number of gas fired thermal power plants will increase in the system with high capacity of wind power plants but the participation of coal power plants will be drastically reduced. Furthermore, the capacity factor of thermal power plants will be reduced that will affect their economy.

It may appear that such development is desirable: a lot of wind, increased or at least unreduced number of thermal power plants on gas and drastic reduction of coal power plants. However, ignoring the coal power plants that are still main facilities in European power industry, can lead to big problems in the future.

Namely, the number of wind power plants, and it is similar with the plants that use biomass, is limited and sooner or later will thermal power plants on coal (with CCS system) be required together with nuclear power plants. Radically focusing on wind power and other renewable sources in the EU can reduce the interest for further progress in the development of modern technological solutions on conventional facilities and undoubtedly jeopardize the technological advantages that now exist in relation to the world competition. In fact, for equipment manufacturers wind power plants are very attractive. Investment in 200 GW wind power plant that will produce annually approximately 500 TWh of electricity is about 250 billion euros. Investment in coal-fired power plants that will annually produce the same amount of electricity is maximum of 150 billion euros. While doing so, it will be necessary to produce 60,000 units of wind power plants, which is much easier job then only 100–150 units of thermal power plants on coal that are also much more site-specific and complicated to design and build.

5. It seems that the wind potential in EU is considerably overrated. Significant incentives have brought to a situation where the wind power plants are profitable even at locations with a low wind potential. Data analysis in Eurostat 2008 for the period 2001–2006 indicates an average capacity factor of wind power plants in 27 Member States of only 1,642 h/year. In energy balances the wind power plants are balanced with factor 0.23 (System Adequacy Forecast 2009) corresponding to an involvement of approximately 2,000 h/year normalized to nominal capacity. As mentioned data indicate, the real situation is much worse. Indeed, there are differences between Member States. The lowest utilization is in Germany (with the highest installed capacity), of only 1,385 h, while the highest utilization is in Great Britain (2,072 h).

It seems that decision on the wind power plants construction depends much more on established tariff system, than on a real wind potential at certain location. However, the incentives are envisaged to last just until the price of construction is reduced due to the technological progress and until the market volume develops to the optimum size. Unfortunately, no technological progress would make the wind power plants profitable at annual utilization of 1,600 h. To increase the utilization factor it would be necessary to implement a major part of new objects at off-shore locations, which, beside significantly higher price, brings additional problems as well, required to be solved in a short time period.

6. Concept with radical focus on renewables costs much more than business as usual case. Cumulative investment cost will be higher for 300 billion euros and yearly additional cost are projected to be 5–10 billion euros yearly in the period 2005–2020.

The question is whether the gains are sufficient compensation for these costs and how will the increased costs be assigned to the member states. All the more that neither the benefits achieved will be uniformly distributed.

Although increased construction of renewable energy sources is often motivated by desirable decentralization in energy production, local employment, ecologically sustainable energy management, reduced dependence on import of energy products, etc., the fact is that the results of concept could be totally different.

Renewable energy sources, particularly wind and solar power plants, represent so called "natural monopoly" based on a fact that it is possible to significantly reduce the electricity price out of these sources only by large capacity increasing (increasing of installed power for 100% reduces the price of generated electricity for 20%, Renewable Energy Technology Roadmap). Necessary consequence is the concentration at the level of development and production of equipment, so many countries will not participate. Only few will profit, while the system development would be paid by all equally. The European solidarity in implementation of the whole project could be significantly shaken.

4. The Status of Renewable Energy in Croatia

Croatia is a small country in the Southeast Europe that is just preparing for full EU membership. With its 56,500 km^2 area and 4,436,000 residents it will be on 19th place according to its area and on 20th according its population in the EU. Total energy consumption (GIC) from 9.0 Mtoe amounts to around 0.5% of all EU spending (21st place), and with energy consumption per capita from 1442 toe/per capita is significantly below the EU average and is on 24th place in the EU.

The Croatian energy sector is of little impact on the EU level, but events in the EU have a large impact on Croatia. If the EU pulls in the right direction that will certainly be useful also for Croatia and the opposite. Even the smaller disturbances may be fatal for Croatia.

The EU invites Croatia to set an appropriately ambitious target for the percentage of electricity produced from renewable energy sources to be achieved by the deadline set in Directive 2001/77/EC, and commensurate with the objective to achieve an increase in the EU's share of renewable energy consumption from around 7% in 2005 to 20% in 2020, as agreed by European Council in March 2007.

The EU furthermore encourages Croatia to establish a longer-term perspective for the development of renewable energy sources in electricity production in Croatia.

The EU takes note of Croatia's indicative national target by 2010 for the share of biofuels for transport, and encourages Croatia to ensure that this target can be attained given the relative low present levels. The EU furthermore encourages Croatia to establish a longer-term perspective for the development of biofuels in Croatia.

The proposal of the action plan for the utilization of renewable energy sources in Croatia is under way (Stručne podloge za izradu Prijedloga akcijskog plana za obnovljive izvore energije u Republici Hrvatskoj 2009).

In a proposal of new energy strategy of the Republic of Croatia (Strategija energetskog razvoja Republike Hrvatske, nacrt Bijele knjige, travanj 2009), GFEC is planned in 2020 of 9.7 Mtoe, and the share of renewable energy sources would be 1.96 Mtoe, so the increase in relation to 2005 would be 1.12 Mtoe. It is intended to be achieved by increased usage of biomass in GFEC (0.70 Mtoe, out of which 0.34 Mtoe for biofuels), wind energy (0.21 Mtoe), additional hydropower (0.084 Mtoe) and electricity generation in other RES (0.13 Mtoe).

Croatia has already implemented support mechanism for electricity generation from renewable energy sources. Feed-in tariffs are approx. 0.10 €/kWh for wind power plants, 0.16 €/kWh for biomass power plants and 0.3 €/kWh for solar power plants. Incentives for cogeneration systems, as well as for other RES, have also been adopted.

Croatia does not have developed technology for RES plants, so quite small or no influence whatsoever of construction and exploitation of RES objects could be expected towards the increasing of local employment and involving of domestic industrial and other capacities. For example, during construction of coal fired power plant or gas power plant, even nuclear power plant, local companies could increase their participation in construction, operation and maintenance of power plants, than in a case of construction of wind or solar power plants. It is unlikely that incentives, which would be paid by Croatian consumers for promoting the construction of RES, would create certain positive effects. That money (on the average of 200 millions of euro/year) will be a Croatian expression of solidarity towards new European energy policy, which with few disadvantages and necessary corrections, still represents an important step forward.

Acknowledgments

The article is a part o the scientific project No. 177-0000000-1931 realised with the support of Ministry of science, education and sports.

References

Directive 2001/77/EC of 27 September 2001 on the promotion of electricity produced from renewable energy sources in the internal electricity market.
Directive 2003/30/EC of 8 May 2003 on the promotion of the use of biofuels or other renewable fuels for transport.
EU energy and transport in figures, Statistical Pocketbook 2009.
Renewable Energy Road Map, Renewable energies in the 21st century: building a more sustainable future, COM(2006) 848 final, Brussels 10.1.2007.
Directive 2009/28/EC of 23 April 2009 on the promotion of the use of energy from renewable sources amending and subsequently repealing Directives 2001/77/EC and 2003/30/EC, Brussels.
Biomass action Plan, COM(2005) 628, Brussels, 7.12.2005.
Renewable Energy Technology Roadmap, 20% by 2020, European Renewable Energy Council, 2008.
Wood resources availability and demands – implications of renewable energy policy, UNECE, FAO, University Hamburg, 2007.
System Adequacy Forecast 2009-2020, UCTE, January 5th 2009.
Stručne podloge za izradu Prijedloga akcijskog plana za obnovljive izvore energije u Republici Hrvatskoj, Rev. 6, Ekonerg 2009.
Strategija energetskog razvoja Republike Hrvatske, nacrt Bijele knjige, travanj 2009.

Additional Literature and Sources of Data

Energy and environment report 2008, EEA Report, No 6/2008.

Green Paper – A European Strategy for Sustainable, Competitive and Secure Energy, COM(2006) 105 final, Brussels, 8.3.2006.

Annex to the Green Paper, A European Strategy for Sustainable, Competitive and Secure Energy, What is at stake – Background document COM(2006) 105 final.

Communication from the Commission to the European parliament, the Council, the European Economic and Social Committee and the Committee of the Regions, 20 20 by 2020, Europe's climate change opportunity, COM(2008) 30 final, Brussels, 23.1.2008.

Green Paper follow-up action – Report on progress in renewable electricity, COM(2006) 849 final, Brussels, 10.1.2007.

Commission staff working document: The support of electricity from renewable energy sources – Accompanying document to the Proposal for a Directive on the promotion of the use of energy from renewable sources, Brussels, 23.1.2008, SEC(2008) 57.

The EU's Target for Renewable Energy: 20% by 2020, Volume I: Report, House of Lords, European Union Committee, 27th Report of Session 2007–08, 24 October 2008.

Action Plan for Energy Efficiency: Realising the Potential, CEC, COM(2006) 545 final, Brussels, 19.10.2006.

Benchmark of Bioenergy Permitting Procedures in the European Union, Ecofys, Golder Associates, 2009.

Directive 2004/8/EC of the 11 February 2004 on the promotion of cogeneration based on a useful heat demand in the integral energy market and amending Directive 92/42/EEC.

Commission decision of 21 December 2006 establishing harmonised efficiency reference values for separate production of electricity and heat in application of Directive 2004/8/EC of the European Parliament and of the Council, 2007/74/EC.

European Energy and Transport, Trends to 2030–Update 2007, EC Directorate-General for Energy and Transport, European Communities, 2008, ISBN 978-92-79-07620-6.

Renewable Energy Made in Germany, Federal Ministry of Economics and Technology, March 2008.

R. Edwards (IES), S. Szekeres (IPTS), F. Neuwahl (IPTS), V. Mahieu (IES), Biofuels in the European Context: Facts, Uncertainties and Recommendations, JRC Scientific and Technical Reports, EUR 23260 EN, European Communities, 2008, ISBN 978-92-79-08393-8.

M. W. Kennedy, 2009. Where is European Electricity supply going?, presentation, Ekonerg, Zagreb, May 2009.

An EU Energy Security and Solidarity Action Plan, Europe's current and future energy position, Demand-resources-investments, COM(2008) 781 final, Brussels, 13.11.2008.

An EU Energy Security and Solidarity Action Plan, Energy Sources, Production Costs and Performance of Technologies for Power Generation, Heating and Transport, COM(2008) 781 final, Brussels, 13.11.2008.

An EU Energy Security and Solidarity Action Plan, The Market for Solid Fuels in the EU in 2004-2006 and Trends in 2007, COM(2008) 781 final, Brussels, 13.11.2008.

An EU Energy Security and Solidarity Action Plan, Second Strategic Energy Review, COM(2008) 781 final, Brussels, 13.11.2008.

EU Energy in figures 2009, Part 2: Energy.

EU Energy in figures 2009, Part 4: Environment.

EU Energy in figures 2009, Greenhouse Gas Emission by Sector.

EU Energy in figures 2009, Greenhouse Gas Emission from Transport by Mode.

EU Energy in figures 2009, Electricity Generation from Renewables.

Energija u Hrvatskoj, Godišnji energetski pregled 2007. Ministarstvo gospodarstva, rada i poduzetništva, ISSN 1847-0602.

Directive 2003/30/EC of 8 May 2003 on the promotion of the use of biofuels or other renewable fuels for transport.

Energy for the Future: Renewable Sources of Energy – White Paper for a Community Strategy and Action Plan, COM(97)599.

Green Paper, Towards a secure, sustainable and competitive European energy network, COM(2008) 782 final, Brussels, 13.11.2008.

An EU Strategy for Biofuels, COM(2006) 34, Brussels, 8.2.2006.

Biofuels Progress Report, Report on the progress made in the use of biofuels and other renewable fuels in the Member States of the EU, COM(2006) 845, Brussels, 10.1.2007.

B. Thomsen, Draft Report on a Roadmap for Renewable Energy in Europe (2007/2090 (INI)), European Parliament, Committee on Industry, Research and Energy, 11.5.2007.

Offshore Wind Energy: Action needed to deliver on the Energy Policy Objectives for 2020 and beyond, COM(2008) 768, Brussels, 13.11.2008.

A European Strategic Energy Technology Plan (SET-Plan), Towards a low carbon future, COM(2007) 723, Brussels, 22.11.2007.

Update of the Nuclear Illustrative Programme in the Context of the Second Strategic Energy Review, COM(2008) 776, Brussels, 13.11.2008.

Impact of large amount of wind power on design and operation of power systems, http://www.ieawind.org/AnnexXXV/Task25_Publications.html.

J. M. Farley, 2009. "Capture-Ready": A global example, Power-Gen Europe, Koeln, Germany.

M. Schuknecht, L. Kirchner, R. Winter, 2009. Third Party Concept – Carbon Capture Readiness, Power-Gen Europe, Koeln, Germany.

J. Kreusel, 2009. ABB AG, Germany, When grids get smart – the way towards the power system oft he future, Power-Gen Europe, Koeln, Germany.

Statistical Yearbook UCTE 2007.

Green Paper on Energy Efficiency or Doing More With Less, COM(2005) 265 final, Brussels, 22.6.2005.

Energy, Yearly Statistics 2006, Eurostat, 2008 edition.

THE KEY ASPECTS OF ELECTRICITY AND NATURAL GAS SECURITY OF SUPPLY IN CROATIA

GORAN MAJSTROVIC[*], NIJAZ DIZDAREVIC,
MARIO TOT, AND DINO NOVOSEL
*Energy Institute Hrvoje Pozar, Savska 163, 10000 Zagreb,
Croatia*

Abstract   The key aspects related to security of electricity and natural gas supply in Croatia are analysed in this paper. First, the paper refers to relevant legal framework, key stakeholders and their responsibilities, including public service obligation. Then, it proceeds to current and future generation and network capacities, operational security and balance, and supply business. Finally, two key vulnerability indicators – *Energy Intensity* and *Energy Dependency* – are calculated for Croatia and eight comparable EU countries. In this way the authors presented basic aspects and figures of security of supply of these two energy sectors and compared them to characteristic values in EU countries. It represents the basis for quantification of short-term risks of energy supply in Croatia.

Keywords: Electricity, natural gas, security of supply, vulnerability indicators

1. Introduction

Despite the opening of the energy markets and consequent development, the State institutions still have to take care of security of supply to protect public interest, particularly in the transient period before competition between market participants can guarantee the security of supply to consumers. Security of electricity and natural gas supply are defined as abilities of the two systems to supply final consumers with electricity and natural gas at acceptable price, sufficient quantity and prescribed quality. Croatian legislation

---
[*] To whom correspondence should be addressed: Goran Majstrovic, Energy Institute Hrvoje Pozar, Savska 163, 10000 Zagreb, Croatia; Email: gmajstro@eihp.hr

F. Barbir and S. Ulgiati (eds.), *Energy Options Impact on Regional Security*,
DOI 10.1007/978-90-481-9565-7_10, © Springer Science + Business Media B.V. 2010

contains provisions for secure energy supply. Further on, it is necessary to define the regulation mechanism which will not slow down market activities, but lead to the satisfactory level of security of supply. In the following chapters the key aspects of electricity and natural gas supply described, particularly referring to legal framework, key stakeholders and their responsibilities, including public service obligation, generation and network capacities, operational security and balance, and supply business. Finally, two key vulnerability indicators – *Energy Intensity* and *Energy Dependency* – are calculated for Croatia and compared to EU countries' ones.

2. Legal Framework

In the process of EU accession, Croatia is obliged to implement *aqcuis communautaire* in energy sector, among others (Directive 2003/54/EC 2003; Regulation 1228/2003/EC 2003; Commission Decision of 9 November 2006; Directive 2003/55/EC 2003; Regulation 1775/2005/EC 2005; Directive 2005/89/EC 2005; Directive 2004/67/EC 2004). Besides, it is also one of the Contracting Parties to the *Treaty establishing Energy Community since* 2005 (Treaty establishing the Energy Community 2005). The *Treaty* refers to extension of EU legal framework in energy sector, competition, environment and renewables to South East Europe. Energy sector assumes electricity and natural gas in particular, including the relevant aspects of security of supply. Relevant provisions from EU legislation in this field (Croatian Energy Law; Croatian Energy Market Law) are transposed into Croatian national legislation. Among others, Croatia is obliged to publish the *Security of Supply Statement* on regular basis, biannually (Statements on monitoring the security of supply 2007). Data provided in this paper cover the key supply aspects from the *Statement*.

3. Electricity Sector

3.1. KEY STAKEHOLDERS AND THEIR RESPONSIBILITIES

Pursuant to the energy legislation package relating to the electricity sector (Energy Law, Electricity Market Law and Energy Regulation Law), the Ministry responsible for energy sector (Ministry of Economy, Labour and Entrepreneurship, hereinafter, Ministry), among others conducts administrative supervision over the implementation of the relevant legislation and regulations, issues approvals for the construction of new production capacities, makes recommendations to the Government concerning tariffs, issues opinions on documents and conducts other expert tasks in the field of the electricity sector.

The Croatian Energy Regulatory Agency (CERA, hereinafter Agency) establishes and conducts the duties of regulation of energy activities. It is obliged to apply regulations to protect market competitiveness relating to energy sector activities and provide technical assistance to the Croatian Competition Agency.

The Croatian Energy Market Operator (HROTE) is responsible for organisation of the electricity market. Transmission system operator (HEP TSO) is, among others, obliged to assure long-term transmission network reliability in order to satisfy reasonable requests for electricity transmission; contribute to security of supply with adequate transmission capacities and transmission network reliability; maintain the electricity quality parameters; prepare explanations and recommendations to the Ministry or the Agency about the need to construct new production units in order to keep adequate security of supply; monitor the security of supply and prepare security of supply reports at least biannually. Distribution system operator (HEP DSO) has similar role for distribution network.

3.2. PUBLIC SERVICE OBLIGATION

All eligible customers from the households category who do not want to excercise their eligibility right to choose supplier or do not manage to find one, contract electricity supply with the carrier of public service obligation of electricity supply. In cases where a supplier of eligible customers ceases to be operational, these customers are also entitled to electricity supply through public service obligation. HEP DSO (as a supplier of last resort) and its parent company HEP are the carriers of public service obligation of electricity supply. Electricity Maket Law foresees a regulatory supervised public tender for procurement of electricity HEP DSO needs to supply households with, starting from 1 January 2011 for the period of 5 years. It implies a procurement of app. 6.4 TWh/year, worth 320 million euros per year or 1.6 billion euros in the 5-year period.

3.3. CURRENT GENERATION CAPACITIES

Total installed generation capacity is equal to 4,226.75 MW with system peak load of about 3,100 MW. The average age of hydro power plants in Croatia is over 35 years. The last one was built in 1989. The average age of thermal power plants is more than 30 years. This total generation capacity does not include production capacities constructed in other neighbouring countries to supply Croatian consumers pursuant to leasing contracts or through ownership shares:

- Bosnia and Herzegovina – TPP Gacko; installed power: 300 MW; fuel: coal; legal basis: share in ownership (1/3 of power over a period of 25 years)
- Serbia – TPP Obrenovac; leasing rights to power based on loan to construct this TPP (installed power: 305 MW; fuel: coal)

Electricity produced in these units is still not available and the status of these plants is still not resolved. Contentious issues in the contracts relating to investments in these units have been left to the duration of contracts, the treatment of investments and the manner of determining the price of electricity supplies.

3.3.1. *Operational security*

In Croatian power system 49% of generation capacity is installed in HPPs. During dry hydrological conditions the production of HPPs is limited, as well as NPP Krško production due to river Sava's cooling water limitations. Accordingly, power import or TPP production is increased.

In the period of limitations in natural gas supply, households supply is of the highest priority, which can result with gas supply restrictions to the industrial consumers and gas-fired TPPs of HEP. Since there are dual-fuel TPPs (natural gas and oil; CHP Osijek: 42 MW, CHP EL-TO Zagreb: 90 MW, CHP TE-TO Zagreb: 135 MW, total: 267 MW), the switch in fuels is applied in the case of natural gas restrictions.[1] Accordingly, HEP-Generation is responsible for oil purchasing for electricity production needs.

3.4. CONSTRUCTION OF NEW PRODUCTION CAPACITIES

The construction of production units is at the discretion of energy under-takings with a licence to produce electricity under the condition that they fulfill criteria determined in the *Authorisation Procedures*. The method of implementing prescribed principles and criteria is determined by the Croatian Government at the recommendation of the Ministry following previously obtained opinion from the Agency. The Ministry issues authorisations for construction. In 2005 and 2006 authorisation was issued to two units (HPP Lešće and CHP Zagreb - L).

If construction of production units based on issued authorisation in addition to demand side management and energy efficiency is not sufficient, then a further decision can be adopted to construct another production

---

[1] Some large TPPs are also designed for dual-fuel operation, but currently can not be operated using natural gas.

capacities via the Tendering Procedures in the interest of security of supply, environmental protection or energy efficiency promotion.

Decisions for the Tendering Procedures for the construction of new production capacities with power up to 50 MW and the selection of the most favourable offer is the responsibility of the Agency; for objects with power higher than 50 MW and the selection of the most favourable offers the responsibility is with the Croatian Government following a recommendation of the Agency.

The Tendering Procedure method, tender conditions, detailed description of contractual provisions and other relevant procedures which each of the parties must satisfy as well as the list of criteria to select the most favourable offer is determined by a regulation set by the Minister. The Agency is responsible to implement the tender procedures. A call for tenders must be published in the National Gazette. The deadline to submit tenders must be at least 6 months and not more than 12 months. Apart from this, the producers and consumers are obliged to obtain approval to connect to the grid or to increase their installed power from the HEP TSO or HEP DSO. Currently, plants under construction or planned for construction are HPP Lešće (42 MW, expected in operation in 2010), CHP Zagreb L (100 MWe, 80 MWt, 2009), various WPPs (up to 360 MW till 2013).

The construction of a new gas (1,200 MW) and coal (1,200 MW) thermal power plants are being contemplated within new Energy Strategy, currently still under preparation. So far, no firm decision has been made on nuclear power plant, rather development of methodology and procedure to eventually lead toward it.

3.5. NETWORK OPERATION AND DEVELOPMENT

Sufficiency of network infrastructure to guarantee a satisfactory level of security of supply in Croatia can be summarized by the fact that the electricity network was initially constructed for consumption dominated by the industrial sector share. Due to 1991–1995 war conflict and destruction in Croatia followed by transitional reconstruction period, the Croatian industrial sector suffered significant degradation. At the same time, there was a significant expansion of the service sector causing different foundations for the network planning and development. Apart from that, in the former Yugoslavia the Croatian power system was not developed on the principle of self-sufficiency. Consequently, today the Croatian transmission network is not topologically determined on the principle of self-sufficiency. For the same reason, upon acquiring independency, many "internal former Yugoslavian" lines then became cross-border interconnecting ones. Having in mind the geographical shape of the country, it is clear that power system of Bosnia and Herzegovina strongly affects operation of the Croatian power system.

Today Croatia is characterized by the high ratio between available cross-border capacity and system peak load (Fig. 1). Theoretically, it is possible to import total power system needs. Installed cross-border capacity is equal to 290% of peak load. Available cross-border import capacity in 2006 was 17% of system peak load and 2.3 times higher than EU average. Additionally, new 2x400 kV interconnection line to Hungary is under construction and it will definitely make Croatia one of the best interconnected systems in Europe.

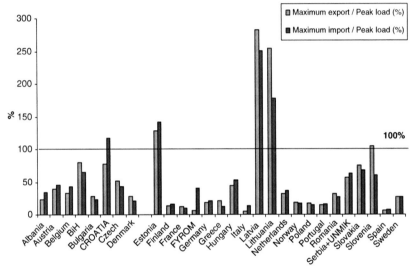

Figure 1. Ratio between available cross-border capacity (import/export) and system peak load in 2006 in European countries (Source: EIHP)

3.5.1. *Network ageing and availability*

Key indicator of the network functionality is its unavailability. The average total transmission lines unavailability in Croatia in the period 1995–2004 by voltage levels ranges from 0.84% to 3.21%. At the same time, the average total transformers unavailability in transmission network in Croatia ranges from 1.10% to 3.58%. Having in mind the difference between planned and forced outages, it can be generally concluded that the level of transmission network availability is satisfactorily high despite high age of the network (more than 30 years in average).

In 2006 electricity supply interruptions data due to distribution network unavailability are given as follows: 3.7 interruptions per consumer or 541 min of average supply interruption duration per consumer (0.1% of the total time) and 146 min of average supply interruption duration per consumer per fault. Despite all maintenance efforts, due to high age of the network elements, certain drop in system availability can be expected in the coming period. It might influence the operational security and security of end users supply.

3.6. ELECTRICITY BALANCE

Even though installed generation capacity (4,227 MW) is higher than the peak load (3,100 MW), Croatia strongly depends on import due to lower prices than domestic generation costs. Available electricity in Croatia for the timeframe 1988–2006 (Fig. 2) shows that import level (of about 25% of total consumption after deducing own production from abroad) strongly depends on hydrology. Moreover, knowing of the existing generation mix (49% HPPs, 44% oil-fired TPPs, 21% gas-fired TPPs, 18% coal-fired TPPs etc.) it is clear that Croatian electricity balance strongly depends on fuel prices. Accordingly, within new Energy Strategy the Croatian Government tries to define very ambitious generation development plan for the next 10 years: among others, 1,200 MW of new gas-fired TPPs, 1,200 MW of new coal-fired TPPs, 300 MW of new large HPPs, 1,200 MW of wind power plants.

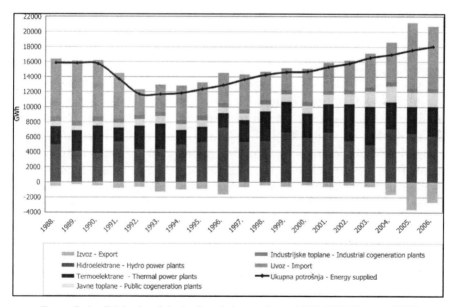

Figure 2. Available electricity in Croatia in timeframe 1988–2006 (Source: EIHP)

3.7. ELECTRICITY SUPPLY

Since 1 July 2008 all 2.2 million customers have been granted an eligibility status. As of today, there are three licensed power suppliers in Croatia: HEP Supply and HEP DSO which are active in the market and Korlea (still passive, only licensed).

So far, there was no customer switching, apart from HEP DSO to HEP Supply for non-households. All customers are still supplied by the incumbent HEP's dependent companies HEP Supply or HEP DSO – the main reason being lower prices (either offered by HEP Supply or regulated to HEP DSO). In the South East Europe region, due to public service obligation, incumbent companies are asked firstly to supply local tariff (regulated) consumers, while free market supply is possible only with an eventual surplus. A public tender for procurement of households consumption (40% of total consumption) is expected in near future. It might dramatically push for further liberalisation of electricity market in Croatia.

4. Natural Gas

4.1. KEY STAKEHOLDERS AND THEIR RESPONSIBILITIES

Key stakeholders in the natural gas market are INA, the Ministry of Economy, Labour and Enterpreneurship, the Agency, TSO (Plinacro) and 38 DSOs. In future HROTE is also expected to facilitate further gas market development.

4.2. PUBLIC SERVICE OBLIGATION

Eligible customers from households category can choose a gas supplier within 6 months from the market opening and conclude a supply contract. Each eligible customer from the households category which has not chosen a supplier within six months from the market opening has a right for gas supply from the supplier who is the carrier of public service obligation. The supplier which is the carrier of public service obligation for gas supply to the households for a period of next 5 years is INA. It is defined as the supplier which on 31 July 2008 carried out gas supply activities for tariff consumers in the households category. This carrier has a right to procure gas from the supplier which is the carrier of public service obligation of gas purchase – INA. Recently, some plans emerged as to transfer the ownership over gas purchase business from INA Company to independent state owned company.

4.3. PRODUCTION CAPACITIES, RESERVES AND STORAGE

Natural gas is produced from 20 on-shore gas fields and two off-shore fields, covering nearly 61% of total demand. The largest quantities come from Molve and Kalinovac, where central gas stations for gas processing and transport preparation were built – Molve I, II and III. Total installed capacity equates to 9 million cubic metres per day.

4.4. STORAGE

During the summer months the storage contains a certain quantity of gas for the Slovenian Geoplin. The daily quantity dispatched for Geoplin varies depending on daily temperatures ranging from 0 to 720,000 m^3/day.

The operational capacity of the underground natural gas storage Okoli (PSP Okoli) is 550 million cubic metre, with 50 million cubic metre, reserved for Geoplin. The maximum daily inlet capacity amounts to 3.8 million cubic metre per day, and the maximum capacity of exhaustion is 5 million cubic metre per day. The maximum hourly capacity amounts to 200,000 m^3. The system consists of 17 production and compression wells. The intensity of lower pressure in the reservoirs is ~186 bar. Cushion gas amounts to about 400 million cubic metre. In the period from 1987 to 2000 the total amount dispatched was 3,536,200,000 m^3 and the total of 3,695,100,000 m^3 was compressed. The total quantity of 7,231,300,000 m^3 passed through the Okoli underground storage facility.

In addition to existing storage capacities at Okoli 1, by 2011 Croatia intends to construct another underground natural gas storage – Okoli 2 with a capacity of 500 million cubic metre, and another one, Beničanci, with an exceptionally large capacity of ~2 billion cubic metre per year. These storage capacities will create conditions to store natural gas for neighbouring countries too. UGS business is being separated from INA, daughter company Podzemno skladište plina d.o.o. is established as of 1 January 2009. INA and Plinacro signed on 30 January 2009 in Zagreb the Sales Agreement.

4.5. CAPACITIES OF GAS TRANSPORT AND DISTRIBUTION NETWORKS

Transport of natural gas is carried out by state owned Plinacro. The natural gas transportation system comprises 1,657 km with diameters ranging from DN 80 to DN 700. The whole system was designed for the working pressure of 50 bar and partly for 75 bar. During peak demand ~510,000 m^3/h are transported, while the maximum quantities of 620,000 m^3/h are delivered to consumers. The total theoretical capacity of the transportation system is 2,000,000 m³/h. The system includes 12 metering and reduction stations with 210 metering points. Gathering gas lines are not included in the transportation system.

There are 36 natural gas distribution companies in Croatia and the total gas pipeline lengths amounts to 15,980 km. Additionally, there are two distribution companies for city gas and LPG/air mixture distribution, with the total network length of 239 km. Thus, the total distribution network in Croatia is 16,219 km long.

4.6. GAS BALANCE

The net gas balance is based on self consumption and degasolisation is omitted and losses are balanced with direct consumption which is divided into consumption for distribution and consumption by direct industrial consumers (Fig. 3).

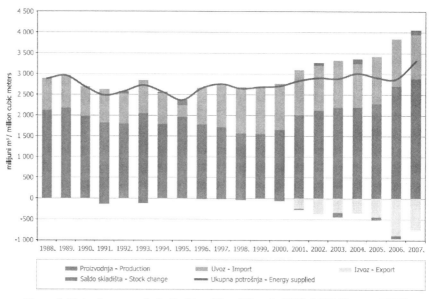

Figure 3. Natural gas supply in the Republic of Croatia 1988–2007 (Source: EIHP)

Figure 4 shows the structure of consumption for all categories of gas consumers in the period from 1988 to 2007.

According to official forecasts for domestic production as well as imports (The Strategic Plan of INA Naftaplin 2002–2011), Fig. 5 presents the available amounts of gas in Croatia in the period 2007–2010.

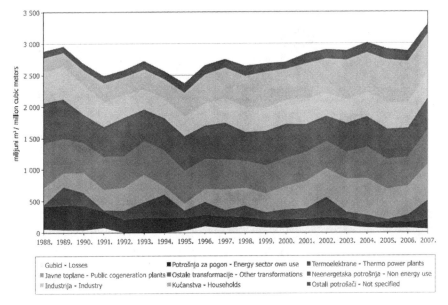

Figure 4. Structure of consumption of natural gas in Croatia 1988–2007 (Source: EIHP)

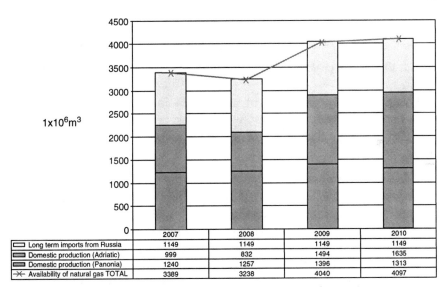

	2007	2008	2009	2010
Long term imports from Russia	1149	1149	1149	1149
Domestic production (Adriatic)	999	832	1494	1635
Domestic production (Panonia)	1240	1257	1396	1313
Availability of natural gas TOTAL	3389	3238	4040	4097

Figure 5. Production structure and supplies of natural gas 2007 – 2010 (Source: INA Naftaplin)

Figure 6 depicts the existing and planned gas system in Croatia with possible supply directions.

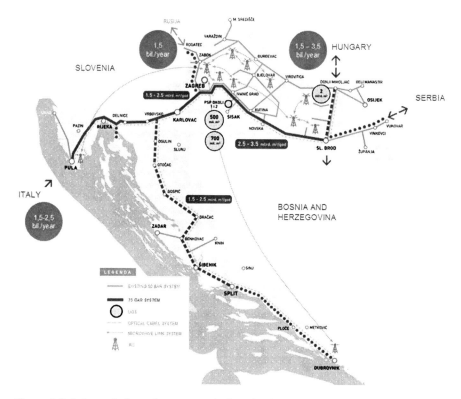

Figure 6. Existing and planned gas system in Croatia with supply routes (Source: EIHP)

Potential supply directions of natural gas in Croatia, that is, possible inter-connections with neighbouring countries, are the following (having in mind that the said schedule is arbitrary and does not represent any scheduled priority):

* Adria LNG terminal.
* Connection to the Hungarian gas system. This option (alongside additional imports of Russian gas, including SEGP) will enable later connections to the Nabucco and Caspian region and Middle East regions.
* Connection to the Romanian system via Serbia. This option, like the previous one, enables imports of additional quantities of Russian gas and later connections to the Nabucco and later Caspian and Middle East regions.

- Expanding capacities of existing import route (Baumgarten/TAG[2]/SOL-Cerŝak-Rogatec). This would enable taking over part of the quantity from the Nabucco pipeline in Baumgarten, and in any case a number of transit countries being led to the least possible measure.
- Supply of quantities of Russian gas via Slovenia through the Volta pipeline.
- Long term connection to the IGI (Italy – Greece Interconnector) or TAP (Trans Adriatic Pipeline) via Montenegro with later extensions for the planned pipeline to Dubrovnik. Alongside the above supplies from the Caspian and Middle East regions this would eventually allow supplies of certain quantities of Russian gas from the Blue Stream pipeline through the Black Sea to Turkey.

4.7. NATURAL GAS SUPPLY

After reviewing available quantities of natural gas and analysing total requirements, a comparative graph was prepared to show available quantities and total requirements of natural gas – these exhibit certain differences so it is necessary to determine new potential sources of supply. Figure 7 shows that there is an estimated shortage in gas supplies amounting to -737 million cubic metre in 2010.

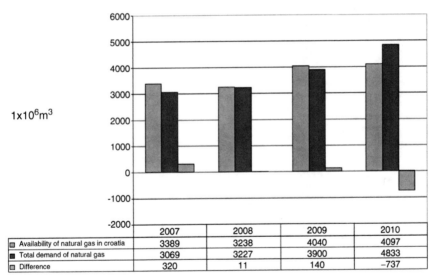

Figure 7. Supplies of natural gas in Croatia (1×106 m^3) (Source: INA Naftaplin)

---

[2] TAG pipeline capacity was increased by 3.2 billion cubic metre per year by installing additional compressors.

The following possible supply options are foreseen (Development and Modernization Plan of Gas Transport System in Croatia 2002–2011):

(a) To settle estimated consumption in that year, a part of the supplies is intended through the Slovenian route (Rogatec – Zabok) which has an excess capacity of 400 million cubic metre of transportation capacity.
(b) New transportation systems from Hungary (Slobodnica – D. Miholjac – Dravaszerdahely) with intended commencement of construction by Plinacro by the end of 2009.
(c) New gas systems Lučko – Zabok – Rogatec with start of construction estimated by the end of 2011 according to that business plan to develop the transportation system.

Long term import contract from Russia for gas supplies amount to 1,149 million cubic metre.

5. Energy Supply Vulnerability Indicators

Energy supply vulnerability can be calculated using different indicators. Detailed study on energy supply vulnerability indicators was recently prepared (Croatian Energy Sector Vulnerability Study, EIHP 2008) to help in detecting and future positioning of various aspects of security of supply. Even though it is not in the main scope of this paper, illustrations of two most common indicators are provided here. *Energy Dependency* is one of the most important indicators – it is defined as ratio between net energy import

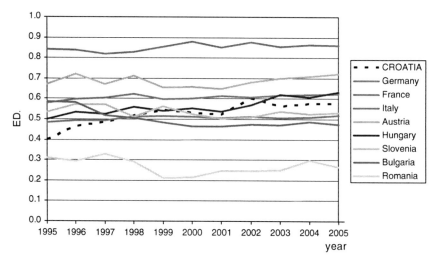

Figure 8. Energy Dependency among different European countries in 1995–2005 (Source: EIHP)

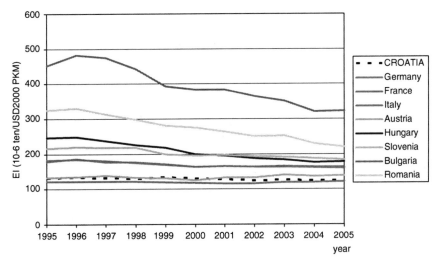

Figure 9. Energy Intensity indicator among several European countries in 1995–2005 (Source: EIHP)

and total energy consumption. Figure 8 shows a position of Croatia between several EU countries, as exhibiting an increasing *Energy Dependency* in the last decade.

On the other side, *Energy Intensity* implies effectiveness of energy use at increasing added value. It is defined as ratio between *Total Primary Energy Supply (Gross Inland Consumption)* and *Gross Domestic Product.* Comparison among several European countries is given in Fig. 9. Croatia has a relatively constant value, while other countries face a decreasing *Energy Intensity.*

6. Conclusions

This paper provides an insight into security of electricity and natural gas supply in Croatia, as mostly relating to the next short-term period. The key electricity market stakeholders and their responsibilities are analysed first as relating to security of supply. This is followed then by the presentation of the regulatory framework for electricity production, transmission, distribution and supply. Public service obligation is also given attention. Further on, existing and planned production capacities are described, as well as electricity network capacities with particular reflection on the cross-border transmission capacities and transmission network access and control. Finally, achievements and plans to satisfy electricity demand needs in the Croatian system are presented.

In the chapter on natural gas, firstly, the key stakeholders in the market and their responsibilities are analysed, as well as the public services obligations. This is followed by a presentation of production capacities, reserves and storage of natural gas, existing and planned capacities of the gas network. In conclusion, energy balance and natural gas supplies data are presented.

Finally, two illustrative energy supply vulnerability indicators are calculated – *Energy Dependency* and *Energy Intensity* – for several European countries as to present actual position of Croatia amongst different EU countries.

Finally, to conclude, the Croatian regulatory framework is rounded in the context of security of supply. The responsibilities of key stakeholders are nominated. Public service obligation of electricity/gas supply is established. Electricity and gas system development plans are continuously verified and upgraded. Croatian electricity system is very well connected to the neighbouring countries, while gas network develops in that direction – these being important for further market development within Croatia and in the South East Europe region.

Putting this together with favourable electricity generation mix (large share of hydro power plants) and diversification of future gas purchase directions, the Croatian electricity and gas sectors generally target high level of supply security.

The vulnerability indicators prove that Croatia has had an increasing *Energy Dependency* in the last decade. Moreover, it exhibits one of the lowest *Energy Intensity* among observed European countries.

References

Treaty establishing the Energy Community, October 2005.
Directive 2003/54/EC of the European Parliament and of the Council of 26 June 2003 concerning common rules for the internal market in electricity, June 2003.
Regulation 1228/2003/EC of the European Parliament and of the Council of 26 June 2003 on conditions for access to the network for cross-border exchanges in electricity, June 2003.
Commission Decision of 9 November 2006 amending the Annex to Regulation (EC) No 1228/2003 on conditions for access to the network for cross-border exchanges in electricity (2006/770/EC), European Commission, November 2006.
Directive 2003/55/EC of the European Parliament and of the Council of 26 June 2003 concerning common rules for the internal market in natural gas, June 2003.
Regulation 1775/2005/EC of the European Parliament and of the Council of 26 June 2003 on conditions for access to the natural gas transmission networks, September 2005.

Directive 2005/89/EC of the European Parliament and of the Council of 18 January 2006 concerning measures to safeguard security of electricity supply and infrastructure investment, January 2005.

Directive 2004/67/EC of the European Parliament and of the Council of 18 January 2006 concerning measures to safeguard security of natural gas supply, April 2004.

Statements on monitoring the security of supply: Explanatory Note of the Secretariat, Albania, Bosnia and Herzegovina, Croatia, The former Yugoslav Republic of Macedonia, Montenegro, Serbia, UNMIK, 5th PHLG Meeting, Becici, Montenegro, June 2007.

Croatian Energy Law (NN 68/01, NN 177/04, NN 76/07, NN 152/08).

Croatian Electricity Market Law (NN 177/04, NN 76/07, NN 152/08).

Croatian Energy Regulation Law (NN 177/04, NN 76/07).

Croatian Gas Market Law (NN 163/08).

The Strategic Plan of INA Naftaplin 2002–2011.

Development and Modernization Plan of Gas Transport System in Croatia 2002–2011.

Croatian Energy Sector Vulnerability Study, EIHP, 2008.

ZERO EMISSION SOURCES OF ELECTRICITY: COST, CAPACITY, ADVANTAGES AND DISADVANTAGES

OKSANA UDOVYK AND OLEG UDOVYK[*]
National Academy of Sciences, Kyiv, Ukraine

Abstract This paper shows how much carbon-free energy might ultimately be available at the global scale. It is based on the international reports and incorporates data from our more recent research. The paper examines cost, capacity, advantages and disadvantages of hydropower, nuclear fission, biomass, wind, geothermal, solar and ocean energy. It highlights which sources make most sense.

Keywords: Hydropower, nuclear fission, biomass, wind, geothermal, solar and ocean energy

1. Introduction

The Sun shines down on Earth with a power of 174,000 TW. It lifts water from the surface of the sea to fall back down on the high mountains of the land; drives ocean currents from the tropics to the poles; spins winds around the world; and powers the green engine of life, producing practically every calorie ever consumed on Earth.

Unfortunately, solar energy, reduced to waste heat, no longer leaves the planet as easily as it once did. Instead it tarries in the atmosphere, thanks to the ever stronger greenhouse effect. This change in the planet's energy balance is a small one — but so great is the flow of energy from the Sun that, over decades and centuries, it is doing terrible damage.

Clearly, society must slow and eventually reverse the rate at which the greenhouse effect is being strengthened. The usual way proposed to achieve this is to increase the cost of emitting carbon dioxide and other greenhouse

---

*To whom correspondence should be addressed: Oleg Udovyk, National Academy of Sciences, Kyiv, Ukraine; E-mail: oleg.udovyk@hotmail.com

gases, through a cap-and-trade scheme, or possibly a direct carbon tax. But for this strategy to work there must be attractive alternative sources of energy.

How much carbon-free energy might ultimately be available at the global scale? What are their cost, capacity, advantages and disadvantages? And which sources make most sense?

2. Hydropower

The world has a lot of dams – 45,000 large ones, according to the World Energy Council, and many more at small scales. Its hydroelectric power plants have a generating capacity of 800 GW and they currently supply almost one-fifth of the electricity consumed worldwide. As a source of electricity, dams are second only to fossil fuels, and generate 10 times more power than geothermal, solar and wind power combined. With a claimed full capacity of 18 GW, the Three Gorges dam in China can generate more or less twice as much power as all the world's solar cells. An additional 120 GW of capacity is under development. One reason for hydropower's success is that it is a widespread resource – 160 countries use hydropower to some extent. In several countries hydropower is the largest contributor to grid electricity – it is not uncommon in developing countries for a large dam to be the main generating source. Nevertheless, it is in large industrialized nations that have big rivers that hydroelectricity is shown in its most dramatic aspect. Brazil, Canada, China, Russia and the United States currently produce more than half of the world's hydropower.

2.1. COST AND CAPACITY

According to the International Hydropower Association (IHA), installation costs are usually in the range of US$1 million to more than $5 million per megawatt of capacity, depending on the site and size of the plant. Dams in lowlands and those with only a short drop between the water level and the turbine tend to be more expensive; large dams are cheaper per watt of capacity than small dams in similar settings. Annual operating costs are low – 0.8–2% of capital costs; electricity costs $0.03–0.10 per kilowatt-hour, which makes dams competitive with coal and gas.

The absolute limit on hydropower is the rate at which water flows downhill through the world's rivers, turning potential energy into kinetic energy as it goes. The amount of power that could theoretically be generated if the entire world's run-off were 'turbaned' down to sea level is more than 10 TW. However, it is rare for 50% of a river's power to be exploitable, and in many cases the figure is below 30%. Those figures still offer considerable opportunity for new capacity, according to the IHA. Europe currently sets a benchmark for hydropower use, with 75% of what is deemed feasible already exploited. For Africa to reach the same level, it would need to increase its hydropower capacity by a factor of 10 to more than 100 GW. Asia, which already has the greatest installed capacity, also has the greatest growth potential. If it were to triple its generating capacity, thus harnessing a near-European fraction of its potential, it would double the world's overall hydroelectric capacity. The IHA says that capacity could triple worldwide with enough investment.

2.2. ADVANTAGES AND DISADVANTAGES

The fact that hydroelectric systems require no fuel means that they also require no fuel-extracting infrastructure and no fuel transport. This means that a gigawatt of hydropower saves the world not just a GW's worth of coal burned at a fossil-fuel plant, but also the carbon costs of mining and transporting that coal. As turning on a tap is easy, dams can respond almost instantaneously to changing electricity demand independent of the time of day or the weather. This ease of turn-on makes them a useful back-up to less reliable renewable sources. That said, variations in use according to need and season mean that dams produce about half of their rated power capacity.

Hydroelectric systems are unique among generating systems in that they can, if correctly engineered, store the energy generated elsewhere, pumping water uphill when energy is abundant. The reservoirs they create can also provide water for irrigation, a way to control floods and create amenities for recreational use.

Not all regions have large hydropower resources – the Middle East, for example, is relatively deficient. And reservoirs take up a lot of space; today the area under man-made lakes is as large as two Italy. The large dams and reservoirs that account for most of that area and for more than 90% of hydro-generated electricity worldwide require lengthy and costly planning and construction, as well as the relocation of people from the reservoir area.

Dams have ecological effects on the ecosystems upstream and down-stream, and present a barrier to migrating fish. Sediment build-up can shorten their operating life, and sediment trapped by the dam is denied to those downstream. Biomass that decomposes in reservoirs releases methane and carbon dioxide, and in some cases these emissions can be of a similar order of magnitude to those avoided by not burning fossil fuels.

Climate change could itself limit the capacity of dams in some areas by altering the amount and pattern of annual run-off from sources such as the glaciers of Tibet. Because hydro is a mature technology, there is little room for improvement in the efficiency of generation.

Also, the more obvious and easy locations have been used, and so the remaining potential can be expected to be harder to exploit. Small (less than 10 MW) 'run-of-river' schemes that produce power from the natural flow of water – as millers have been doing for four millennia – are appealing, as they have naturally lower impacts. However, they are about five times more expensive and harder to scale than larger schemes.

So, hydropower – a cheap and mature technology, but with substantial environmental costs; roughly a terawatt of capacity could be added.

3. Nuclear Fission

When reactor 4 at the Chernobyl nuclear power plant melted down on 26 April 1986, the fallout contaminated large parts of Europe. That disaster, and the earlier incident at Three Mile Island in Pennsylvania, blighted the nuclear industry in the West for a generation. Worldwide, though, the picture did not change quite as dramatically. In 2008, 35 nuclear plants were under construction, almost all in Asia. The 439 reactors already in operation had an overall capacity of 370 GW, and contributed around 15% of the electricity generated worldwide, according to the most recent figures from the International Atomic Energy Agency (IAEA), which serves as the world's nuclear inspectorate.

3.1. COSTS AND CAPACITY

Depending on the design of the reactor, the site requirements and the rate of capital depreciation, the light-water reactors that make up most of the world's nuclear capacity produce electricity at costs of between US$0.025 and $0.07 per kilowatt-hour. The technology that makes this possible has benefited from decades of expensive research, development and purchases subsidized by governments; without that boost it is hard to imagine that nuclear power would currently be in use.

Because nuclear power requires fuel, it is constrained by fuel stocks. There are some 5.5 million tones of uranium in known reserves that could profitably be extracted at a cost of US$130 per kilogram or less, according to the latest edition of the 'Red Book', in which the IAEA and the Organization for Economic Co-operation and Development (OECD) assess uranium resources. At the current use of 66,500 t $year^{-1}$, that is about 80 years' worth of fuel. The current price of uranium is over that $130 threshold. Geologically similar ore deposits that are as yet unproven – 'undiscovered reserves' – are thought to amount to roughly double the proven reserves, and lower-grade ores offer considerably more. Uranium is not a particularly rare element – it is about as common a constituent of Earth's crust as zinc. Estimates of the ultimate recoverable resource vary greatly, but 35 million tones might be considered available. Nor is uranium the only naturally occurring element that can be made into nuclear fuel. Although they have not yet been developed, thorium-fuelled reactors are a possibility; bringing thorium into play would double the available fuel reserves. Furthermore, although current reactor designs use their fuel only once, this could be changed. Breeder reactors, which make plutonium from uranium isotopes that are not themselves useful for power production, can effectively create more fuel than they use. A system built on such reactors might get 60 times more energy out for every kilogram of natural uranium put in, although lower multiples might be more realistic. With breeder reactors, which have yet to be proven on a commercial basis, the world could in principle go 100% nuclear. Without them, it is still plausible for the amount of nuclear capacity to grow by a factor of 2 or 3, and to operate at that level for a century or more.

3.2. ADVANTAGES, DISADVANTAGES

Nuclear power has relatively low fuel costs and can run at full blast almost constantly – US plants deliver 90% of their rated capacity. This makes them well suited to providing always-on 'base load' power to national grids. Uranium is sufficiently widespread that the world's nuclear-fuel supply is unlikely to be threatened by political factors.

There is no agreed solution to the problem of how to deal with the nuclear waste that has been generated in nuclear plants over the past 50 years. Without long-term solutions, which are more demanding politically than technically, growth in nuclear power is an understandably hard sell. A further problem is that the spread of nuclear power is difficult to disentangle from the proliferation of nuclear weapons capabilities. Fuel cycles that involve recycling, and which thus necessarily produce plutonium, are particularly worrying. Even without proliferation worries, nuclear power stations may make tempting targets for terrorists or enemy forces (although in the latter case the same is true of hydroelectric plants). A long-term commitment to greatly increased use of nuclear power would require public acceptance not just of existing technologies but of new ones, too – thorium and breeder reactors, for instance. These technologies would also have to win over investors and regulators. Nuclear power is also extremely capital intensive; power costs over the life of the plant are comparatively low only because the plants are long lived. Nuclear power is thus an expensive option in the short term. Another constraint is a lack of skilled workers. Building and operating nuclear plants requires a great many highly trained professionals, and enlarging this pool of talent enough to double the rate at which new plants are brought online might prove very challenging. The engineering capacity for making key components would also need enlarging. In light of these obstacles, predictions of the future role of nuclear power vary considerably. The European Commission's Outlook up to 2050 contains a bullish scenario that assumes that, with public acceptance and the development of new reactor technologies, nuclear power could provide about 1.7 TW by 2050. The IAEA's analysts are more cautious. They see capacity rising to not more than 1,200 GW by 2050. Massachusetts Institute of Technology described a concrete scenario for tripling capacity to 1,000 GW by 2050, a scenario predicated on US leadership, continued commitment by Japan and renewed activity by Europe. This scenario relied only on improved versions of today's reactors rather than on any radically different or improved design.

Reaching a capacity in the terawatt range is technically possible over the next few decades, but it may be difficult politically. A climate of opinion that came to accept nuclear power might well be highly vulnerable to adverse events such as another Chernobyl-scale accident or a terrorist attack.

4. Biomass

Biomass was humanity's first source of energy, and until the twentieth century it remained the largest; even today it comes second only to fossil fuels. Wood, crop residues and other biological sources are an important energy source for more than two billion people. Mostly, this fuel is burned in fires and cooking stoves, but over recent years biomass has become a source of fossil-fuel-free electricity. As of 2008, the World Energy Council estimates biomass generating capacity to be at least 40 GW. Biomass can supplement coal or in some cases gas in conventional power plants. Biomass is also used in many co-generation plants that can capture 85–90% of the available energy by making use of waste heat as well as electric power.

4.1. COSTS AND CAPACITY

The price of biomass electricity varies widely depending on the availability and type of the fuel and the cost of transporting it. Capital costs are similar to those for fossil-fuel plants. Power costs can be as little as $0.02 per kilowatt-hour when biomass is burned with coal in a conventional power plant, but increase to $0.03–0.05 per kilowatt-hour from a dedicated biomass power plant. Costs increase to $0.04–0.09 per kilowatt-hour for a cogeneration plant, but recovery and use of the waste heat makes the process much more efficient. The biggest problem for new biomass power plants is finding a reliable and concentrated feedstock that is available locally; keeping down transportation costs means keeping biomass power plants tied to locally available fuel and quite small, which increases the capital cost per megawatt.

Biomass is limited by the available land surface, the efficiency of photosynthesis, and the supply of water. An OECD round table in 2008 estimated that there is perhaps half a billion hectares of land not in agricultural use that would be suitable for rain-fed biomass production, and suggested that by 2050 this land, plus crop residues, forest residues and organic waste might provide enough burnable material each year to provide 68,000 TW-h. Converted to electricity at an efficiency of 40%, that could provide a maximum of 3 TW. The Intergovernmental Panel on Climate Change pegs the potential at roughly 120,000 TW-h in 2050, which equates to slightly more than 5 TW on the basis of a larger estimate of available land. These projections involve some fairly extreme assumptions about converting land to the production of energy crops. And even to the extent that these assumptions prove viable, electricity is not the only potential use for such plantations. By storing solar energy in the form of chemical bonds, biomass lends itself better than other renewable energy resources to the production of fuel for transportation. Although turning biomass to biofuel is not as efficient as just burning, it can produce a higher-value product. Biofuels might easily beat electricity generation as a use for biomass in most settings.

4.2. ADVANTAGES, DISADVANTAGES

Plants are by nature carbon-neutral and renewable, although agriculture does use up resources, especially if it requires large amounts of fertilizer. The technologies needed to burn biomass are mature and efficient, especially in the case of co-generation. Small systems using crop residues can minimize transportation costs. If burned in power plants fitted with carbon-capture-and-storage hardware, biomass goes from being carbon neutral to carbon negative, effectively sucking carbon dioxide out of the atmosphere and storing it in the ground. This makes it the only energy technology that can actually reduce carbon dioxide levels in the atmosphere. As with coal, however, there are costs involved in carbon capture, both in terms of capital set-up and in terms of efficiency.

There is only so much land in the world, and much of it will be needed to provide food for the growing global population. It is not clear whether letting market mechanisms drive the allocation of land between fuel and food is desirable or politically feasible. Changing climate is diminishing the productivity and availability of suitable land. There is likely to be opposition to increased and increasingly intense cultivation of energy crops. Use of waste and residues may remove carbon and nutrients from the land that would otherwise have enriched the soil; long-term sustainability may not be achievable. Bioenergy dependence could also open the doors to energy crises caused by drought or pestilence, and land-use changes can have climate effects of their own: clearing land for energy crops produces emissions at a rate the crops themselves cannot offset.

If a large increase in energy crops proves acceptable and sustainable, much of it may be used up in the fuel sector. However, small-scale systems may be desirable in an increasing number of settings and the possibility of carbon-negative systems – which are plausible for electricity generation but not for biofuels – is a unique and attractive capability.

5. Wind

Wind power is expanding faster than even its fiercest advocates could have wished a few years ago. Globally, capacity has risen by nearly 25% in each of the past 5 years, according to the Global Wind Energy Council. Wind Power Monthly estimates that the world's installed capacity for wind as of 2008 was 94 GW. If growth continued at 21%, that figure would triple over 6 years. Despite this, the numbers remain small on a global scale, especially given that wind farms have historically generated just 20% of their capacity.

5.1. COSTS AND CAPACITY

Installation costs for wind power are around US$1.8 million per megawatt for onshore developments and between $2.4 million and $3 million for offshore projects. That translates to $0.05–0.09 per kilowatt-hour, making wind competitive with coal at the lower end of the range. With subsidies, as enjoyed in many countries, the costs come in well below those for coal – hence the boom. The main limit on wind-power installation at the moment is how fast manufacturers can make turbines. These costs represent significant improvements in the technology. In 1988, a wind farm might have consisted of an array of 50-kW turbines that produced power for roughly $0.40 per kilowatt-hour. Today's turbines can produce 30 times as much power at one-fifth the price with much less down time.

The amount of energy generated by the movement of Earth's atmosphere is vast – hundreds of terawatts. Researchers calculated that at least 72 TW could be effectively generated using 2.5 million of today's larger turbines placed at the 13% of locations around the world that have wind speeds of at least 6.9 m s^{-1} and are thus practical sites.

5.2. ADVANTAGES, DISADVANTAGES

The main advantage of wind is that, like hydropower, it doesn't need fuel in operation. The only costs therefore come from building, maintaining and eventual decommissioning of the turbines and power lines. Turbines are getting bigger and more reliable. The development of technologies for capturing wind at high altitudes could provide sources with small footprints capable of generating power in a much more sustained way.

Wind's ultimate limitation might be its intermittency. Providing up to 20% of a grid's capacity from wind is not too difficult. Beyond that, utilities and grid operators need to take extra steps to deal with the variability. Another grid issue, and one that is definitely limiting in the near term, is that the windiest places are seldom the most populous, and so electricity from the wind needs infrastructure development – especially for offshore settings. The recoupable energy is some two orders of magnitude lower because of turbine spacing and engineering constraints. As well as being intermittent, wind power is, like other renewable energy sources, inherently quite low density. A large wind farm typically generates a few watts per square meter – 10 W is very high. Wind power thus depends on cheap land, or on land being used for other things at the same time, or both. It is also hard to deploy in an area where the population sets great store by the value of a turbine-free landscape. Wind power is also unequally distributed: it favors nations with access to windy seas and their onshore breezes or great empty plains. Germany has covered much of its windiest land with turbines, but despite these pioneering efforts, its combined capacity of 22 GW supplies less than 7% of the country's electricity needs. Britain, which has been much slower to adopt wind power, has by far the largest offshore potential in Europe – enough to meet its electricity needs three times over, according to the British Wind Energy Association. Industry estimates suggest that the European Union could meet 25% of its current electricity needs by developing less than 5% of the North Sea. Such truly large-scale deployment of wind-power schemes could affect local, and potentially global, climate by altering wind patterns. Wind tends to cool things down, so temperatures around a very large wind farm could rise as turbines slow the wind to extract its energy. Some experts suggest that 2 TW of wind capacity could affect temperatures by about 0.5°C, with warming at mid-latitudes and cooling at the poles – perhaps in that respect offsetting the effect of global warming.

With large deployments on the plains of the United States and China, and cheaper access to offshore, a wind-power capacity of a terawatt or more is plausible.

6. Geothermal

Earth's interior contains vast amounts of heat, some of it left over from the planet's original coalescence, some of it generated by the decay of radio-active elements. Because rock conducts heat poorly, the rate at which this heat flows to the surface is very slow; if it were quicker, Earth's core would have frozen and its continents ceased to drift long ago. The slow flow of Earth's heat makes it a hard resource to use for electricity generation except in a few specific places, such as those with abundant hot springs. Only a couple of dozen countries produce geothermal electricity, and only five of those – Costa Rica, El Salvador, Iceland, Kenya and the Philippines – generate more than 15% of their electricity this way. The world's installed geothermal electricity capacity is about 10 GW, and is growing only slowly – about 3% per year in the first half of this decade. A decade ago, geothermal capacity was greater than wind capacity; now it is almost a factor of 10 less. Earth's heat can also be used directly. Indeed, small geothermal heat pumps that warm houses and businesses directly may represent the greatest contribution that Earth's warmth can make to the world's energy budget.

6.1. COSTS AND CAPACITY

The cost of a geothermal system depends on the geological setting. The situation as being similar to mineral resources. There is a continuum of resource grades – from shallow, high-temperature regions of high-porosity rock, to deeper low-porosity regions that are more challenging to exploit. The cost of exploiting the best sites – those with a lot of hot water circulating close to the surface – at about US$0.05 per kilowatt-hour. Much more abundant low-grade resources are exploitable with current technology only at much higher prices.

Earth loses heat at between 40 and 50 TW a year, which works out at an average of a bit less than a tenth of a watt per square meter. For comparison, sunlight comes in at an average of 200 W m^{-1}. With today's technology,

70 GW of the global heat flux is seen as exploitable. With more advanced technologies, at least twice that could be used. Using enhanced systems that inject water at depth using sophisticated drilling systems it would be possible to set up 100 GW of geothermal electricity in the United States alone. With similar assumptions a global figure of a terawatt or so can be reached, suggesting that geothermal could, with a great deal of investment, provide as much electricity as dams do today.

6.2. ADVANTAGES, DISADVANTAGES

Geothermal resources require no fuel in use. They are ideally suited to supplying base-load electricity, because they are driven by a very regular energy supply. At 75%, geothermal sources boast a higher capacity factor than any other renewable. Low-grade heat left over after electricity generation can be used for domestic heating or for industrial processes. Surveying and drilling previously unexploited geothermal resources has become much easier thanks to mapping technology and drilling equipment designed by the oil industry. A significant technology development programmed – Tester suggests $1 billion over 10 years – could greatly expand the achievable capacity as lower-grade resources are opened up.

High-grade resources are quite rare, and even low-grade resources are not evenly distributed. Carbon dioxide can leak out of some geothermal fields, and there can be contamination issues; the water that brings the heat to the surface can carry compounds that shouldn't be released into aquifers. In dry regions, water availability can be a constraint. Large-scale exploitation requires technologies that, although plausible, have not been demonstrated in the form of robust, working systems.

Capacity might be increased by more than an order of magnitude. Without spectacular improvements, it is unlikely to outstrip hydro and wind and reach a terawatt.

7. Solar

Not to take anything away from the miracle of photosynthesis, but even under the best conditions plants can only turn about 1% of the solar radiation that hits their surfaces into energy that anyone else can use. For comparison, a standard commercial solar photovoltaic panel can convert 12–18% of the energy of sunlight into useable electricity; high-end models come in above 20% efficiency. Increasing manufacturing capacity and decreasing costs have led to remarkable growth in the industry over the past 5 years: in 2002, 550 MW of cells were shipped worldwide; in 2008 the figure was six times that. Total installed solar-cell capacity is estimated at 9 GW or so. The actual amount of electricity generated, though, is considerably less, as night and clouds decrease the power available. Of all renewable, solar currently has the lowest capacity factor, at about 14%. Solar cells are not the only technology by which sunlight can be turned into electricity. Concentrated solar thermal systems use mirrors to focus the Sun's heat, typically heating up a working fluid that in turn drives a turbine. The mirrors can be set in troughs, in parabolas that track the Sun, or in arrays that focus the heat on a central tower. As yet, the installed capacity is quite small, and the technology will always remain limited to places where there are a lot of cloud-free days – it needs direct sun, whereas photovoltaic can make do with more diffuse light.

7.1. COSTS, CAPACITY

The manufacturing cost of solar cells is currently US$1.50–2.50 for a watt's worth of generating capacity, and prices are in the $2.50–3.50 per watt range. Installation costs are extra; the price of a full system is normally about twice the price of the cells. What this means in terms of cost per kilowatt-hour over the life of an installation varies according to the location, but it comes out at around $0.25–0.40. Manufacturing costs are dropping, and installation costs will also fall as photovoltaic cells integrated into building materials replace free-standing panels for domestic applications. Current technologies should be manufacturing at less than $1 per watt within a few years. The cost per kilowatt-hour of concentrated solar thermal power is estimated by the US National Renewable Energy Laboratory (NREL) in Golden, Colorado, at about $0.17.

Earth receives about 100,000 TW of solar power at its surface – enough energy every hour to supply humanity's energy needs for a year. There are parts of the Sahara Desert, the Gobi Desert in central Asia, the Atacama in Peru or the Great Basin in the United States where a GW of electricity could be generated using today's photovoltaic cells in an array 7 or 8 km across. Theoretically, the world's entire primary energy needs could be served by less than a tenth of the area of the Sahara. Advocates of solar cells point to a calculation by the NREL claiming that solar panels on all usable residential and commercial roof surfaces could provide the United States with as much electricity per annum as the country used in 2005. In more temperate climes things are not so promising: in Britain one might expect an annual insulation of about 1,000 kW-h m^{-1} on a south-facing panel tilted to take account of latitude: at 10% efficiency, that means more than 60 m^2 per person would be needed to meet current UK electricity consumption.

7.2. ADVANTAGES, DISADVANTAGES

The Sun represents an effectively unlimited supply of fuel at no cost during operation, which is widely distributed and leaves no residue. The public accepts solar technology and in most places approves of it – it is subject to less geopolitical, environmental and aesthetic concern than nuclear, wind or hydro, although extremely large desert installations might elicit protests. Photovoltaic can often be installed piecemeal – house by house and business by business. In these settings, the cost of generation has to compete with the retail price of electricity, rather than the cost of generating it by other means, which gives solar a considerable boost. The technology is also obviously well suited to off-grid generation and thus to areas without well developed infrastructure. Both photovoltaic and concentrated solar thermal technologies have clear room for improvement. It is not unreasonable to imagine that in a decade or 2 new technologies could lower the cost per watt for photovoltaic by a factor of ten, something that is almost unimaginable for any other non-carbon electricity source.

The ultimate limitation on solar power is darkness. Solar cells do not generate electricity at night, and in places with frequent and extensive cloud cover, generation fluctuates unpredictably during the day. Some concentrated solar thermal systems get around this by storing up heat during the day for use at night (molten salt is one possible storage medium), which is one of the reasons they might be preferred over photovoltaic for large installations. Another possibility is distributed storage, perhaps in the batteries of electric and hybrid cars. Another problem is that large installations will usually be in deserts, and so the distribution of the electricity generated will pose problems. German Aerospace Centre proposed that by 2050 Europe could be importing 100 GW from an assortment of photovoltaic and solar thermal plants across the Middle East and North Africa. They also noted that this would require new direct-current high-voltage electricity distribution systems. A possible drawback of some advanced photovoltaic cells is that they use rare elements that might be subject to increases in cost and restriction in supply. It is not clear, however, whether any of these elements is either truly constrained – more reserves might be made economically viable if demand were higher – or irreplaceable.

In the middle to long run, the size of the resource and the potential for further technological development make it hard not to see solar power as the most promising carbon-free technology. But without significantly enhanced storage options it cannot solve the problem in its entirety.

8. Ocean Energy

The oceans offer two sorts of available kinetic energy — that of the tides and that of the waves. Neither currently makes a significant contribution to world electricity generation, but this has not stopped enthusiasts from developing schemes to make use of them. There are undoubtedly some places where, thanks to peculiarities of geography, tides offer a powerful resource. In some situations that potential would best be harnessed by a barrage that creates a reservoir not unlike that of a hydroelectric dam, except that it is refilled regularly by the pull of the Moon and the Sun, rather than being topped up slowly by the runoff of falling rain. But although there are various schemes for tidal barrages under discussion – most notably the Severn Barrage between England and Wales, which proponents claim could offer as much as 8 GW – the plant on the Rance estuary in Brittany, rated at 240 MW, remains the world's largest tidal-power plant more than 40 years after it came into use. There are also locations well suited to tidal-stream systems – submerged turbines that spin in the flowing tide like windmills in the air. The 1.2 MW turbine installed in the mouth of Stanford Lough, Northern Ireland, is the largest such system so far installed. Most technologies for capturing wave power remain firmly in the testing phase. Individual companies are working through an array of potential designs, including machines that undulate on waves like a snake, bob up and down as water passes over them, or nestle on the coastline to be regularly overtopped by waves that power turbines as the water drains off.

8.1. COSTS AND CAPACITY

Barrage costs differ markedly from site to site, but are broadly comparable to costs for hydropower. At an estimated cost of £15 billion or more, the capital costs of the Severn Barrage would be about $4 million per megawatt. British Carbon Trust which spurs investment in non-carbon energy puts the costs of tidal-stream electricity in the $0.20–0.40 per kilowatt-hour range, with wave systems running up to $0.90 per kilowatt-hour. Neither technology is anywhere close to the large-scale production needed to significantly drive such costs down.

The interaction of Earth's mass with the gravitational fields of the Moon and the Sun is estimated to produce about 3 TW of tidal energy – rather modest for such an astronomical source. Of this, perhaps 1 TW is in shallow enough waters to be easily exploited, and only a small part of that is realistically available. EDF, a French power company developing tidal power off Brittany, says that the tidal-stream potential off France is 80% of that available all round Europe, and yet it is still little more than a GW. The power of ocean waves is estimated at more than 100 TW. The European Ocean Energy Association estimates that the accessible global resource is between 1 and 10 TW, but sees much less than that as recoverable with current technologies. An analysis in the MRS Bulletin in April 2008 holds that about 2% of the world's coastline has waves with an energy density of 30 kW m^{-1}, which would offer a technical potential of about 500 GW for devices working at 40% efficiency. Thus even with a huge amount of development, wave power would be unlikely to get close to the current installed hydroelectric capacity.

8.2. ADVANTAGES, DISADVANTAGES

Tides are eminently predictable, and in some places barrages really do offer the potential for large-scale generation that would be significant on a countrywide scale. Barrages also offer some built-in storage potential. Waves are not constant – but they are more reliable than winds.

The available resource varies wildly with geography; not every country has a coastline, and not every coastline has strong tides or tidal streams, or particularly impressive waves. The particularly hot wave sites include Australia's west coast, South Africa, the western coast of North America and western European coastlines. Building turbines that can survive for decades at sea in violent conditions is tough. Barrages have environmental impacts, typically flooding previously intertidal wetlands, and wave systems that flank long stretches of dramatic coastline might be hard for the public to accept. Tides and waves tend by their nature to be found at the far end of electricity grids, so bringing back the energy represents an extra difficulty. Surfers have also been known to object.

Therefore, ocean energy is marginal on the global scale.

9. Conclusion

The tendency among governments and traditional utilities to see renewable energy sources as oddities or add-ons is thus deeply misplaced. These sources are the future; the only question is how fast we can harness their potential. For the most part they are driven by the Sun – either directly, or via the indirect means of wind, water and plants – and it is striking to see how much they have to offer. Many of these technologies offer more than a terawatt of additional global capacity, and all have ample room for growth. Today's largest, hydropower and nuclear power could grow by a factor of 2 or more – albeit with potentially troublesome non-climate consequences.

But the greatest proportional capacity for growth lies with the current small-fry contributors. Wind turbines, for example, offer much greater possibilities for improvement than hydroelectric plants; the technologies for harnessing the wind far out to sea – or, for that matter, in the high currents of the jet-stream – are still comparatively immature. And even wind does not have the ultimate potential promised by the direct use of sunlight. In the future, new manufacturing methods that use alternative materials should make the capture of sunlight at least an order of magnitude cheaper than it is now.

The challenge is how to scale these new technologies up for a global market. A significant part of the answer is investment in focused research and development. Sadly, in most of the world, government spending on research in the energy sector has been stagnant or worse for decades. That has to change. Also crucial is an entrepreneurial culture that can get new ideas to mass markets. That requires both a social network geared to innovation and a supply of ready capital (Udovyk 2007).

Still, enthusiastic venture capitalists will not be enough. Firms with new energy technologies need to be able to access the capital of the stock markets if they are to grow large enough to change the world. One way to make such firms attractive to investors is to guarantee them sales. This can be done through regulation – insisting on a certain amount of renewable energy in all new developments, say – through subsidy, or through direct intervention; there is nothing to stop governments from acting as bulk buyers of new capacity. Governments would be wise, though, not to indulge protectionist tendencies in the process. For investors in some place to know that they can sell in any another place, and to be sure of a market in any place, is a great benefit.

Finding economically efficient ways to stimulate the growth of a particular industry is not easy. The types of subsidy that have worked to date – such as the feed-in inducements that have built Germany its solar-cell industry – are expensive even at the megawatt scale. At the scale of tens and then hundreds of gigawatts they are likely to be unsustainable. Perhaps worse than the cost, though, is the risk that poorly designed subsidies will damage markets and thwart the development they seek to encourage. However, this is a challenge to encourage technologies in a smart and flexible way, not a reason to leave the markets alone. New whole-system approaches to market transformation are available, offering opportunities for change on the scale of what needs to be done (Greyson 2010).

The challenges are daunting. But the wind keeps blowing, the grass keeps growing, and the Sun keeps shining. The energy is there.

References

O. Udovyk, 2007. "Prospects for sustainable development of Ukrainian energy sector" in "Assessment of Hydrogen Energy for Sustainable Development", Eds. J.W. Sheffild et al., Springer, pp. 249–256.

J. Greyson, 2010. "Seven policy switches for global security" in this book, pp. 69–92.

ANALYSIS OF ENERGY OPTIONS IMPACT ON REGIONAL SECURITY: THE CASE OF EASTERN EUROPE

OLEG UDOVYK[*] AND OKSANA UDOVYK

National Academy of Sciences, Kyiv, Ukraine

Abstract   This paper examines energy security in three East European countries – Belarus, Moldova, and Ukraine. It is based on the report of the Environment and Security Initiative and incorporates analysis of recent energy security events in the region. The paper highlights the importance of recognizing the region's geopolitical positioning between the European Union and the Russian Federation, improving energy security without jeopardizing the environment, addressing long-standing frozen conflicts, and strengthening cooperation.

Keywords: Ukraine; energy dilemma; Chernobyl legacy; frozen conflicts, European Union, Russian Federation, Gazprom, Nabucco project

1. Introduction

The impact of energy on economy and security is clear and profound, and this is why in recent years energy security has become a source of concern to most countries. However, energy security means different things to different countries based on their geographic location, their endowment of resources, and their strategic and economic conditions (Korin and Luff 2009).

NATO leaders recognize that the disruption of the flow of vital resources could affect Alliance security interests. They have declared their support for a coordinated, international effort to assess risks to energy infrastructure and to promote energy infrastructure security. At the Bucharest Summit in April 2008, the Allies noted a report on "NATO's Role in

---

[*] To whom correspondence should be addressed: Oleg Udovyk, National Academy of Sciences, Kyiv, Ukraine; E-mail: oleg.udovyk@hotmail.com

F. Barbir and S. Ulgiati (eds.), *Energy Options Impact on Regional Security*,
DOI 10.1007/978-90-481-9565-7_12, © Springer Science + Business Media B.V. 2010

Energy Security" (www.nato/int) which identifies guiding principles and outlines options and recommendations for further activities. Those principles were reiterated at the Strasbourg/Kehl Summit in April 2009. The report identifies five key areas where NATO can provide added value: information and intelligence fusion and sharing; projecting stability; advancing international and regional cooperation; supporting consequence management; and supporting the protection of critical infrastructure. Consultations are ongoing as to the depth and range of NATO's involvement in this issue.

This paper examines energy security in East European nations. Eastern Europe extends from the northern shore of the Black Sea in Ukraine up to the Baltic Sea basin in Belarus. It covers 845,000 km^2 and is home to almost 60 million people. These nations share common borders, watersheds, and infrastructure and have many similarities in their geography, history, culture, and economy. Belarus, Moldova, and Ukraine are nations with recent sovereign statehood. They are positioned between an enlarging EU and a historically influential Russia. The area's unique position and history have played a large part in the overlapping of security issues, which have evolved over three distinct periods: the Soviet years of intensive industrialization, a difficult period of political and economic transition, and the most recent period with its new challenges (Udovyk 2008a).

2. The Regional Context

The region has negotiated the difficult transition years without suffering violent conflict of the kind that paralyzed the Balkans, the Caucasus, and Central Asia. Eastern Europe gained much sympathy by deciding not to preserve military nuclear capacity and transferring weapons inherited from the Ukraine to Russia. Furthermore, disagreements between Russia and Ukraine regarding the status of the Soviet Black Sea fleet have been satisfactorily managed and largely resolved, sparing Europe a major security risk.

However, there are plenty of regional security issues reaching beyond the borders of Eastern Europe to feature on the security agenda of the whole continent. The Transnistrian conflict in Moldova is one example. There are also difficult issues of supply and transit of Russian fuel. The key challenge for the three countries is still to strengthen contemporary state institutions to fully address economic, social, demographic, environmental, and security problems.

The legacy of the Chernobyl disaster – almost synonymous for the outside world with environmental problems in Eastern Europe – epitomizes the difficulties involved in dealing with all these problems at the same time.

26 April 1986 a violent explosion at the Chernobyl nuclear power plant destroyed the reactor and started a large fire that lasted 10 days. During the explosion and the fire a huge amount of radioactivity was released into the environment, spreading into Belarus, Ukraine, and beyond (Udovyk 2007a). The disaster also clearly demonstrated that an accident in one country may threaten human lives and health all over a continent (Udovyk 2009).

Given this legacy, the recent announcements of plans by the governments of Belarus and Ukraine to expand the use of nuclear power reflect the dramatic challenges facing these countries. Their current dependence on energy imports is seen as a key security concern. The region does not have sufficient energy resources of its own, but energy is critically important for both social stability and economic development, particularly with such energy-intense economies. The energy issue is all the more important because Eastern Europe stands at the crossroads of east–west and north–south energy corridors linking Russia to Western Europe, and the Black Sea to the Baltic (Udovyk 2008b).

2.1. THE GEOPOLITICAL POSITION

Despite common borders and many similarities, the three countries of Eastern Europe do not constitute a region in the sense of political community. Belarus, Moldova, and Ukraine have not yet developed visible capacity and projects for regional integration. On the contrary, Eastern Europe is a zone of geopolitical attraction among major powers, including the Russian Federation to the east and the European Union to the west. Eastern Europe's pivotal location at the intersection of strategic transport corridors, such as between Russian and Caspian producers of fuel and European energy consumers, further amplifies such influence.

After expanding eastward over the last decade, the EU seems to be experiencing "enlargement fatigue." Its capacity to absorb additional members was compromised. The EU is also the most important trade partner for all three countries. It is therefore still important for the EU to have friendly, politically stable, and economically prosperous countries on its doorstep, forming a solid bulwark against unwanted migration, terrorism, and other threats such as drug, arms, and human trafficking. Ukraine and Moldova are the only two European countries among the "top ten" sources of illegal migrants to the EU. EU's most comprehensive attempt to deal with Eastern Europe is through its Neighborhood Policy, which aims at strengthening stability in the region and cross-border cooperation.

On the eastern side, Eastern European countries must forge new relations with Russia, with which they share strong historic, cultural, and social ties.

Russia is keen to maintain secure transit routes through Eastern Europe while retaining the ties of the past and developing political and economic cooperation. Russia remains a key market for Eastern European products and the most important energy supplier for all three countries. As is the case with the EU, this economic cooperation makes relations with Russia extremely important and political disagreements – for example regarding the settlement of the Transnistrian conflict in Moldova – very painful. Russian security interests are also related to the presence of its military facilities in Moldova (Transnistria) and Ukraine (Crimea).

2.2. INTERNAL SECURITY CHALLENGES

Internal problems and tensions are no less important than geopolitical challenges. Not only may they weaken young states and increase their vulnerability to external factors, but they may also present security challenges in their own right. Not surprisingly, such internal security factors feature prominently in the national security doctrines of all three countries.

In the past Belarus, Ukraine, and Moldova were intricately linked to the rest of the Soviet economy. The collapse of the USSR and economic liberalization opened up local markets, increased competition, and severed some of the ties with former Soviet republics. However, access to Western markets, especially in the EU, has been very limited and often conditional on political or further economic reform. Moreover, the new patterns of trade with Europe have increasingly consisted of exports of raw materials in exchange for imports of manufactured goods. Finally, it has proven difficult to restructure the old heavy industry that was often the mainstay of the Soviet-era economy.

Economic restructuring has consequently not delivered on its promise of universally higher living standards and political stability. The decline in agricultural production contributed to increased poverty and further deterioration in the basic infrastructure of rural areas in all three countries. Social problems have also become more acute in some heavily industrialized regions. In certain cases this has coincided with tension and conflict. Here again, the most striking example is Transnistria, home to almost all Moldovan industry with traditionally strong ties to the former Soviet economic space. Another example of a region suffering from economic restructuring is the heavily industrialized Donbas region in Ukraine, where economic and social problems mesh with issues of environmental and energy security.

The economic and social problems of rural and heavily industrialized areas are aggravated by demographic trends, severely affected by the declining birth rates that are now below the replacement level in all three countries. The populations of Ukraine and Belarus will shrink significantly. Outgoing labor migration makes the situation even worse.

Coping with these difficulties requires effective, resourceful, and committed state government. However, government bodies in the region are not always able to implement reform of social welfare, health care, and education. They themselves are often in need of reform, to effectively deal with public sector corruption, for example.

2.3. INTERPLAY OF INTERNAL AND EXTERNAL SECURITY CHALLENGES

Internal and external security challenges are closely linked (Fig. 1).

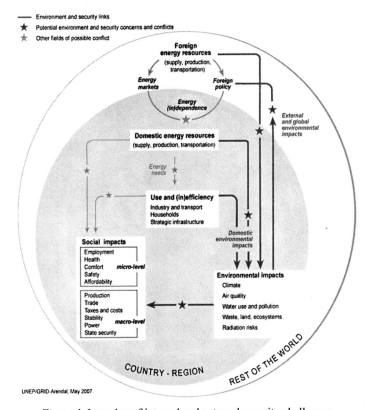

Figure 1. Interplay of internal and external security challenges

On the one hand, internal weaknesses increase vulnerability to external threats; on the other hand, external pressures often shape economic and political reforms with their social, environmental, and other security repercussions. Energy, among other issues, is at the core of both internal and external security challenges in the region.

3. The Energy Dilemma and Chernobyl Legacy

Given the tragic Chernobyl legacy (Fig. 2), why are both Belarus and Ukraine currently considering expanding their nuclear energy generating capability? The answer lies in the special role played by energy and energy security in Eastern Europe.

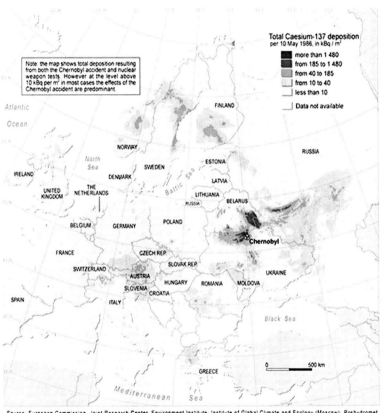

Source: European Commission, Joint Research Center; Environment Institute; Institute of Global Climate and Ecology (Moscow); Roshydromet (Russia); Minchernobyl (Ukraine); Belhydromet (Belarus). *Atlas of Caesium Deposition on Europe after the Chernobyl Accident.* 1998.

Figure 2. Chernobyl legacy

Energy is vital for the internal and external security of all three countries. A secure, affordable domestic energy supply is critical to economic development, particularly in energy-hungry industrial sectors. It is also essential to meet social needs (heating, transportation, etc.), especially for vulnerable groups. Since the region's own energy resources and production capacities, especially in Moldova and Belarus, are insufficient, a significant proportion of energy has to be imported, primarily from Russia. This is, in turn, a major factor in the external security of Eastern Europe. Another factor is the location of the region at the crossroads of major energy transport corridors linking producers in Russia and the Caspian region with consumers in Central, Western, and Northern Europe. The stability of oil and gas transportation routes is becoming increasingly important for Russia, the EU and other countries.

A good illustration of the external aspect of energy security was the heated debate over arrangements for the supply of Russian natural gas to Belarus and Ukraine, tariffs for transporting gas across these countries, and ownership of gas transportation facilities. In the context of price negotiations, Belarus also agreed to sell 50% of shares of Beltransgaz – the Belarus national gas distribution and transportation company – to Russia's state-owned Gazprom. The dispute between Russia and Ukraine over gas prices resulted in disruption of gas supplies to Western Europe and sparked a strong reaction from the EU that had worldwide resonance. Most observers considered that Russia was exerting political pressure by increasing gas prices. A similar dispute over tariffs on the export of Russian oil and its products to and through Belarus resulted in a brief disruption of oil supplies.

Imported energy is important to fuel economic development, particularly energy-hungry heavy industry, such as machine building and steel production in Ukraine and fertilizer and chemical production in Belarus. Refining of oil products in Mozyr and Novopolotsk used to be a key sector in the Belarus economy, but profits may drop substantially after Russia imposed tariffs on the export of oil to Belarus. The survival of much of the metallurgy and machine-building industry in the Donbas depends directly on a cheap, secure supply of natural gas currently imported from Russia, or on finding an alternative such as electricity from Ukraine's domestic power sources. Most of heavy industries were inherited from the Soviet Union and are often located in environmentally and socially stressed areas, while forming the mainstay of the existing economy. It may not be economically feasible to restructure them to improve energy security. Moreover, these industries are socially (and politically) important, as they constitute the main source of employment in densely populated areas with a poorly diversified economy. There are many other ways in which energy is linked to social and ultimately political issues.

The need is so pressing that Belarus and Ukraine are turning to nuclear power try to solve their energy problems. Belarus plans to build a domestic nuclear power plant by 2015, while the Energy Strategy adopted by Ukraine proposes building new nuclear reactors and extending the service life of existing ones. But the deployment of nuclear power is also associated with various security challenges ranging from enforcement of nonproliferation to concerns about terrorism, the operation of reactors, and radioactive waste disposal.

On the other hand, Ukraine and Belarus are determined to increase energy efficiency and implement cleaner energy technologies (Udovyk 2007b). The need to increase energy independence has focused fresh attention on the coal sector, which currently provides up to half of all energy and fuels up to a quarter of the electricity production in Ukraine. Belarus also has substantial deposits of brown coal. The importance of coal to the region could potentially increase, but would require major capital investment. Much as nuclear power, it could result in significant environmental risks, although new technologies may ensure cleaner (albeit more expensive) coal-based energy generation. Other domestic energy supply options, such as hydropower or using wood and other bio-fuels, are associated with environmental, social, and security impacts as well (e.g., the impact of newly-built hydropower facilities on downstream areas).

Whatever the strategic choices, restructuring of the energy sector in Eastern Europe will continue, and will have a major impact on the economy and social stability as well as the state of the environment. As long as the key lessons of Chernobyl remain on the agenda, these impacts need to be fully understood and integrated into policy making.

4. Resent Energy Security Events

The Russian economy, largely based on its oil and gas industry, has been hard hit by the current economic crisis. Gazprom's revenue on the European market is expected to shrink to $40 billio from $73 billion in 2008, and production to drop by 10–15% in 2009 – a production level likely to last 4–5 years owing to lack of future investments in production facilities. Gazprom's investment program will be reduced by more than 50% to RUB400 billion in 2009. Simultaneously, export prices have dropped to around half the record level of the last quarter of 2008. Despite this slump, the Putin regime and Russian oil and gas companies – the regime's extended arm – are continuing an aggressive strategy towards the European market – what could even be called a crisis-exploiting strategy. In April 2009 Gazprom in cooperation with Credit Suisse issued Euro-bonds worth $2.25 billion; a move that enhances the number of European stakeholders with a distinct interest in Gazprom's success. Concurrently, with Gaz de France's apparent readiness to join in, the Nord Stream pipeline project might be facing a critical breakthrough that could lead the Swedes to drop their procrastination on the project, which is destined to increase Russian political leverage in Europe. Nord Stream, however, also requires permission from Finland, which continues to withhold authorization perhaps reflecting the Finnish objection to Russian restrictions on timber trade.

Other aspects of this strategy are seen in Ukraine, Azerbaijan, Turkey and Hungary. Kiev and Brussels in March brokered a deal in which the EU will finance the modernization of Ukraine's pipeline infrastructure, which transports 80% of Russian gas exports to Europe. The Kremlin considers this to be a hostile move which is why Gazprom threatened Ukraine with fines for buying less gas in the first quarter of 2009 than agreed – a step Gazprom had hitherto repudiated. As a consequence of this agreement, and the EU's recently adopted Third Energy Packet, which enables European energy companies to buy Russian gas at the Russia-Ukraine border, the Ukrainian pipeline system could come under control of European firms; this situation would not only lessen Russian political clout but pose a financial threat to Russian gas companies as the transit fees paid are likely to almost double. Such developments led President Medvedev to suggest scrapping the Energy Charter, which Russia has signed but not ratified and thus does not feel bound to, in favor of a new structure for energy security in which producers and consumers jointly supervise the maintenance of pipeline infrastructure in third countries. This scenario is aimed at undermining the role of transit countries such as Ukraine and Poland. Medvedev is unlikely to succeed with this proposal; it remains unclear, however, if Russia will play a role in the upgrade of the Ukrainian gas pipeline system. With the Ukrainian economy in serious trouble, Prime Minister Yulia Tymoschenko recently invited Russia to participate in the modernization work in exchange for Gazprom forgoing its right to levy a fine upon her country. Gazprom and the Kremlin upheld strong pressure on Ukraine until Naftogaz on June 2009 managed to pay Gazprom $657 million, including $500 million for gas placed in underground storage. How the Russians responds to this remains to be seen.

Also, in Azerbaijan Gazprom has been making inroads. In late March 2009 Gazprom and SOCAR, the Azeri state oil company, signed a memorandum of understanding to start negotiations on exporting Azeri gas through Russia from January 2010. No doubt this – in part – reflects Gazprom's interest in spoiling the Nabucco project by drawing on Azeri gas reserves. However, it is unlikely to mean Azeri deflection from the Nabucco project. The issued joint memorandum is vaguely worded implying that Azerbaijan is perhaps using the negotiations in order to pressure its European partners into firmly committing to the Nabucco project. With the recent progress in negotiations on the Nabucco project it is increasingly unlikely that Azerbaijan would choose the Russian alternative. President Aliyev is intent on securing a transit agreement rather than the delivery-at-frontier (DAF) agreement Moscow is offering; an outcome far more likely with transit via Turkey who has, nevertheless, also been insisting on DAF.

Meanwhile, Gazprom and Turkey reached a preliminary agreement on constructing a second branch of the Blue Stream pipeline. As Turkey's gas market is saturated, the construction of a second Blue Stream trunk-line would be for export, possibly to Israel or Southern Europe – a region also targeted by the Nabucco project. Thus clearly, Gazprom is exploiting the current risk-averse environment to blatantly challenge the realization of Nabucco. On May 2009 Bulgaria, Greece, Italy, Serbia and Russia signed a deal on the South Stream pipeline, which designates 2015 as the year in which the pipeline will become operational, and sets the end of this year as deadline for a final decision on the pipeline's route. Furthermore, it was announced that the capacity of South Stream will be expanded from 31 billion cubic metres to 63 billion cubic metres. Since South Stream is a project aimed at debilitating the Southern Corridor, this can be seen as a decisive moment for European energy security. Though the Nabucco project, the quintessential branch of the Southern Corridor, has seen progress recently there are still severe obstacles to realization, namely uncertainty regarding both financing and supplies. These obstacles are well illustrated by the recent postponement of production from phase two of the Azeri Shah Deniz gas field, which is designated for the Nabucco pipeline, and under-lines the necessity of a firm European response.

The perhaps most aggressive Russian move in spring of 2009 was, however, not made by Gazprom but by Surgut Neftegaz when the company acquired Austrian OMV's 21.2% stake in Hungary's MOL oil and gas company under nebulous circumstances, making Surgut the single largest shareholder in MOL. Whereas the Hungarian government has signed on to the South Stream project – directly aimed at undermining Nabucco – MOL has stayed loyal to the Nabucco project; a stance which the Russians might seek to reverse. Surgut Neftegaz has a commercial interest in MOL's refining capacities, which are considered the most efficient in Central Europe. Yet, with MOL being a partner in the Nabucco project and in a LNG project on the Croatian Adriatic coast – both projects aimed at lessening European energy dependence on Russia – the bargain has strategic implications allowing Russia influence into European decision making on energy security. It seems particularly disturbing that Russians are gaining a say in MOL's initiative for enhancing Central European energy security the New Europe Transmissions System (NETS), which is intended as a mechanism for sharing gas supplies in case of emergencies by creating a single regional market. In other words, an initiative aimed at counteracting consequences of Russian actions, such as the cut-off of January 2009, will now come under Russian influence. Russia is effectively jumping past transit countries into the heart Europe thereby increasing its leverage over these very transit countries, particularly Ukraine.

Surgut Neftegaz is – even measured against Russian standards – an opaque business with an obscure shareholding structure. Putin is believed to be the major shareholder owning, directly or indirectly, a 37% stake in the business. The company is also headed by a long-time confidant of Putin's, Vladimir Bogdanov. The transaction followed OMV's unsuccessful attempt at a hostile takeover of MOL in 2008. Already then, OMV was rumoured to be fronting for a Russian takeover attempt. Brussels and Budapest – as well as MOL – were kept in the dark until OMV and Surgut publicized their deal. As late as a week before the transaction was revealed, on March 2009, OMV CEO Wolfgang Ruttenstorfer declared that OMV had no current plans of selling its stake in MOL, though it was likely to do so eventually. It can only be speculated what happened in between but it appears that the price paid by Surgut – 1.4 billion euros, twice the market price – exactly covers the costs held by OMV for buying and holding the shares in MOL; a fact supporting the suspicion that OMV was fronting for a Russian takeover attempt.

Surgut's long term intensions are unclear. It is assumed that the company will be tied up with Rosneft – a story supported by Bogdanov's recent nomination to sit on Rosneft's board. Another rumour has it that Surgut will in time resell its shares to either Rosneft, Lukoil or Gazpromneft, which are all in financially weak positions at the moment. Such a resale could double the Russian voting power in MOL. Surgut Neftegaz is estimated to have cash reserves of $20 billion. These reserves, along with its new stake in MOL, put the company in a good position for future takeover attempts. As demonstrated with the recent deal, Surgut can – unlike companies account-able to their shareholders – easily overpay for shares. Furthermore, Surgut can with its cash reserves tempt the Hungarian government with recapitalizing the country's energy sector; the company can rely on Putin supporting their agenda through foreign policy. MOL is dependent on Russian oil supplies from the Druzhba pipeline, which could be critical if a takeover battle emerges as Russia has been known to cut off supplies through Druzhba.

Another ongoing offensive towards the Hungarian energy sector under-scores this point. In late April 2009 Emfesz KMT – the second largest gas distributor in Hungary, controlling 20% of the market – announced that it would cease importing gas from RosUkrEnergo in favour of RosGas AG. Less than 10 days later, Emfesz was bought by RosGas, which is believed to be within Gazprom's network of interests but has an unintelligible ownership structure. The company has so far failed to provide information on this matter to the Hungarian Energy Office, which has nevertheless approved of the sale. According to Global Witness, the agency shows no sign of withdrawing such approval in the case that RosGas does not present the required information on deadline. Furthermore, Rosgas is believed to be the future target of takeover by Bulgarian Overgas. The latter, likewise, has an opaque ownership structure, and is thought to be a proxy of Gazprom. Thus, Russian leverage over the Hungarian energy sector is only set to increase.

While pursuing an aggressive strategy towards Europe, Russia seems to be losing ground in Central Asia. This highlights how the crisis-exploiting strategy pursued by Russia is untenable. Putin, and Russian oil and gas companies, are seeking to recreate a situation of excessively expensive energy. At the same time, it is questionable whether Gazprom is able to follow through on its many present commitments. With Gazprom's own production set to shrink and stay at a lower level for years to come, losing access to the gas reserves in Central Asia would effectively undermine the strategy.

The pipeline blast on the Davletbat–Dariyalyk pipeline (CAC-4) near the Turkmen–Uzbek border is a pertinent example of the troubled state in Russian–Turkmen relations. With the explosion, virtually no gas is flowing from Turkmenistan to Russia, as the only other pipeline route – the western branch of the Central-Asia Centre (CAC) pipeline – carries only 2 billion centimetre/year. In 2007 Kazakhstan, Russia and Turkmenistan agreed on the re-construction of the western branch, the Caspian Coastal Pipeline (CCP). However, since then construction has been postponed (though Kazakhstan recently ratified the agreement), and Ashgabat has sought other bidders for the construction as Turkmen President Berdimukhamedov doubts Gazprom's ability to get the job done.

The explosion followed a somewhat unsuccessful meeting in Moscow between Berdimukhamedov and Medvedev at which the two failed to agree on the construction of an East–West pipeline across Turkmenistan. This partly reflects Moscow's fear that the Trans-Caspian pipeline project will resurface in which case Gazprom investment into the East–West pipeline would damage the company by facilitating easier export of gas from eastern Turkmenistan to Europe, circumventing Russia.

The blame game following the explosion has seen the Kremlin remain silent while Russian experts attribute the explosion to a combination of worn-out Soviet-era infrastructure and Turkmen negligence. The Turkmen authorities on the other hand blame Gazpromexport for decreasing the amount of gas drawn from the pipeline without warning to the Turkmen authorities, thereby causing the explosion. Whatever the cause, gas has stopped flowing and repairs – anticipated to take 2–3 days – have not begun while the blame game continues. Both sides could be speculating that the spat will position them better in upcoming price negotiations, which will settle the price paid by Russia for Turkmen gas from 2010. Clearly, it reflects Gazprom's position in that it has become less dependent on Turkmen gas owing to declining demand, and in which the current "European" price it pays for Turkmen gas is hurting the company's bottom-line. Indeed, Gazprom is attempting to bully Turkmenistan into accepting a lower gas price on with the deputy chief executive, Valery Gobulev, on 2 June 2009 threatening to buy less gas off Turkmenistan unless a lower price is agreed on.

Turkmenistan, on the other hand, is using the row to signal to the West that it is not a Russian satellite state, and that the country will be open to countervailing proposals from the West. As recent as the first week of June of 2009, a high-level Turkmen delegation led by Foreign Minister Rashid Meredov visited Brussels. While Gazprom might find it difficult to finance an upgrade of Turkmen infrastructure, Turkmenistan certainly does not have the funds to do so itself. In that sense, the cessation of gas supply probably suits both sides well.

In mid-April of 2009 Turkmenistan and the German Rheinisch-Westfälische Elektricitätswerk AG (RWE) signed an agreement to have RWE develop the Turkmen offshore gas field, Bloc 23; included in the agreement was the objective of RWE eventually exporting Turkmen natural gas. This deal could prove a breakthrough for the Nabucco project; however, the subsequent summit on the Southern Corridor in Prague on 7 May 2009 did not see Turkmenistan, nor Kazakhstan or Uzbekistan, sign on to the project. This highlights the precarious situation Central Asian states are in. Continuing on the case of Turkmenistan, the country is hugely dependent on stable relations with Russia and Gazprom to fill the state coffers. This income could, in principle, be supplanted by the EU but a definite, unmistakable and long-term commitment from Europe is requisite for the Turkmens to make such a risky policy shift. Considering further, that the prospect of exporting via a Southern Corridor continues to be some years away, the possibility of Turkmenistan returning to Moscow's flock is great.

Nevertheless, at present Europe would be well-advised to beef up its relations with the Central Asian states, particularly Turkmenistan, in an effort to break the Russian stronghold on the European energy market. The financial crisis, and the resulting cracks in the Russian strategy, call for urgent European action – now is the time, so to speak. "The EU must, concurrently, commit strongly and unequivocally to the Nabucco project in order to achieve the tantalizing prospect of bringing both Azerbaijan and

Turkmenistan firmly onside thereby securing the Southern Corridor. Given both countries present signalling for Western engagement, the chance to lessen dependence on Russia certainly exists" (Hagelund 2009).

5. Conclusion

The most important factor shaping the future security of Eastern Europe is the interplay of political and economic interests in the pan-European region. Many in Eastern Europe are attracted by Western models, but drawn East by historic, cultural, and linguistic affinities, and, last but not least, by close trade and energy links. Most probably the three countries of Eastern Europe will continue to search for a balance between the two poles. However, the three states themselves are not passive objects in a geopolitical game, but active players, and much of the regional security architecture will depend on the ability of Chisinau, Kyiv, Minsk, and other capitals to seek mutual understanding and reach compromises.

References

A. Korin, G. Luff, 2009. "Energy security challenges for the 21st century: a reference handbook (contemporary military, strategic and security issues), Praeger, Santa Barbaea, CA, 200 pp.

"NATO's Role in Energy Security" www.nato/int

O. Udovyk, 2007a. "Learning from Chernobyl: past and present responses" in "Multiple Stressors: a challenge for the Future", Ed C. Mathersill et al., Springer, The Netherlands, pp. 449–454.

O. Udovyk, 2007b. "Prospects for sustainable development of Ukrainian energy sector" in "Assessment of Hydrogen Energy for Sustainable Development", Eds J.W. Sheffield et al., Springer, The Netherlands, pp. 249–256.

O. Udovyk, 2008a. "Energy, environment and security in Eastern Europe" in "Sustainable Energy Production and Consumption – Benefits, Strategies and Environmental Costing", Eds F. Barbir and S. Ulgiati, Springer, The Netherlands, pp. 359–372.

O. Udovyk, 2008b. "Risk assessment for coupled critical infrastructures" in "Advanced Technologies and Methodologies for Risk Management in the Global Transport", Eds C. Bersani et al., IOS Press, Amsterdam, The Netherlands, pp. 239–248.

O. Udovyk, 2009. "Working together for nuclear safety" in "New Techniques for Detection of Nuclear and Radioactive Agents", Eds G.A. Aycik et al., Springer, The Netherlands, pp. 295–305.

O. Udovyk, 2008. "Risk assessment for coupled critical infrastructures" in "Advanced Technologies and Methodologies for Risk Management in the Global Transport", Eds C. Bersani et al., IOS Press, Amsterdam, The Netherlands, pp. 239–248.

C. Hagelund, 2009. "Europe's chance to face up to Russia's energy bullying", "Energy Security", June 2009.

SOCIO-ECONOMIC AND ENERGY DEVELOPMENT IN UKRAINE – PROBLEMS AND SOLUTIONS

ALEXANDER GOROBETS*
*Sevastopol National Technical University,
Management Department, Streletskaya Bay,
Sevastopol 99053, Ukraine*

Abstract   In this paper the development of socio-economic and energy sectors in Ukraine is analyzed since its independence and the major roots of systemic crisis in these sectors are identified. The specific policy regulations, i.e. institutional (integrated accounting and auditing, eco-cities, environmental regulation, training programmes), economic ("green" tax, correct pricing, investments in science and education) and technological (eco-efficiency, renewable energy) are suggested to approach energy security and general sustainability in Ukraine. The core social (crime and unemployment rate, gap between rich and poor, number of volunteers, quality of education), economic (resource efficiency and quality of goods/services) and environmental (biodiversity state) indicators are proposed to measure the progress in development. Proportion of population in full health is proposed as the principal new indicator of sustainable development.

Keywords: Ukraine, development, socio-economic indicators, public health, energy, policy regulations

1. Introduction

On 16th of September 2009, Ukraine' Cabinet of Ministry established a National Council of Sustainable Development in Ukraine that is a very important step towards an implementation of international agreements on

---
* To whom correspondence should be addressed: Alexander Gorobets, Sevastopol National Technical University, Management Department, Streletskaya Bay, Sevastopol   99053, Ukraine; E-mail: alex-gorobets@mail.ru

F. Barbir and S. Ulgiati (eds.), *Energy Options Impact on Regional Security*,
DOI 10.1007/978-90-481-9565-7_13, © Springer Science + Business Media B.V. 2010

sustainable development signed in Rio de Janeiro in 1992 and in Johannesburg in 2002. Indeed, the following problems remain sharp in Ukraine since its independence:

- Social – poverty, epidemics (tuberculosis, HIV), cardiovascular diseases, the general public health crisis (UKRSTAT 2009)
- Economic – increasing gap between rich and poor and unemployment rate, enormously high material and energy consumption per unit of gross domestic product
- Environmental – air, water and land pollution, wastes accumulation, deforestation and land erosion (MENR 2009; UNDP 2009)

The main reasons for this are as follows:

1. The general moral crisis of society (e.g. high level of corruption), and passive public position
2. An extremely high material and energy intensity economy based on the old, environmentally polluting and health risky industries (heavy, chemical, etc.) and technologies (e.g. drinking water chlorine purification) since Soviet Union time
3. Inadequacy of institutions to the current needs of society, especially in the fields of environmental management and education, self-governance and civil control
4. Weak understanding of sustainable development principles both by the government authorities and by public due to a lack of appropriate education
5. The inability of the political elite to adapt to multi-party democracy and as a result the lack of internal, within Ukraine, and external, among New Independent States, cooperation
6. Incompetent management at all levels and psychological barriers of taking measures in a new, dynamic and complex environment
7. Absence of sustainable development policy regulations and consistent sustainability indicators

Therefore the goal of this research is to analyze the development of socio-economic and energy sectors in Ukraine since its independence, to develop the appropriate policy regulations and the major indicators of sustainable development that will be consonant with the national strategies of sustainable development and millennium development goals: poverty reduction, quality life-long education, environmental sustainability, improved health and reduced HIV/AIDS and other diseases, gender equality, global partnership (UN 2009).

2. Energy Development in Ukraine

Currently Ukraine has one of the highest material and energy-intensive economies in the world, i.e. it is still three times more energy intensive than an average level in European Union despite a gradual improvement in energy efficiency during the last years (EIA 2009). Although energy consumption in Ukraine has decreased since the country's independence due to the economic fall, its dependence on energy imports has declined only marginally. At present, the most of Ukraine's oil and gas and all of its nuclear fuel comes from or through Russia. This dependence increases the risks of energy supply and makes Ukraine' economy highly vulnerable to international price shifts. Because of its geographic location, Ukraine does not have many available supply alternatives besides of developing its own reserves of oil, coal and gas or big investments in renewable energy (solar and wind).

Therefore, as it is shown on Fig. 1 (EIA 2009), presently the total primary energy consumption in Ukraine is almost two times higher than energy production.

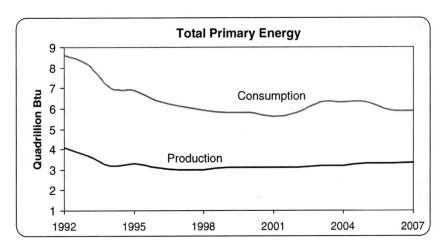

Figure 1. Total primary energy in Ukraine

Another up-to-date economic problem in Ukraine is that the most of domestic energy prices (except oil products) are much below international levels and they only cover operational costs. Such conditions limit invest-ments in infrastructure, and incentives for efficiency. The highest risk comes from nuclear energy, where the nuclear tariff does not fully cover capital expenditures and nuclear waste disposal (IEA 2006). Despite recent increases in import prices, retail natural gas prices remain several times lower than prices in Western Europe and they are also lower than prices in neighbours like Russia. As a result, Ukrainian natural gas industry which belongs to the state-owned firm Naftogaz is very close to financial default (MFE 2009). Coal prices do not cover even production costs and therefore coal mines are in financial crisis.

Moreover, despite privatization of some energy assets (i.e. coal and electricity sectors) the government still maintains a strong role in owning and regulating energy assets that minimizes competition and reduces efficiency (IEA 2006).

Today, natural gas is the most important energy source in Ukraine (41%), followed by coal (30%), nuclear (17%) and oil (11%). Renewable energy sources (not including hydro power stations) produce only about 1% (MFE 2009).

The dynamics of natural gas production and consumption in Ukraine is shown on Fig. 2 (EIA 2009). According to the Ministry of fuel and energy, Ukraine has roughly 1,104 trillion cubic meters of natural gas reserves (MFE 2009). Since 1992, Ukraine's consumption of natural gas has decreased by approximately 35% in 2008 but still remains three times higher than its production (Fig. 2). The deficit of natural gas is imported from Russian Federation.

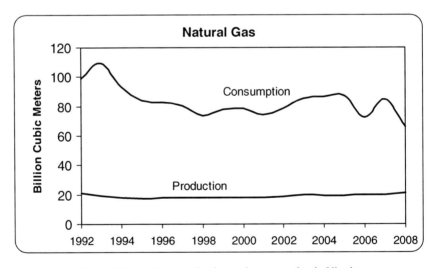

Figure 2. Natural gas production and consumption in Ukraine

The dynamics of oil production and consumption in independent Ukraine is shown on Fig. 3 (EIA 2009).

Although Ukraine has explored some oil reserves, particularly in its sector of Black Sea and Azov Sea, oil production has remained relatively stable since independence (Fig. 3). According to the Ministry of Fuel and

Energy, Ukraine has 395 million barrels of proven oil reserves in 2008, the majority of these are located in the eastern Dnieper–Donetsk basin (MFE 2009). Although oil consumption has dropped dramatically from 819,000 barrels per day in 1992 to around 353,000 barrels/day in 2008 (Fig. 3), Ukraine remains highly dependent on imported oil, most of which comes from Russia (EIA 2009).

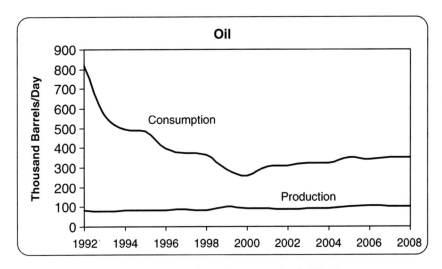

Figure 3. Oil production and consumption in Ukraine

Traditionally, industrial growth in Ukraine in the twentieth century was based on coal. But, after the collapse of the Soviet Union, coal production dropped almost twice and now fluctuating around this level as it is shown on Fig. 4 (EIA 2009).

Despite the significant coal resources, i.e. Ukraine has 37.6 billion short tons in proven coal reserves (about 15% of the former Soviet Union's total reserves and the seventh largest in the world) (EIA 2009), the country is a net coal importer due to a lack of investment in coal industry and incompetent management (e.g. big governmental subsidies in this sector).

The most of the mines are still in state hands and they are not profitable. Therefore some of them have already been shut down or privatized. Most of Ukraine's coal mines are located in the eastern region of the country (Donetsk basin). The country's coal industry, which employs about 500,000 people, is badly managed by state organizations and has serious problems, such as unsafe working conditions, inefficiency, low productivity and labor strikes. Ukraine has the second most dangerous mining industry in the world (after China) (EIA 2009).

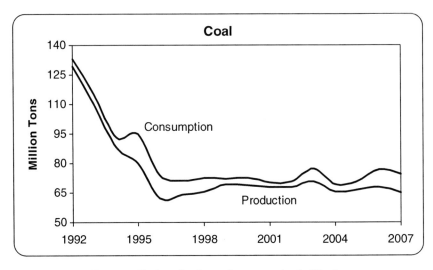

Figure 4. Coal production and consumption in Ukraine

Electricity generation and consumption in Ukraine is shown on Fig. 5 (EIA 2009).

Since generation of electricity in Ukraine exceeds its internal consumption there is a capacity to sell it abroad. Ukraine's power sector has the 12th-largest installed capacity in the world, i.e. 54 GW (EIA 2009). Although generation and consumption fell sharply since independence, they have started to increase again since 2000 (Fig. 5). According to the Ministry of Fuel and Energy, Ukraine generated 184 billion kilowatt hours of electricity in 2008 that is 25% higher than its own consumption (148 billion kilowatt hours) in this year (MFE 2009).

Ukraine has sufficient generating capacity to provide more than twice its electricity needs, but the country's ageing infrastructure (particularly transmission and distribution systems) needs an investment and maintenance (IEA 2006).

Therefore, improving energy efficiency, correct pricing and concentration on its own renewable energy production has to be a key strategy in Ukraine to increase energy security and economic competitiveness and reduce its environmental footprint.

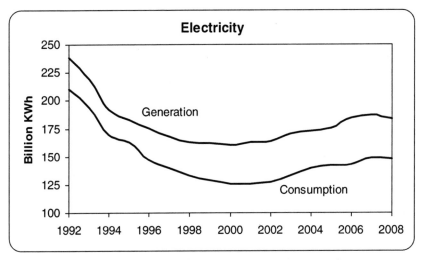

Figure 5. Electricity generation and consumption in Ukraine

3. Socio-economic Development in Ukraine

At the present conditions of the world economic crisis, the general socio-economic situation in Ukraine is becoming even more dangerous than it was in the last years, that is confirmed by such indicators as increasing poverty – at least 29% of the population is living below the poverty line, with 3% in extreme poverty, increasing gap between rich and poor (UNICEF 2007; UNDP 2009) and very short average lifespan – 62 years for men and 72 years for women (UKRSTAT 2009).

According to a forecast by the State Employment Center unemployment in Ukraine will increase from 6.4% at the end of 2008 (ILO 2009) to 9% in 2009 (SEC 2009). The World Bank predicts that Ukraine's economy will fall by 15% in 2009 with inflation rate at about 13.4% (World Bank 2009). In the same time, devaluation of the national currency, i.e. grivna to euro and American dollar is more than 50% in 2009 that is causing an appropriate growth in prices for imported goods and services. The dynamics of the gross domestic product in Ukraine in percent to previous year is shown on Fig. 6 (UKRSTAT 2009; World Bank 2009).

The current demographic situation in Ukraine and its trend of development can be characterized as catastrophic, that is reflected in Table 1, showing the major causes of population decline as a percentage of the total (Ukrstat 2009).

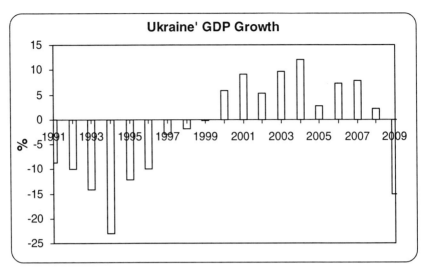

Figure 6. Gross domestic product growth in Ukraine in percent to previous year

TABLE 1. Ukraine' demography from 1990 to 2008

Years	Population (million)	Birth rate	Death rate	Blood circulation (%)	Tumors (%)	Respiratory (%)	External (%)
1990	51.94	12.7	12.1	52.9	16.2	5.9	8.8
1993	52.11	10.7	14.2	55.0	14.1	5.7	9.2
1996	50.82	9.2	15.3	57.5	12.8	5.7	10.4
1999	49.43	7.9	15.0	60.7	13.3	5.0	9.6
2002	48.00	8.1	15.7	61.6	12.6	4.2	10.1
2005	46.93	9.1	16.7	62.5	11.7	3.6	9.0
2008	46.14	11.0	16.3	63.6	11.8	3.1	8.1

According to the State Statistics Committee of Ukraine after 1992 there is a steady decline in population (Ukrstat 2009). The basic explanation could be connected with a general economic crisis which had place until 1999 (Ukrstat 2009), but the following economic growth has not affected the demographic trend because this economic benefit was not used efficiently (e.g. by investing in health care and human development system) and distributed equally among different groups of population, i.e. the gap between rich and poor has been increasing. According to the estimated trend equation (Gorobets 2007, 2008) Ukraine' population will decrease to about 45 million by 2010 if the present inhuman socio-economic policy is not changed. This trend is based mostly on a natural population decrease (number of deaths is more than number of births), where the most of deaths is caused by blood circulation disease (Ukrstat 2009). An analysis of the above table shows that, despite the improving birth rate situation since 2002, deaths have outstripped births during almost all of the transition period in independent Ukraine. The major cause of mortality (blood circulation diseases) shows a 10% growth over the last 18 years, although tumors and respiratory diseases have declined during this time. The danger sign is a relatively high and stable proportion of external causes (about 10%) – such as alcohol poisoning and other substance use, traffic accidents, suicides and injuries as a result of crime.

Furthermore, there are also alarming statistics (not shown in the table) on AIDS (the number of reported cases of HIV has increased 20 times in the past 5 years), tuberculosis, smoking, alcohol consumption, and physical health, especially among youth and infants (UNICEF 2007). These can be accounted for by the deep social and moral crisis of Ukrainian society, despite the sharp growth of religious organizations (from 15,784 in 1995 to 31,227 in 2007), public associations (from 1,047 in 1996 to 2,595 in 2007) and higher education students (from 1,638 million in 1990 to 2,764 million in 2008) (UKRSTAT 2009).

4. Policy Regulations and Development Indicators

4.1. POLICY REGULATIONS

To improve energy efficiency and general socio-economic and ecological situation in Ukraine the following policy regulations are proposed:

4.1.1. *Economic responses*

(a) Elimination of subsidies in the energy sector and setting of tariffs that cover full, long term costs
(b) Increasing of competition and transparency in the energy sector to promote corporate efficiency
(c) Conduct a cost-benefit analysis of renewable energy options for different regions in Ukraine
(d) Financial support for potential developers and users of renewable energy sources
(e) "Green" tax – heavy taxation of natural resources use (incl. fossil fuels) and no taxation on intellectual products (Brown 2001)
(f) Priority of investments in science, eco-efficient technologies, renewable energy (e.g. solar and wind) to reduce dependence on energy imports and eco-centric and social infrastructure (e.g. bicycle' tracks and sport' fields) to support human development and healthy lifestyles

(g) Restructuring the economy from industrial to knowledge based socio-ecological economics (Daly and Farley 2003; Brown 2001), based on "green" technologies, information technologies, bicycle' infrastructure, eco-villages, organic agriculture
(h) Transition to a service-sector oriented economy with a low material-energy throughput

4.1.2. *Energy and technology responses*

(a) Adoption of eco-efficient technologies (minimization of resources used and wastes) in industry, reusing and recycling materials
(b) Efficient use of information technology (e.g. work and study at home, distance education programs, internet conferences, electronic documents) to reduce consumption of energy and material resources
(c) Renewable energy – solar, wind, hydrogen, biofuel use in all economic sectors, e.g.
(d) "Solar" buildings and houses using low energy alternatives, accumulation of human' motion energy for domestic use
(e) Development of "green" transport system (fueled by biogas, hydrogen, ethanol, electricity)

4.1.3. *Institutional responses*

(a) Scientific verification of decision-making for socio-ecological and economic security, including preventive measures, precautionary and individual responsibility principles
(b) Development of cooperation between governmental institutions in energy, environmental and socio-economic issues
(c) Development of a credible greenhouse emissions inventory
(d) Obligation for all energy consumers to have heat, electricity and gas meters
(e) Public awareness campaigns and training programmes to increase energy efficiency at households
(f) Energy-efficiency standards for equipment and buildings
(g) Environmental auditing of industry and transport to encourage efficiency and reduce emissions
(h) Adoption of System for Integrated Environmental and Economic Accounting, aimed at accounting of ecological factors in national economic statistics (UNSTAT 2006)
(i) Internalization of the environmental costs of energy production into energy prices

(j) Changing the production and consumption patterns through the legal regulation, price controls (setting right prices, incl. environmental costs), eco-taxation, adoption of the 'polluter/consumer pays' principle

(k) Development of the appropriate eco-centric institutions to integrate human activity within natural cycles – eco-villages and eco-cities (e.g. Auroville in India) (Kapoor 2007)

(l) Free transition of eco-efficient technologies and appropriate knowledge base (e.g. via open access electronic journals) to all interesting economic agents, instead of financial donations, to stimulate their own development

(m)Set the priority to physical indicators of socio-economic and environmental development

4.2. DEVELOPMENT INDICATORS

Because monetary indicators (e.g. gross domestic product) are artificial and misleading (Common 2007), i.e. they do not reflect all environmental and social externalities, there is a need to use natural – physical and biological indicators for monitoring progress in socio-economic and environmental development. Therefore,

- Social development of country can be measured by – crime rate, corruption level, unemployment rate, trust in institutions, number of volunteers, quality of education, job satisfaction, gap between rich and poor.
- Economic development can be measured by ecological footprint, resource-efficiency, i.e. time, energy, water and material use per unit of output, amount of wastes and emissions per unit of output (life cycle assessment); quality of goods/services (e.g. functionality and useful life span).
- Environmental development can be measured by biodiversity state (e.g. living planet index) as an indicator of environmental health and by the quality/pollution of natural resources (fresh water, air, land).

Since public health depends on all dimensions of human life: genetic factors, natural and built environment, socio-economic development and lifestyle, its integrated characteristic, i.e. – proportion of population in full health (no mental and physical diseases) is proposed as the principal new indicator of sustainable development in Ukraine and worldwide:

$$SDI = \text{Population in full health/total population} \qquad (1)$$

Therefore, theoretical values of SDI belong to the interval (0; 1] and its optimal value should approach the upper limit, i.e. 1, which means nil proportion of sick population.

The advantage of this indicator in comparison with many other sustainability indices (Bohringer and Jochem 2007) is that it has an objective and integrated character (depends on all dimensions of human life) and relatively simple methodology for its calculation (e.g. medical records, sampling, inferential statistics) and therefore can be used widely for international comparison between different countries.

5. Conclusions

In this paper, the state of energy sector and socio-economic development in Ukraine are analyzed. The specific policy regulations, i.e. institutional (integrated accounting and auditing, environmental regulation, price controls, shifting in consumption and production patterns, eco-centric cities, training programmes), economic ("green" tax, full costs reflective pricing, investments in science, education and service sector) and technological (eco-efficiency, renewable energy, information technology) are suggested to approach energy security and general sustainability in Ukraine. The core social (crime and unemployment rate, gap between rich and poor, education quality), economic (resource efficiency, quality of goods and services) and environmental (biodiversity state) indicators are proposed to measure the progress in development. Proportion of population in full health (with no mental and physical diseases) is proposed as the principal new indicator of sustainable development in Ukraine and worldwide.

To start transition to energy efficiency and sustainable development in Ukraine, scientists, educators and all concerned people must take initiative and responsibility in their own hands and actively participate in political processes by all possible ways: joining political parties, participating in different campaigns, writing short influential articles in popular newspapers and magazines, working with authorities, business leaders and the producers of media programmes (Gorobets 2006).

Integration of Ukraine in the highly competitive globalizing world economy should be based on using its key advantages such as a great human capital, strategic geographical location (transit routes) between Europe and Russian Federation, a vast amount of productive agricultural land, and a well developed from Soviet time aerospace, defense and shipbuilding industry.

References

Brown, L.R., 2001, *Eco-Economy: Building an Economy for the Earth*. W.W. Norton & Company, New York: 352p http://www.earth-policy.org/Books/Eco_contents.htm

Böhringer, C., Jochem, P.E.P., 2007. Measuring the immeasurable — A survey of sustainability indices. Ecological Economics 63, 1–8.

Common, M., 2007. Measuring national economic performance without using prices. Ecological Economics 64, 92–102.

Daly, H.E. and Farley, J. (Eds.), 2003. *Ecological Economics: Principles and Applications*. Island, Washington, DC.

EIA, 2009. Energy Information Administration's primary report of international energy statistics. International Energy Annual. Official energy statistics from the U.S. government. http://www.eia.doe.gov/iea/

Gorobets, A., 2006. An eco-centric approach to sustainable community development. Community Development Journal 41 (1), 104–108.

Gorobets, A., 2007. Sustainability and environmental security management tools. In: A. Morris and S. Kokhan (Eds.), *Geographic Uncertainty in Environmental Security. NATO Science for Peace and Security Series C: Environmental Security*. Springer, The Netherlands

Gorobets, A., 2008. An independent Ukraine: sustainable or unsustainable development? Communist and Post-Communist Studies 41 (1), 93–103.

IEA, 2006. Ukraine, energy policy review. International Energy Agency, Paris. www.iea.org

ILO, 2009. International Labor Organization. Database of labour statistics. Geneva. http://laborsta.ilo.org/

Kapoor, R., 2007, Auroville: A spiritual-social experiment in human unity and evolution. Futures 39, 632–643.

MENR, 2009. Ministry of Environment and Natural Resources of Ukraine, Kiev. http://www.menr.gov.ua

MFE, 2009. Ministry of Fuel and Energy of Ukraine, Kiev. http://mpe.kmu.gov.ua

SEC, 2009. State Employment Center of Ukraine, Kiev. www.dcz.gov.ua

UKRSTAT, 2009. State Statistics Committee of Ukraine, Kiev. http://www.ukrstat.gov.ua/

UN, 2009. The Millennium Development Goals Report 2009. United Nations, New York. http://www.un.org/millenniumgoals/

UNDP, 2009. UN Development Programme in Ukraine, Kiev. http://www.undp.org.ua/

UNICEF, 2007. Key challenges for children in Ukraine. Article on website of UNICEF Ukraine. http://www.unicef.org/ukraine/children.html

World Bank, 2009. Ukraine economic update, July 2009. http://www.worldbank.org.ua/

ADVANCES IN SMALL HYDROPOWER ENERGY PRODUCTION IN UKRAINE

IGOR WINKLER*
*Yu. Fedkovich Chernivtsi National University,
Chernivtsi, Ukraine*

Abstract Traditional sources of energy are becoming shorter and more expensive. Current global financial crisis can temporally postpone raising in the energy price but nothing can stop shortening of the fossil energy sources, which currently supply a significant part of the global energy production. Wider involvement of various renewable energy sources is unavoidable for maintaining development of any national economy. Wide network of the small hydropower plants (SHPs) can ensure stable energy production in Ukraine especially for small and medium size consumers. These plants can be installed at numerous small rivers, industrial water/wastewater pipes and require comparatively low initial/renovation investment. Maintenance fee of SHPs is very moderate while working term and prospects are quite long. Unlikely to regular hydropower plants SHPs cause minimal negative environmental effects and should occupy more significant place in the nearest energy production development plans. The Ukrainian SHPs currently produce about 2% of the total electrical energy output from the national hydropower industry. Rebuilding of old Ukrainian SHPs and construction of all prospective plants can at least double this value.

Keywords: Small hydropower plants, sustainable energy production, replacing fossil fuel sources

---

* To whom correspondence should be addressed: Igor Winkler, Yu. Fedkovich Chernivtsi National University, Kotsiubinsky St. 2, Chernivtsi, Ukranine; E-mail: igorw@ukrpost.ua

F. Barbir and S. Ulgiati (eds.), *Energy Options Impact on Regional Security*,
DOI 10.1007/978-90-481-9565-7_14, © Springer Science + Business Media B.V. 2010

1. Introduction

1.1. RECENT TRENDS IN THE ENERGY PRODUCTION

Renewable energy production (REP) is one of the fastest growing branches in the EU countries. REP supplies ~ 8–10% in the total energy production and this part is constantly growing. Current energy production strategy plans to reach at least 12% of REP (excluding regular hydropower energy plants) by 2010. EU candidate countries are required to reach at least 6% of REP in their national energy production balance before joining the EU.

Biomass energy production is the most promising branch in REP and it is planned that it should supply about 75% of the total REP in the EU (see Table 1) (Geletuha and Kudrya 2005). Hydropower energy production is considered as the second important renewable energy source (about 17% of the total REP), which can save up to 30 million tons of oil equivalent (TOE).

TABLE 1. Renewable energy production in the EU countries

| Renewable energy source | Energy production | | | |
| | 1995 | | 2010 (Plan) | |
	Million TOE	Percent	Million TOE	Percent
Wind	0.35	0.5	6.9	3.8
Hydropower	**26.4**	**35.5**	**30.55**	**16.8**
Photovoltaic	0.002	0.003	0.26	0.1
Biomass	**44.8**	**60.2**	**135**	**74.2**
Geothermal	2.5	3.4	5.2	2.9
Solar water heater	0.26	0.4	4	2.2
TOTAL	74.3	100	182	100

REP in Ukraine is lower and reaches only 2.5% from the total national energy production. However, potential of REP in Ukraine is much higher and can save up to 75,000 TOE/year.

Ukrainian hydropower energy production is one of the promising REP sources since Ukraine has very significant total hydropower potential, which is not completely used. Potential hydropower potential of Ukraine is about 44.7 million kilowatt-hour. However, economically profitable potential is lower than the third part of the total. Current total hydro-energy plants power in Ukraine reaches about 5,443 MW, which is about 8% in the gross domestic energy plants power. This hydropower energy is mostly produced at regular hydropower energy plants while total electrical power of SHPs is about 100 MW (about 2% from the total hydropower plants power)

(Tsymbalenko and Vikhorev 2005). But potential electric capacity of the SHPs in Ukraine is much higher and can substitute the third part of the current electric power of the regular hydropower plants.

1.2. CURRENT CONDITIONS OF THE SHPS IN EUROPE

A hydropower plant is considered as small if it does not have any water storage reservoir (a plant uses natural waterfall or just a river stream) or if the reservoir's surface area is smaller than 1 ha. Such power plants cause the lowest negative environmental effect.

Considering an integral environmental effect of the energy production the author of (Syrotiuk 2008) has proposed an integral environmental penalty coefficient. This coefficient involves assessment of the global warming, ozone layer depletion, soils acidulation, euthrofication, heavy metals contamination, winter and summer smog formation, disposal of wastes (including radioactive materials) and potential exhaustion of the energy sources. Table 2 represents the penalty coefficients for the various energy production branches.

TABLE 2. Integral environmental penalty coefficients for the various energy production branches

Energy production branch	Penalty coefficient
Brown coal	1,735
Oil	1,398
Coal	1,356
Nuclear	672
Photovoltaic	461
Natural gas	267
Traditional hydropower plants	118
Wind	65
SHP	5

It can be seen that SHPs cause the lowest anthropogenic load and can be considered as ecologically safe power source. Even traditional hydropower plants cause numerous negative effects (Winkler and Tevtul 2006):

• Changes in the hydrological regime lead to changes in the water temperature, oxygen, mineral and organic compounds content, which result in significant changes in the river's bioproductivity. This leads to vanishing of some living species and increase in the number of the others (sometimes less valuable).

- Flooding of wide bank areas, coastline changes, increase in the water surface area. This leads to raise of the groundwater level and dramatic climate changes in the nearby area. The climate becomes softer and more humid.
- Large water reservoirs increase local seismity and become very dangerous technological objects. Serious failure or a terrorist attack on the dam can cause catastrophic consequences.
- Hydropower plant's dam seriously influences the normal spawning fish migration even though many of them are equipped with special spawning run locks.

Most of the abovementioned negative effects are inapplicable to the SHPs since they do not have artificial water reservoirs (or have very small ones). That is why this branch attracts more and more attention in many European countries. Traditional big hydropower resources are almost completely involved in the energy production in most European countries including Ukraine, which warms up this interest towards SHPs.

Table 3 represents some parameters of the SHP industry in the EU-15 countries. It can be seen that the highest number of SHPs is in operation in Germany, France, Austria and Italy. Mountain and hill areas are usually reach of small rivers and streams and significant part of these countries' territory is occupied by such areas, which makes them leaders in the SHPs number. Germany operates mostly low-power SHPs and occupies only third position in the total power of national SHP's after Italy and France.

The SHP percentage in the total hydropower energy production is higher than 5% in most EU-15 countries (it is interesting that all the Denmark's hydropower energy is produced at its SHP's only). However, the SHP percentage in the total national energy production is still insufficient and this value is higher than 3% only in Austria, Luxembourg, Sweden and Spain.

All data in the Table 3 are related to the years 2000–2004 but this branch is constantly growing and current situation is even better.

Many new members and candidates to EU (EU-13 countries) also operate SHPs but this branch is weakly developed comparing to the "old" EU members. Similar characteristics of the SHPs in EU-13 countries are represented in the Table 4.

Data of the Table 4 shows that the Czech Republic, Poland and Slovenia operate the highest number of the SHPs. But their electrical power is much lower comparing to the leaders from the Table 3. Percentage of the SHPs electricity in the total hydropower energy production seems very significant for many countries but in fact this is because of the low part of the hydropower energy in the total national energy production. For example, the third part of the SHPs energy in the Czech Republic makes only 1% to the national energy balance. Comparing data of Tables 3 and 4 we can conclude that the average percentage of the SHPs energy production in the national energy production in the EU-13 countries is significantly lower than in the EU-15.

TABLE 3. Some parameters of SHPs in the EU-15 countries (Syrotiuk 2008)

Country	Number of SHP	Total power (MW)	Part of the SHP production in total hydropower energy production	Part of the SHP production in total national energy production
Belgium	82	96	6.86	0.61
Denmark	40	11	100	0.09
Germany	6,200	1,500	16.67	1.27
Greece	40	69	2.3	0.63
Spain	1,106	1,607	9.08	3.06
France	1,730	2,000	7.81	1.73
Ireland	45	23	4.51	0.48
Italy	1,510	2,229	10.98	3.12
Luxembourg	29	39	3.55	3.25
Netherlands	3	2	2.22	0.01
Austria	1,700	866	7.53	4.89
Portugal	74	286	6.36	2.62
Finland	204	320	11.03	1.96
Sweden	1,615	1,050	6.40	3.20
UK	10	162	3.77	0.21
TOTAL	14,488	10,260	8.67	1.77

Both EU-15 and EU13 countries declare priority of the renewable energy production. However, "old" members mostly operate SHPs, which are in service for the decades. That is why renovation of the existing SHPs is the main direction of development for the EU-15 countries while "new" members have to combine renovation and building of new SHPs. Total amount of the

SHPs energy production is raising in all the countries but its percentage is lowering in some countries because of more advanced development of other renewable energy productions.

TABLE 4. Some parameters of SHPs in the EU-13 countries (Syrotiuk 2008)

Country	Number of SHP	Total power, MW	Part of the SHP production in total hydropower energy production	Part of the SHP production in total national energy production
Cyprus	1	0.5	< 0.1	< 0.1
Czech Rep.	1,136	250	32.3	1.0
Estonia	13	3.0	100	0.1
Hungary	35	8.6	25.4	0.1
Latvia	57	1.7	0.5	0.3
Lithuania	29	9.3	3.6	0.2
Malta	0	0	0	0
Poland	472	127	31.0	0.5
Slovak Rep.	180	31	3.8	0.7
Slovenia	413	77	7.6	2.3
Bulgaria	64	141	22.3	1.0
Romania	9	44	1.8	0.5
Turkey	67	138	1.7	0.4

Spain puts more efforts and attention to development the wind energy and PV, which become its national priority. SHPs are also developing in this country but the percentage of their energy production is decreasing.

Austria is very reach with the water energy and SHPs branch is its long-term national priority. This country shows how to use effectively even comparatively small and unusual water energy sources.

The Vienna municipal water supply uses mostly karst water, which runs from the mountain springs for about 180 km before reaching the city water taps [http://www.wien.gv.at/english/environment/watersupply/supply/index.html]. Only one pumping station works at the initial part of one of the water supply lines and the water runs mostly because of the surface slope. The water supplying company has installed 12 SHPs along the water lines and now they generate ~65 million kilowatt-hours per year [http://www.wien.gv. at/english/environment/watersupply/energy.html]. Part of this energy is used for the own needs and other energy is being supplied to the local customers. Total energy amount would be enough to supply a 50,000 capital of Lower Austria St. Pölten. This example shows how SHPs can be effectively operated even in rather unusual conditions.

2. SHPs in Ukraine: Current Condition and Future Prospects

Small hydropower energy production started in Ukraine years ago together with traditional hydropower production. Small hydropower plants were constructed at numerous small rivers, streams, rapids and waterfalls all over Ukraine. They supplied energy for the local consumers mostly in remote and mountain areas and often were not connected to the national electric circuit network (Winkler 2009). The total number of SHPs reached almost 1,000 in the late 1950s. Somewhere they were constructed even in the site of old watermills.

However, the price of electricity produced at the SHPs was reasonably higher than the price of the traditional hydropower energy. The SHPs were gradually ousted by the traditional hydropower plants and the total number of the Ukrainian SHPs was cut down to 42 in the 1980s.

This branch got the 'second wind' in Ukraine in the mid-late 1990s after a series of 'oil crises' resulted in the steady increase of the energy price. On other hand, many negative results of the traditional hydropower plants functioning become very obvious, which also promoted rebirth of idea of the SHPs.

Total number of working SHPs in Ukraine exceeded 85 in 2008 and is still increasing. Some old SHPs are being modified in order to increase their power and flexibility of the energy production. Table 5 represents the closest prospects of the SHP energy production in Ukraine (Tsymbalenko and Vikhorev 2005).

As seen from the Table 5, the Transkarpathian and Lviv regions are very rich with the small hydropower resource and it is needful to pay special attention to development of the SHPs in these regions. Total profitable potential of the SHPs energy production is 3,747 million kilowatt-hours per year while all traditional Ukrainian hydropower plants produce about 11,000 million kilowatt-hours per year. 'Big' hydropower energy sources have been almost completely involved in the energy production in Ukraine therefore, further increase of the hydropower energy production is possible mostly with new and modified SHPs. In some regions they can even play the leading role in the electricity supply.

TABLE 5. Prospects of the SHP development in Ukraine

Region of Ukraine	Number of working SHPs (2007)	Prospective number of SHPs	Installed power of SHPs (MW)	Potential energy production (mln. kWh/ year)	Economically profitable potential (mln kWh/ year)
Vinnitsa	12	17	18.5	238	108
Zhytomyr	4	21	1.3	222	101
Transcarpathian	3	11	29.0	2,991	1,357
Lviv	0	6	0	1,197	544
Ternopil	6	11	8.2	282	128
Khmelnitsky	8	7	5.0	200	91
Chernivtsi	0	5	0	583	265
Other regions	37	19	40	2,539	1,153
TOTAL	70	97	102	8,252	3,747

An energy potential of the small rivers of the Transcarpathian region of Ukraine is estimated as about 400 MW (www.zik.com.ua/ua/news/2006/05/29/40649). It is more than installed power of all existing SHPs of Ukraine. This potential would be enough to supply electricity to all consumers of this region. Same situation is in Chernivtsi region – local small rivers can completely supply local consumers with electricity.

SHPs should not be forgotten in other regions too since they are very important for the energy supply to consumers in remote and mountain areas. There are no big industrial consumers in such regions and existing subscribers are scattered over wide area. Low-power and steady working SHP is the best energy source for such low-power local network.

3. Conclusion

Small hydropower plants are experiencing fast growth in all European (and not only European) countries now. The can generate electrical energy to local low-power consumers and also can supply national electrical network.

Economically profitable potential of SHPs in some countries or regions is high enough to supply energy to all local consumers. Taking into account minimal negative environmental effect of the SHPs they can be installed even in the natural reserved areas.

However, prospects of the small hydropower energy production should not be overestimated. This branch would not be able to substitute traditional high-power energy sources especially in highly industrialized areas. On other hand, even 'insignificant' energy production at the SHPs saves about 75,000 TOE in Ukraine and can save from 69 to 100 million TOE worldwide in 2010.

References

Geletuha G, Kudrya S. Ukraine, 2005. Non-traditional and Renewable Energy Sources. Green Energy 19, 8–10.

http://www.wien.gv.at/english/environment/watersupply/energy.html, accessed 10 June 2009.

http://www.wien.gv.at/english/environment/watersupply/supply/index.html, accessed 10 June 2009.

http:// www.zik.com.ua/ua/news/2006/05/29/40649, accessed 11 June 2009.

Syrotiuk M., 2008. Renewable Energy Sources. Lviv University Publishing Centre, Lviv.

Tsymbalenko O, Vikhorev Yu., 2005. Small Hydroenergy: Raise Market Prospects. Green Energy 19, 12–19.

Winkler I., 2009. Small Hydropower Resources and Prospects of Small Hydropower Electric Plants in the Near-Border Regions of Ukraine. Energy and Environmental Challenges to Security (eds. Stephen Stec and Besnik Baraj). Springer, 371–378.

Winkler I. and Tevtul' Ya., 2006. Ecologically Safe Energy Sources. Part 2. Renewable Energy Sources. Ruta, Chernivtsi.

SECURING ENERGY SUPPLY AT THE REGIONAL LEVEL – THE CASE OF WIND FARMING IN GERMANY: A COMPARISON OF TWO CASE STUDIES FROM NORTH HESSE AND WEST SAXONY

JAN MONSEES[1*], MARCUS EICHHORN[2], AND CORNELLA OHL[1]

[1]*Helmholtz Centre for Environmental Research – UFZ, Department of Economics, Permoserstrasse 15, 04318 Leipzig, Germany*
[2]*Helmholtz Centre for Environmental Research – UFZ, Department of Ecological Modelling, Permoserstrasse 15, 04318 Leipzig, Germany*

Abstract Wind power is one way we can reduce our dependency on fossil fuel imports and mitigate climate change. However, wind power can play this important role only if sufficient space for wind farming is made available off-shore as well as on-shore including sites far away from the seashores. Against this background this paper presents a comparative analysis of two hinterland case studies from Germany. Applying GIS and official wind speed data we evaluate the effectiveness of designated wind farming areas in the regions of West Saxony and North Hesse in terms of their expected wind energy yields and potential for repowering. We show that, in this respect, the current spatial allocation for wind power generation in both study regions is not as effective as it could be, but for different reasons. We contrast this finding with an alternative proposal which not only meets the legal requirements for wind power generation but also yields better results in terms of expected energy output. This proposal takes into account the availability of different turbine types and a spatial re-allocation of wind farming areas within the study regions.

---
* To whom correspondence should be addressed: Dr. Jan Monsees, Department of Economics, Helmholtz Centre for Environmental Research – UFZ, Permoserstrasse 15, 04318 Leipzig, Germany; E-mail: jan.monsees@ufz.de

F. Barbir and S. Ulgiati (eds.), *Energy Options Impact on Regional Security*,
DOI 10.1007/978-90-481-9565-7_15, © Springer Science + Business Media B.V. 2010

Keywords: Case study, Germany, Hesse, land use, regional planning, renewable energy, repowering, Saxony, spatial allocation, wind farming, wind power, wind turbines

1. Introduction

Wind power is contributing considerably to replacing fossil fuels and to reducing CO_2 emissions. Of all the renewable energies, it is wind power that has experienced the most impressive growth in the last 2 decades. Globally, the installed capacity has multiplied by a factor of 15 between 1996 and 2007, from 6.1 to 93.9 GW rated power. By the end of 2007 Germany was in the lead in absolute terms with 22.3 GW, followed by the USA (16.8 GW), Spain (15.1 GW), India (7.8 GW), and China (5.9 GW), whereas Denmark was leading in relative terms with a 21% wind power share in the country's electricity supply (GWEC and Greenpeace 2008). However, to achieve more ambitious climate policy targets that are now being set, for example, by the European Union or the German Federal Government (BMU 2007), it is necessary for wind power to maintain its growth trend in the coming decades. In terms of on-shore expansion, the installation of wind turbines requires suitable land which, in Germany, needs to be designated by regional planning agencies.

Our paper examines if and how the regional planning agencies of West Saxony and North Hesse manage to secure suitable space for wind power generation by designating land for this purpose in their regional plans. The paper is organized as follows. Section 2 briefly refers to wind farming regulations in Germany insofar as they are relevant for our analysis, i.e. the planning system with a focus on the regional level and the Renewable Energy Sources Act. In Section 3 we describe our calculation method for analysing the wind energy potentials at the regional level. Sections 4 and 5 feature the two case studies by introducing the study regions and their regional planning as regards wind farming, applying our method and presenting and discussing the results. Section 6 compares the main findings from the two case studies and, finally, Section 7 draws conclusions with regard to secure wind energy supply at the regional level.

2. Case-Relevant Wind Farming Regulations in Germany

Within the German multi-level governance system the regional level forms the link between the sub-national level, which is represented by the *Bundeslaender*, i.e. state level, and the county level. Overriding spatial planning guidelines and tools are provided by the Federal Regional Planning Act which is enacted at the federal and national level respectively.

The law defines a framework which leaves ample scope for the *Bundeslaender* to design long-term oriented state development plans, including principles and objectives for regional planning in their domain. Based on this, central responsibility for the designation of land for wind power is assigned to regional planning agencies. The state (*Bundesland*) of Hesse, for example, is comprised of three planning agencies called *Regionalversammlung*, whereas the state of Saxony has four which are called *Regionaler Planungsverband*. These agencies establish legally binding regional plans.

On the regional level, regional planning agencies make use of specific spatial predeterminations in order to secure space for wind farming as well as other certain types of land use. In legal terms, the agencies either designate priority areas (*Vorranggebiete*), or suitability areas (*Eignungsgbiete*), or a combination, i.e. priority areas with the impact of suitability areas (*Vorrang- und Eignungsgebiete*; here referred to as VE areas). In so far as wind power is concerned, priority areas are where the installation and operation of wind turbines is prioritized over other land-use. In contrast, suitability areas are considered appropriate for wind power generation, however priority is not given for it. At the same time, suitability areas also preclude any other sites in the planning region from being used for wind power generation. The combination in the form of VE areas, assures that in areas designated as such wind farming is given priority over other land-use on the one hand, while at the same time prevent any other sites in the planning region from being used for wind farming (Koeck and Bovet 2008).

The number, dimensions and energy potential of such areas on the regional level obviously play a crucial role in achieving national energy and climate policy targets in terms of lowering the dependence on fossil fuel imports and reducing greenhouse gas emissions. Having said this, priority, suitability and VE areas serve as a means of controlling the further expansion of wind farming, which otherwise could lead to an unwanted spreading of wind turbines across landscapes throughout a whole region. The Renewable Energy Sources Act (EEG), an instrument that promotes renewables from 1991[1] onwards, provides a second means to counteract uncontrolled growth of wind farming in Germany since its 2004 amendment. The EEG now stipulates that grid system operators shall pay a tariff for electricity from a wind turbine only under the condition that this turbine generates a minimum yield according to a technical definition which is given in Appendix 5 of the law. Thus, a wind turbine operator is obliged to prove that his/her turbine at

---

[1] In fact, the EEG itself came into force in the year 2000; however, its predecessor, the electricity-feed-in-law had already become effective in 1991.

the site under consideration will produce at least 60% of the so-called reference yield for that turbine type (EEG 2009). But the EEG also rewards repowering, i.e. the replacement of existing small turbines by tall state-of-the-art turbines which reduce the specific land demand for wind farming (ha/MW).

The designation of VE areas usually is the outcome of a complex, criteria-based selection process, but does not follow a nationwide standardised procedure. Rather, it is partly based on binding legal provisions, and partly on the discretion of the regional planning agency (RPW 2008). However, case law established at the highest judicial level bars negative planning i.e., designating VE areas for wind farming which are unfeasible for economical reasons and in fact impede wind power generation instead of fostering it (Koeck and Bovet 2008). A systematic and stepwise approach is typical whereby a set of exclusion criteria is applied one after another to disqualify those parts of a planning region which are not suitable for wind farming for whatever reason. In addition to a number of mandatory criteria which are derived from the Federal Immission Control Act (noise protection etc.) and the Federal Nature Conservation Act, there are many flexible criteria regarding, for example, the visual impact of wind turbines on the landscape or the dimension of buffer zones to specific types of land use. Furthermore, many stakeholders and the general public are invited to participate in the planning process.

3.   Method Applied to Evaluate Potential Sites for Wind Farming

The method applied in this paper is based on Ohl and Eichhorn (2009).[2] It draws on regional planning criteria, EEG requirements, wind potential data, technical wind turbine parameters and a geographic information system (GIS). In order to find appropriate land for wind farms, we follow a selection procedure as described in detail by Ohl and Eichhorn (2009). Using GIS analysis we begin by excluding physically unsuitable parts of the landscape, which mainly comprise settlement areas, infrastructure facilities, water bodies and forests. Then, we eliminate those areas from the available space which are not legally qualified for the erection and operation of wind turbines, e.g. nature conservation areas and the like. Accordingly, we obtain a number of potential sites that are legally and physically suitable for wind farming.

---

[2] A similar approach, but on a smaller spatial scale, has been applied by Krewitt and Nitsch (2003).

Next, sites selected in this way are overlaid with the grid data on annual energy yield to identify the most promising sites. The database for the calculation of the wind energy potential in the two study regions is provided by the national meteorological agency of Germany (DWD 2007). We employ grid-based frequency distributions of wind speeds as Weibull parameters for form and scale with a horizontal resolution of 1 by 1 km. To determine the energy yield, we utilize the power curves of state-of-the-art wind turbines which are able to deliver the highest amount of electric energy possible at a certain wind speed. The grid cells are the focal units for the computation of energy yields. For every cell there is a certain predicted energy potential per turbine that is ready for exploitation. Grid cells may have one or more turbines subject to the conditions of their optimal spatial arrangement.[3] We consider two different turbine types: at first a turbine of 2 MW rated power, 82 m rotor diameter, 80 m hub height and 121 m total height (type I) and, afterwards, a turbine of 3 MW rated power, 90 m rotor diameter, 105 m hub height and 150 m total height (type II).

Special attention is paid to the priority, suitability and VE areas designated in the regional plans of both study regions. Applying GIS data as to their location and size, we evaluate the suitability of these areas for the operation of turbine type I, and II respectively under the constraints given by the regional plans. Both turbine types can be assumed to be proper means for repowering in the future. This is of particular importance if existing priority, suitability and VE areas are already occupied with low-performing small turbines which then may block more ambitious regional renewable energy and CO_2 reduction targets. Our assessment takes into account the land demand of the two turbine types under consideration, their required minimum distances to built-up areas and, as the case may be, also height limitations for wind turbines. In a final step, we then aim to identify possible reallocation areas which may better fit the purpose of repowering than the designated priority, suitability and VE areas in the regional plans.

4. Case Study West Saxony

4.1. STUDY REGION

West Saxony is one of a total of four planning regions in the German state (*Bundesland*) of Saxony. By the end of 2007 the region had a population of nearly one million and a surface area of 4,388 km² which equals a density

---

[3] Cf. park layout recommendations of the Danish Wind Industry Association (2009).

of 245/km². 221 wind turbines with a total capacity of 235 MW rated power were in operation in the region, generating around 345 GWh of electricity and averting roughly 296,000 t CO_2 per year (Ohl and Eichhorn 2009). The regional plan for West Saxony, in force since 2008, designates 22 VE areas covering a total of 1,145 ha (11.45 km²) which represents 0.26% of the region's total surface area (RPW, 2008).[4] The location of the VE areas within the planning region West Saxony is depicted in Fig. 1 (no. 1–22). The size of individual VE areas ranges between 5 and 240 ha. Since priority, suitability and VE areas are recent planning devices, a number of older wind turbines in West Saxony had previously been erected outside of the 22 VE areas and are now operating under a continuation permit (RPW 2008). Notwithstanding, the bulk of wind power in West Saxony today is generated in VE areas which are already used to capacity with turbines that have been installed between 1994 and 2007. Therefore, an increase of wind power generation in the future will largely depend on repowering. Since the EEG repowering bonus requires that the turbines that are replaced be in operation for at least 10 years, it would take at least until the end of 2017 to repower all existing turbines.

The designation of the 22 VE areas in West Saxony is based on a similar approach as described in the previous section. It follows the objectives and principles for the planning region which in turn are required to be in accordance with the state development plan of Saxony. Neither the state development plan, nor the regional plan quantify targets for wind power in terms of electricity consumption quotas or overall installed rated power. However, the regional planning agency is committed to, and achieves, designating at least 0.25% of the region's surface area as VE areas for wind farming (RPW 2008). Accordingly, a total of 20 exclusion criteria have been applied during the selection process. The criteria are described in detail in the regional plan. Most of these deal with nature conservation and landscape protection matters. Several others prescribe minimum distances of wind turbines to residential areas and infrastructure facilities. In addition, a minimum distance of 5 km between two wind farms is required. Also height restrictions are set, some of which apply to all VE areas and a few are defined specifically for certain VE areas (RPW 2008).

---

[4] The data provided here originates prior to a territorial reform in the state of Saxony as from 1 August 2008. In consequence of a consolidation of counties and planning regions the planning region West Saxony lost one of its former member counties and was scaled down to a surface area of 3,964 km². Nevertheless, the Regional Plan West Saxony effective since 25 July 2008 has remained in force unchanged.

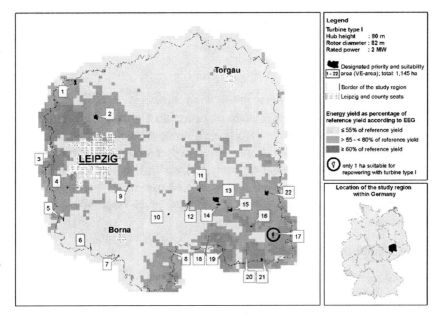

Figure 1. Study region of West Saxony with designated wind farming areas and expected energy yields for turbine type I

4.2. RESULTS

We have to first mention that the results presented here are calculated for state-of-the-art turbines of 2 MW rated power (here called type I) which are qualified to replace most of the existing turbines in the future. As these turbines will fall under the new EEG it will be crucial to meet the reference yield criterion (cf. Section 2). The reference yield for our turbine type I is 5.7 GWh/a, 60% of the amount required by the EEG equates to 3.4 GWh/a. Existing turbines installed prior to the 2004 EEG amendment receive tariffs regardless of their performance, i.e. for these turbines the reference yield is insignificant.

4.2.1. *Land demand*

In order to assess the theoretical potential for repowering we now need to thoroughly apply the requirements of the West Saxony regional plan to some specific parameters of the turbine type I considered here, including land demand, height and noise level. In terms of the land demand of wind turbines, there is no unique method at hand to provide an accurate measure but rather a range of calculation methods depending on context (Ohl and Monsees 2008). A good estimate in the context of our investigation is the toppling distance circular area which measures the hub height plus the rotor blade width in every direction. It amounts to 4.6 ha for every single type I turbine.

4.2.2. *Minimum distance to built-up areas*

The absolute minimum distances of wind turbines to built-up areas in Germany are stipulated by the Federal Emission Control Act (BImSchG). They are based on the spread of a turbine's noise emissions which is mainly influenced by the wind speed, the turbine height and the generator size. The allowed maximum noise level according to the BImSchG is 35 dB dB(A) in residential-only areas and 40 dB(A) in mixed-use areas. For turbine type I, this requirement corresponds to a distance of at least 800 or 500 m, respectively.[5]

4.2.3. *Combined distance and height regulations*

In addition to the universal provisions of the BImSchG, the West Saxony regional plan sets even more restrictive distance regulations linked to turbine heights. According to this, the total turbine height is limited to 100 m in a zone up to 750 m from residential areas. In a zone from 750 to 1,000 m from residential areas, a turbine must be at least ten times its hub height (RPW 2008) away. As a result, type I turbines (hub height: 80 m) may only be erected 800 m or more away from residential areas. Taking into account the location of the 22 VE areas only three (nos. 2, 13 and 22) do have substantial space available beyond 800 m away, while most others are either closer to residential areas or relatively small. In all, just 323 ha (28%) out of the total of 1,145 ha designated for wind farming are sited beyond 800 m.

4.2.4. *Height limitation in some VE areas*

Apart from the aforementioned height regulations that are applicable for the entire planning region, in two VE areas (no. 1 and 2) located close to the Leipzig–Halle airport, the total height of a wind turbine is strictly limited to 100 m. Thus, these areas are also completely unavailable for repowering with type I turbines (total height: 121 m). A further 109 ha have to be subtracted from the VE sectors potentially suitable for repowering so that only 214 ha (19%) remain.

---

[5] According to a wind power expert from Saxony, personal communication, 22 July 2008.

4.2.5. *Expected energy yields*

Figure 1 illustrates the grid data set representing the annual energy yield to be expected for turbine type I. The darker the colour of the grid cells, the higher the expected energy yield. Of particular importance are the dark grey sectors because they indicate that turbine type I achieves 60% or more of its reference yield. Only wind farmers in these sectors will be paid a tariff for wind powered electricity by grid system operators according to the EEG. Figure 1 clearly indicates that only few parts in the northwest and southeast of West Saxony fulfil this criterion. This means that wind farming with turbine type I is unprofitable for most parts of this planning region. With regard to the 22 VE areas designated in the West Saxony regional plan (black spots in Fig. 1) obviously only one VE area (no. 17, circled in black) is partly located in a dark grey sector. This is the only place that a wind farmer has an incentive to repower, i.e. to replace his/her existing turbines with type I turbines.

4.2.6. *Repowering potential in VE areas*

Due to the distance and height regulations referred to above, only 214 ha (19%) out of a total of 1,145 ha designated for wind farming in West Saxony are actually available for repowering with turbine type I. The suitable sectors are located in 11 of the 22 VE areas. Assuming a land demand of 4.6 ha for a single turbine (cf. Section 4.2.1), it is possible to fill this space with 46 type I turbines which would total 92 MW rated power. However, the land available is essential but it is not sufficient. For wind farmers, the key to repowering is achieving the reference yield criterion. As mentioned above (cf. Section 4.2.5) just one too small portion of 1 ha in just one VE area (no. 17) can be expected to meet this criterion, i.e. there is nearly no incentive to repower at all. This unsatisfactory result gives reason to reconsider the selection of VE areas as well as the technical repowering option and will be addressed in the following section.

4.3. DISCUSSION

The analysis carried out in Section 4.2 has revealed that from the total VE area designated for wind power generation only a fraction, 19%, is legally suitable for turbine type I and that an even lesser portion (1%) is economically viable. This considerable discrepancy raises questions as to whether and what alternatives may exist to reconcile it. Ohl and Eichhorn (2009) have primarily identified two options which will be reproduced below – the utilisation of another type of wind turbine, and the spatial re-allocation of VE areas.

4.3.1. *Utilisation of a different type of wind turbine*

Because wind force is stronger and more regular the higher the altitude, it can be expected that a taller turbine would yield more energy than type I at every given location. Therefore, we now reproduce the energy appraisal for turbine type II which is characterised by 3 MW rated power, 90 m rotor diameter, 105 m hub height, 150 m total height and a toppling distance circular area of 7.1 ha to represent its land demand. The EEG reference yield for turbine type II is fixed at 6.9 GWh/a, which results in at least 4.1 GWh/a for EEG tariff eligibility (60% of reference yield). Thus, it requires a 20% higher energy yield for turbine type II to pass the reference yield criterion compared with type I. The expected energy yields for turbine type II are shown in Fig. 2. Despite the more demanding performance standard, EEG tariffs can be expected for turbine type II in more parts of the study region than for type I which makes, in principle, type II the economically superior choice. To a large extent, this can be attributed to the 30% increase in hub height.

However, because of the above-mentioned combined distance and height regulations (cf. Section 4.2.3.) the legally suitable VE area for type II turbines is considerably reduced since these are higher than type I. Only sites of at least 1,000 m distance from built-up areas can be taken into account this time. As a result, just 82 ha or 7% of the total designated space for wind farming remain which are situated in three (no. 2, 13, 22) out of a total of 22 VE areas. Considering the particular height restriction around the airport (cf. Section 4.2.4.) leads to the omission of VE area no. 2, so that just 47 ha (4%) are left for turbine type II compared to 214 ha (19%) for turbine type I. But finally applying the EEG reference yield criterion to the remainder only yields 23 ha (2%) located in two VE areas (no. 13, 22) which are useful from an economic point of view. This leaves space to operate three turbines of 9 MW rated power in total which may potentially yield 12.6 GWh/year. Since the use of other state-of-the-art turbines instead of type II would also face limitations in terms of distance and height, it is obvious that the disparity cannot be resolved by simply altering the turbine type.

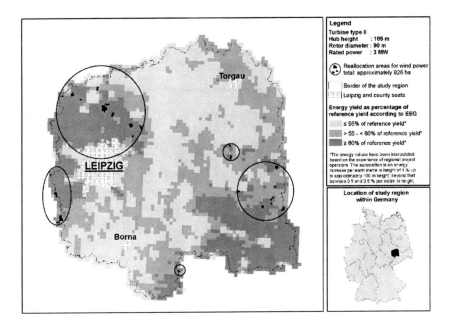

Figure 2. Expected energy yields and potential re-allocation areas in West Saxony for turbine type II

4.3.2. *Spatial re-allocation of wind farming areas*

As a second means to reconcile the disparity between the currently designated wind farming space and that which is economically viable for repowering, we now consider the spatial re-allocation of VE sites. In doing so, we particularly emphasise the energy potential per unit area to economically select more feasible areas for wind farming than those designated in the West Saxony regional plan. We carry out this analysis for both turbine types. The results for turbine type I are shown in Fig. 3. A total of 188 ha partitioned into 11 individual plots (depicted in Fig. 3 inside the six circles) have been identified which satisfy the legally binding selection criteria and the specific distance and height regulations in West Saxony as well as the EEG reference yield criterion. The identified area equates to 16% of the total surface area of the currently designated 22 VE areas and allows for the operation of 40 type I turbines of 80 MW rated power in total which would yield 140 GWh and save 119,840 t CO_2 per year. The equivalent results for turbine type II are illustrated in Fig. 2. Overall, 926 ha partitioned in 21 individual plots (within the five circles in Fig. 2) have been identified in this case which equals 81% of the currently designated VE areas. This enables 130 type II turbines of 390 MW rated power in total to be operated which would yield 546 GWh and save 467,376 t CO_2 per year. Thus, a re-allocation of the VE areas designated at present could bring about more economically viable space for repowering, a larger energy yield and higher CO_2 savings (Ohl and Eichhorn 2009).

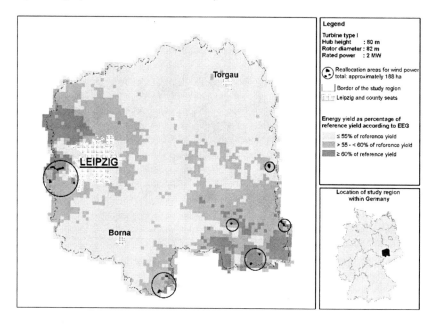

Figure 3. Potential re-allocation areas in West Saxony for wind farming with turbine type I

5. Case Study North Hesse

5.1. STUDY REGION

North Hesse is one of a total of three planning regions[6] in the German state (*Bundesland*) of Hesse. By the end of 2007 the region had a population of little more than 1.2 million and a surface area of 8,289 km². Because this surface area is almost twice as large as West Saxony, we decided to consider only a part of North Hesse of approximately the same size as West Saxony. The selected part comprises the three counties of Kassel, Schwalm-Eder and Waldeck-Frankenberg and the city of Kassel. These will be considered below as our study region North Hesse. It has roughly 785,000 inhabitants and a total surface area of 4,786 km². With a density of 164/km² this study region

---

[6] Although the official term in Hesse is administrative district (*Regierungsbezirk*), we use the term planning region here as well for the sake of comparison with Saxony.

is much less populated than the study region West Saxony. 223 wind turbines with a total capacity of 183 MW rated power were in operation in the region by the end of 2007, generating around 270 GWh of electricity per year (RVN 2009) and averting roughly 230,000 t CO_2.[7] About two-thirds of these turbines have been installed outside VE areas – which were designated at a later date – and are now operating under a continuation permit (RVN 2009). This represents a much larger fraction compared to West Saxony.

The regional plan for North Hesse has been revised lately and waits now for the approval of the government of Hesse to become effective. The plan designates areas for wind farming as the West Saxony regional plan does, but it names these differently. While West Saxony designates VE areas, North Hesse designates priority areas for wind power generation (cf. Sections 2 and 4.). However, referring to the state planning law (*Landesplanungsgesetz*) of Hesse, Section 6 (3), the North Hesse regional plan defines priority area in such a way that the installation of wind turbines outside these areas is not allowed. That is to say, the legal impact of priority areas in North Hesse is exactly the same as that of VE areas in West Saxony (RVN 2009). Hence, for the sake of comparison, we will phrase the priority areas for wind power generation in North Hesse as VE areas as well. Furthermore, unlike West Saxony, North Hesse distinguishes between VE areas already in operation (*Bestand*), i.e. partly occupied with wind turbines, and VE areas not yet in operation (*Planung*), i.e. still without wind turbines (RVN 2009). Eighteen VE areas covering 1,073 ha that are already operating and 17 VE areas covering 1,025 ha that are not yet operating are designated for wind farming. The location of VE areas of both categories is depicted in Fig. 4 (no. 1–18 and 19–35 respectively). The size of individual VE areas ranges from 5 to 114 ha. Overall, 35 VE areas covering 2,098 ha or 20.98 km² are designated for wind farming (RVN 2009). This represents 0.44% of the region's total surface area, a proportion that is almost twice as high as in West Saxony.

The selection procedure of the VE areas in North Hesse also resembles the approach described in Section 3. It follows the objectives and principles as laid down in the regional plan taking the state planning law and the state development plan of Hesse into consideration. As in West Saxony, neither the state development plan nor the regional plan give quantifiable targets for wind power. However, the regional plan conveys that the designation of further VE areas for wind farming in North Hesse is to support the ambitious renewable energy policy targets of the German Federal Government (RVN

---

[7] CO_2 savings are set at 856 t/GWh according to Ragwitz and Klobasa (2005), cited in Ohl and Eichhorn (2009).

2009). The set of exclusion criteria to designate VE areas is similar to West Saxony and exemplified in detail in the regional plan. However, in contrast to West Saxony a few exclusion criteria in North Hesse discriminate between VE areas already in operation and those not yet in operation. First of all, this concerns differing minimum distances between VE areas and built-up areas. Moreover, VE areas that are not yet operating require a minimum surface area of 20 ha unless they are attached to a VE area that is already in operation (RVN 2009). A large portion in the north of the study region is situated within a military radar range, where it is necessary that turbines are staggered.[8] The impact of these requirements on the repowering potential will be analysed in Section 5.2.

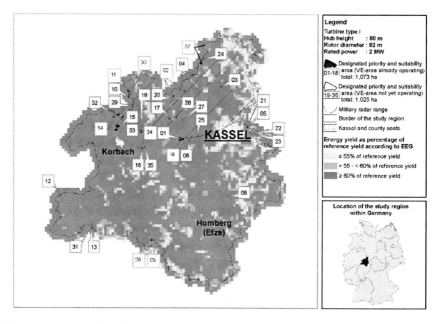

Figure 4. Study region of North Hesse with designated wind farming areas and expected energy yields for turbine type I

---

[8] This condition derives from appeals of the military district administration in the course of building permit procedures on the local level for two repowering sites. Although not a requirement imposed by the regional plan, it has to be considered when it comes to repowering; personal communication: *Regierungspräsidium* (district government) Kassel, 1 October 2009.

5.2. RESULTS

The results presented here again are calculated for state-of-the-art turbines of 2 MW rated power which qualify to replace existing turbines in the years to come and for which it will be crucial to meet the EEG reference yield criterion (cf. Sections 2. and 4.2). In order to make fair comparisons we use the same approach to assess the theoretical repowering potential in North Hesse as in West Saxony. Accordingly, we now apply the requirements of the North Hesse regional plan to turbine type I.

5.2.1. *Land demand*

The land demand per wind turbine is again set at 4.6 ha which represents the toppling distance circular area of type I turbines (cf. Section 4.2.1.).

5.2.2. *Distance requirements*

As already explained in Section 4.2.2, type I turbines need minimum distances of approximately 800 m from residential-only areas and 500 m from mixed-use areas to be in accordance with the BImSchG. Apart from that, the North Hesse regional plan requires that VE areas not yet in operation are at least 1,000 m away from residential areas and 500 m from commercial areas. In contrast, VE areas already in operation only require distances of 750 and 300 m respectively (RVN 2009). The difference can be attributed to the characteristics of the smaller turbines which were common by the time the VE areas already in operation were designated. Compared to West Saxony, the North Hesse region plan does not specify combined distance and height regulations. However, the northern portion within a 40 km military radar range (shaded sector in Fig. 4) requires a distance of at least 750 m between two turbines. This has a significant impact on the utilisation level of VE areas located within this range and affects 901 ha or 43% of the total of designated VE areas. This requirement leads to a virtual increase in the land demand per turbine from 4.6 to 44 ha and, therefore, to a reduction of the number of turbines allowed per affected VE area. Consequently, the energy yields achievable in the affected VE areas are also considerably reduced.

5.2.3. *No height limitation*

Unlike West Saxony, the North Hesse regional plan does not explicitly limit turbine heights. Thus, the repowering potential is not restricted in this respect.

5.2.4. *Expected energy yields*

Figure 4 illustrates the grid data set representing the annual energy yield to be expected for turbine type I. The darker the colour of the grid cells, the higher the expected energy yield. As in Fig. 1, again the dark grey sectors represent those areas where turbine type I is expected to achieve 60% or more of its reference yield and thus qualifies for EEG tariffs. In sharp contrast to West Saxony (cf. Fig. 1), Fig. 4 shows that in North Hesse every single VE area designated for wind farming regardless whether already in operation or not does meet the reference yield criterion. That is to say, from this point of view, turbine type I can be profitably operated without reservation throughout the 35 VE areas or 2,098 ha respectively. However, the effective energy yields are considerably reduced because of the separation requirements within the said radar range. The magnitude of this impact and possible solutions are discussed in the following subsections.

5.2.5. *Repowering potential in VE areas*

Due to less restrictive distance and height regulations, turbine type I can be legally deployed and EEG eligibly operated in every VE area in North Hesse without limitation (cf. Section 5.2.4). Thus, compared to West Saxony, the repowering of existing turbines in VE areas is not hampered in that respect. Yet on the other hand, the above-mentioned distance requirement between turbines has a negative effect on the repowering possibilities in 18 VE areas which are situated within the 40 km radar range in the northern part of the study region. Compared to an unrestricted use of these areas this would lead to a loss of 146 type I turbines with a total capacity of 292 MW, which implies a wind power loss of some 700 GWh/a. However, this requirement would affect repowering by other turbine types in a similar manner.

5.3. DISCUSSION

Although the results obtained so far have shown that the currently designated VE areas in North Hesse are entirely suitable for type I turbines we now discuss the options raised in the West Saxony case study (cf. Section 4.3.) for North Hesse as well, to deal with the loss of space due the military radar range.

5.3.1. *Utilisation of a different type of wind turbine*

The first option we analyse is the use of the taller turbine type II. The energy appraisal is depicted in Fig. 5. It resembles the picture for type I, i.e.

the study region North Hesse in general qualifies for the most part for EEG tariffs and type II could be operated in every designated VE area above the reference yield criterion. Hence, in terms of EEG eligibility there is no difference between type I and II turbines. But turbine type II requires a minimum distance of 1,000 m to residential areas for sound diffusion reasons, which exceeds the requirement for type I by 200 m. However, since VE areas not yet in operation are planned 1,000 m away from human settlements, they are not impacted at all. Of the possibly affected 18 VE areas already in operation, only a few are located closer than 1,000 m to settlements. Therefore, this point can be ignored in our analysis without distorting the results. Apart from that, the use of type II is constrained as well by the said radar range (shaded sector in Fig. 5). In contrast to an unrestricted use of the affected VE areas this would lead to a loss of 77 type II turbines with a total capacity of 231 MW, which implies a wind power loss of approximately 470 GWh/a or 67% in terms of losses for type I.

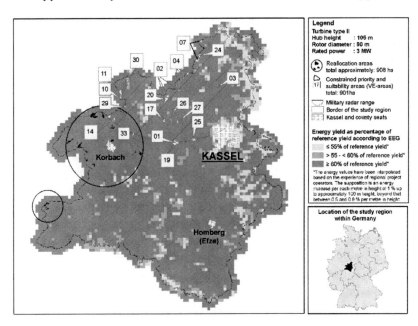

Figure 5. Expected energy yields for turbine type II and potential re-allocation areas in North Hesse

5.3.2. *Spatial re-allocation*

In North Hesse, the need for re-allocation does not arise, as in West Saxony, from turbine height limitations or an underperformance in terms of the EEG, but from the significant utilisation constraint in VE areas within the radar range. We therefore identify a number of spatial re-allocation areas in North Hesse which also qualify for wind farming in terms of legal and physical exclusion criteria and EEG eligibility, but are located outside the radar range. Potential re-allocation areas for turbine type II are illustrated in Fig. 5 (within the black circled sectors). By using such areas instead of the constrained ones, it would be possible, without increasing the total of land designated for wind farming, to regain space for 129 type II turbines with a total capacity of 387 MW rated power which could yield around 970 GWh/a. For turbine type I, in principle, the same sectors of the study region can be considered for a re-allocation. In that case, they could accommodate 199 type I turbines with a total capacity of 398 MW generating approximately 1,270 GWh/a.

6. Comparative Evaluation of the Two Case Studies

When we started our research on wind farming in West Saxony we were somewhat surprised about the initial findings that the designated VE areas to a large extent do not allow for a profitable operation or even an installation of state-of-the-art turbine types which normally can be considered as the logical alternatives for repowering in the next years. This motivated us to investigate another hinterland region, North Hesse, in order to verify whether the results obtained from the West Saxony case study can be viewed as typical for a hinterland in general or not.

First of all, we have to mention two main differences between the two study regions which most likely have a major influence on the outcome. Firstly, the average wind speeds in North Hesse are constantly higher than in West Saxony which is obvious from a comparison of Figs 1–5. This enables the operation of both considered turbine types above the reference yield criterion almost throughout North Hesse, whereas this is guaranteed only at a few sites in West Saxony. Secondly, West Saxony is much more densely populated than North Hesse which, under the given distance requirements towards built-up areas, considerably restricts the search-space for VE areas in West Saxony. The coincidence of these two differences, an unfavourable wind regime and a higher population density makes it even more difficult for the regional planning agency in West Saxony to designate VE areas for wind farming which not only qualify with regard to the applied exclusion criteria but also for a profitable management of wind farms. Hence, we can ascertain that we have no evidence that our findings from West Saxony are representative of German hinterland regions in general.

A detailed comparison of the two case studies reveals some other interesting findings which relate to the two options discussed: the utilisation of another turbine type and a spatial re-allocation of wind farm areas. The effect of the switch from turbine type I to type II, for instance, is more noticeable in West Saxony than in North Hesse. This is because type II can be expected to operate in more sectors of West Saxony above the reference yield criterion than type I (cf. s. 1 and 2). In contrast, in North Hesse there is no such difference regarding this criterion. Although we have found that turbine type II would allow for a better utilisation of VE areas in West Saxony than type I, this option has not proven to be a solution due to the combined distance and height regulations that must be observed (cf. Section 4.2.3.).

That has led us to the consideration of a spatial re-allocation which has revealed that it is possible to find sites other than the designated VE areas which would yield better results from an energetic point of view without compromising the legal requirements for wind farming. In this respect, we can conclude that, subject to our chosen framework, the current spatial allocation of wind farm areas in West Saxony is not as effective as it could be. In North Hesse, on the other hand, the need for a spatial re-allocation of wind farm areas may arise from the constraint imposed by a military radar range. 43% of all the land designated for wind farming is affected by the turbine separation requirement. This implies a significant loss in potential wind power, whereby the potential loss is considerably higher in case of turbine type I than for type II because, in the absence of such a constraint, type I could be arranged more densely than type II. Since we have shown that potential re-allocation areas are available instead, we can conclude that, again according to our chosen framework, the current spatial allocation in North Hesse is also not effective as it could be.

7. Conclusions

What general conclusions for securing wind energy supply at the regional level can now be drawn from the comparative evaluation of the two case studies? In the first place, our investigation clearly shows that even for hinterland regions it should be possible in the near future to produce more wind power than today. Our proposals illustrate that through a repowering with state-of-the-art turbines and a spatial re-allocation of wind farm areas, the current regional wind power quotas could nearly double (in West Saxony) or even more than double (in North Hesse). However, to achieve this ambitious goal it would be necessary to overcome a number of constraints. Distance and height regulations turned out to be major obstacles for a full utilisation of existing VE areas. It should be scrutinised whether such requirements are justified in each particular case and whether technical devices could be made available to mitigate adverse effects on humans and nature. Of course, regional plans are devised for long periods of time and need to take into account numerous criteria and balance many interests. Therefore, a re-allocation of VE areas cannot be expected to take place shortly. Just as well as the wind turbines operating today are usually scheduled to run for 20 years or more. Hence, economic incentives for wind farmers are required if repowering is to start earlier. A step in this direction has been made in Germany with the recent amendment of the EEG (2009). To sum up, the findings from our comparative evaluation of two German case studies clearly indicate that wind farming can play a major role in securing energy supply at the regional level. However, wind power generation in the hinterland will not have reached its limit for a long time to come. We suppose that this holds true not only for Germany but for many other European countries as well.

References

BMU, 2007, Bundesministerium für Umwelt, Naturschutz und Reaktorsicherheit (Federal Ministry for the Environment, Nature Conservation and Nuclear Safety), Klimaagenda 2020: der Umbau der Industriegesellschaft, Hintergrundpapier (Climate Agenda 2020: Reconstructing the Industrial Society, Background Paper), Berlin, Germany. http://www.bmu.de/files/pdfs/allgemein/application/pdf/hintergrund_klimaagenda.pdf.

Danish Wind Industry Association, 2009, Park Effect – Park Layout. http://guidedtour.windpower.org/en/tour/wres/park.htm.

DWD, 2007, Deutscher Wetterdienst (German Meteorological Service), Winddaten für Deutschland Bezugszeitraum 1981-2000, mittlere jährliche Windgeschwindigkeit und Weibull-Parameter 80 m über Grund (Wind Data for Germany from 1981-2000, average annual wind speed and Weibull distribution 80 m above ground level). Abteilung Klima und Umweltberatung, Zentrales Gutachterbüro (Department of Climate and Environment Counselling: Central Assessment Office.) Offenbach, Germany.

EEG, 2009, Gesetz für den Vorrang Erneuerbarer Energien (Renewable Energy Sources Act) (EEG), Bundesgesetzblatt (Federal Law Gazette) Jahrgang (Volume) 2008 Teil I Nr. 49 (Part I No. 49), Bonn. http://www.bgblportal.de/BGBL/bgbl1f/bgbl108s2074.pdf.

GWEC, and Greenpeace, 2008, Global Wind Energy Council and Greenpeace International, Global Wind Energy Outlook 2008, Brussels and Amsterdam. http://www.gwec.net/fileadmin/documents/Publications/GWEO_2008_final.pdf.

Koeck, W., and Bovet, J., 2008, Windenergieanlagen und Freiraumschutz (Wind turbines and free space protection), NuR 30:529–534.

Krewitt, W., and Nitsch, J., 2003, The potential for electricity generation from on-shore wind energy under the constraints of nature conservation: a case study for two regions in Germany, Renewable Energy 28:1645–1655.

Ohl, C., and Eichhorn, M., 2009, The Mismatch between Regional Spatial Planning for Wind Power development in Germany and National Eligibility Criteria for Feed-in Tariffs – A case study in West Saxony, Land Use Policy: article in press, corrected proof, available online 18 July 2009, doi:10.1016/j.landusepol.2009.06.004.

Ohl, C., and Monsees, J., 2008, Sustainable Land Use against the Background of a Growing Wind Power Industry, UFZ-Discussion Paper 16/2008, Helmholtz Centre for Environmental Research – UFZ, Grimma, Germany.

RPW, 2008, Regionaler Planungsverband Westsachsen (Regional Planning Agency West Saxony), Regionalplan Westsachsen 2008 (Regional Plan West Saxony 2008), Leipzig, Germany. http://www.rpv-westsachsen.de/

RVN, 2009, Regierungspräsidium Kassel (District Government Kassel), Geschäftsstelle der Regionalversammlung Nordhessen (Agency of the Regional Assembly North Hesse), Regionalplan Nordhessen 2009 – Genehmigungsentwurf (Regional Plan North Hesse 2009 – On-Approval-Draft) Kassel, Germany. http://www.rp-kassel.hessen.de/irj/ RPKS_Internet?uid=3231993b-5869-0111-0104-3765bee5c948.

RENEWABLE ENERGIES TO PROVIDE SUSTAINABLE DEVELOPMENT PERSPECTIVES FOR NORTH AFRICA: THE SAHARA WIND PROJECT

KHALID BENHAMOU[*]

Sahara Wind Inc., Rabat, Morocco

Abstract   The trade winds that blow along the Atlantic coast from Morocco to Senegal represent one of the largest and most productive wind potentials available on earth. Because of the extremely harsh climatic conditions, populations in these areas are concentrated in a few remote cities where economic activities such as mining or fishing can be sustained. Although growing, the local electricity demand remains very low, and unless this vast renewable energy resource can be utilized in a broader context to supply regional electricity markets, economic development alternatives within the region will be limited. This by itself can be considered a threat to regional security, as a lack of economic development over vast desert areas increases the exposure of the region to illegal activities such as trafficking which may have a potential to grow into broader transnational security threats. While local urban centers are witnessing very high unemployment rates, the region is also under pressure from sub-Saharan migrant population fluxes. Ideally, the supplying of regional electricity markets could pave the way for an integrated development of the region, as trade wind generated electricity can be transferred via High Voltage Direct Current infrastructures at relatively low costs. Such energy options are likely to reduce the reliance on fossil fuel generated electricity for Spain and Portugal which boast Europe's highest CO_2 emission increases from Kyoto's 1990 base reference levels. In creating an integrated, job generating, socially acceptable renewable energy industry, such option could also eliminate the need for countries like Morocco or Portugal to consider Nuclear Energy as a viable option for meeting their growing electricity needs, knowing that Spain has suspended the construction of new nuclear power plants for over 2 decades. This paper aims at describing the renewable energy options of countries in North Africa integrated to the

___
[*] To whom correspondence should be addressed: Khalid Benhamou, Sahara Wind Inc., 32 Av. Lalla Meryem Souissi Rabat, 10100 – Morocco; E-mail: kb@saharawind.com

F. Barbir and S. Ulgiati (eds.), *Energy Options Impact on Regional Security*,
DOI 10.1007/978-90-481-9565-7_16, © Springer Science + Business Media B.V. 2010

Euro-Mediterranean power market, and their potential impact on the region's security.

Keywords: Security threats, energy security, nuclear energy, distributed energy, capacity building, Sahara trade winds, climate change, sustainable development, energy options, High Voltage Direct Current transmission technology, Union for the Mediterranean, Solar Plan

1. Current Status – Security and Development Challenges in the Saharan Regions

1.1. GEOGRAPHIC AND HISTORIC BACKGROUND

The Sahara Desert is the world's largest desert. Agricultural resources for subsistence are very scarce. Furthermore, these areas have been administratively divided (rather artificially) during decolonization processes into independent countries that currently dispose of very limited resources, or possibilities to access them. These are sorely needed for the states to be able to secure or assert their own authorities within their own territories.

As has been seen during the colonization process, low population densities and the remoteness of these areas make it difficult even for the most developed economies to be able to secure territories of that size. A glimpse of this phenomenon is currently highlighted in parts of the South Western United States of America, where federal authorities are trying to support States in asserting more control over international border areas.

As a result of historical, geographical and also conjectural challenges, central authorities in the Saharan countries are either weak -enabling thereby the emergence of security threats- or relying on some form of traditional allegiances for delegating the central authority. The latter option can be either seen as an effective, legitimate form of governance or, in some cases, as a very controversial tool used to dilute responsibilities of a government in case of a conflict where exactions are being committed.

The fact that most countries in the Sahel region have had artificial border delimitations made with limited historical considerations; where economics and territorial control by previous colonizing power justified their rationale, these often times do not reflect the countries' own ethnic groups and thereby senses of identities. As borders are by nature incompatible with the regions traditional and nomadic ways of life, conflicts between neighboring countries are often times the norm rather than the exception in this vast area.

1.2. LIVING CONDITIONS

While quality of life may be improving in Morocco, living conditions in Mauritania and overall in the Sahara desert remains very difficult. Access to electricity, water and other services is significantly limited by the natural conditions under which these regions have been subjected to. The extreme dryness and high solar radiation prevents any vegetation from growing, making human presence and traditional economic activities such as agriculture hardly sustainable. Because of the extremely harsh climatic conditions, the population densities of these areas are in the order of less than one person/km². Most of the population is concentrated around water supply sources that constitute attraction poles, shaping communities into a few cities spread within a very large region.

1.3. ECONOMY

As the local economy can hardly sustain itself through agriculture, the main economic activities are based on mining or fishing and fish processing industries where possible. It is important to note that fishing and agriculture, North West Africa's main economic drivers, are under severe constraints. The first one, due to high demographics is the threat of overexploitation of the fishing resources the second one being the effects of climate change which contributed to furthering the erosion and desertification of the region's few agricultural areas. In this context, and as most mineral resources remain largely untapped in this vast region, mining and mine processing activities are likely to become the sector bearing the most promising economic development perspectives.

1.4. DEMOGRAPHY

Due to improved living conditions, particularly in the cities, the region's population has been witnessing significant growth rates. Boasting one the world's highest demographic growth, the population densities increased in urban areas, and as a consequence, cities have grown tremendously. This is bringing new challenges, as infrastructures, access to electricity and water supplies have to accommodate these changes. As an example, the lack in power generation capacities and the construction of a pipeline project to bring water from the Senegal River to supply Mauritania's capital city Nouakchott from many hundreds of kilometers away, is quite edifying.

1.5. SECURITY CHALLENGES

Besides infrastructure problems created by the concentration of populations, there are significant other challenges that need to be addressed within the growing cities of the Saharan/Sahelian Countries. Being mostly nomadic by tradition, the living conditions of populations concentrated in urban centers require significant social, cultural and behavioral adjustments. In such contexts, the necessity to maintain economic activities and employment figures at higher levels may be a critical element that should be taken in consideration with respect to the region's social and political stability. In order to foster a stable environment with long term political stability, economic opportunities for the creation of jobs remains very important. Providing decent incomes, sometimes rather artificially as in some oil rich North African or Gulf States may not necessarily contribute to the security of the region in this particular context, as has been seen in recent years. These local security issues can have global security ramifications as well, which constitutes a significant global security threat that needs to be addressed.

1.5.1. *Terrorism (safe heavens)*

Current transnational security threats are hardly better characterized then with the existence of areas with very little governmental control or authority, where subsistence can only be guaranteed through some form of illegal activity. The impingement of these activities on the notion of human dignity and self esteem with the psychological frustrations they induce can be very easily exploited to shatter one's own identity, in the radicalization of minds and becoming part of a global ideological struggle. This is particularly true when fundamentals or the moral grounds of an activity (such as trafficking) may be questioned. Providing a sense of purpose is important to all human beings. Absent the legitimacy of accepting the natural struggle of ensuring ones survival in difficult circumstances, as by tradition in these areas, any lack of purpose will be particularly effective in the radicalization of minds. It unfortunately must be mentioned that these are not exclusively confined to isolated or remote environments as in the Saharan desert. The main problem there lies in the fact that these areas are much more difficult to access which makes such threat very difficult to contain or even assess. This makes it therefore a very serious hazard.

1.5.2. *Drug Trafficking*

Government resources for territorial oversight are fairly limited in the Sahel region, hence more lucrative contraband and smuggling activities have recently superseded century old traditions of Trans-Saharan commerce.

When considering the broader context of the extension of Latin American drug-trafficking smuggling routes, which have established solid bases in several sub-Saharan countries, taking almost political control in some of them, it may be expected that some level of lawlessness in this vast Sahara desert is likely to prevail.

1.5.3. *Illegal immigration*

Tackling the global consequences of climate change, environmental degradations and rampant desertification on largely agricultural based societies currently under high demographic pressure is a key social priority, as they do generate economic distress leading to mass migration. Being net exporters of immigrants, Morocco and Mauritania for instance are located on the main routes of migrant populations from sub-Saharan Africa which constitutes a significant security threat to the stability of the region and their European neighbors. The North African continent is currently a "protection zone" for southern European borders (Lahlou) and migrant populations have already altered the workforce and demographics of entire economic sectors. The construction sector in Morocco or the fishing industry in Mauritania already employ significant amounts of sub-Saharan migrants, and in the context of the wider illegal immigration trends that the region is witnessing, human trafficking is likely to remain a lucrative market niche. Hence, if nothing is being done to alter the current situation, the prospects for peace and stability within the region may be affected. This in turn can have dangerous consequences on the broader, global security scheme.

2.   Energy Supply, Energy Access a Development Imperative

With a 96% energy dependency from fossil fuel imports absorbing most of Morocco's export revenues, the impact of such dependency on budgetary spending is quite significant. Since close to 30% of National budgets are dedicated to education in the region, one can easily understand how critical the development of sustainable energy consumption schemes can be. While Mauritania enjoys a slightly improved situation regarding its energy dependency, its scarce population is distributed over a vast territory in which access to electricity is virtually impossible to grant through conventional grid infrastructures. Within such context, granting basic access to energy services such as electricity is essential to develop local, sustainable economic activities capable of preventing and fixing migrant populations.

3. Renewable Energy Resources and the New Energy Economy

Since this region is located on the edge of one of the largest electricity grids (EU grid), its large renewable energy potential could produce significant amounts of cheap renewable energies that could be gradually developed and ultimately end up supplying larger electricity markets. The trade winds that blow along the Atlantic coast from Morocco to Senegal represent one of the largest and most productive wind energy potentials available on earth (Fig. 1). As renewable energies have a strong social component, they tend to generate local industrial activities (green jobs), and developing mechanisms to initially firm this energy locally is very important as they lie on the critical path of major alternative, sustainable energy developments.

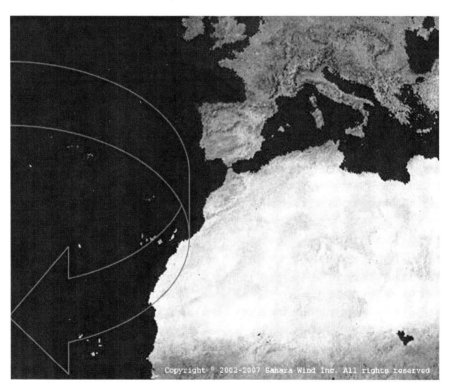

Figure 1. A vast renewable resource potential: global Trade Winds over North West Africa

Due to the geographical dimensions of the areas involved, only isolated distributed grid infrastructures are capable of supplying the cities of the Sahel region. These are very difficult to integrate in a renewable energy system, as their larger size tends to exceed smaller wind or photovoltaic distributed generation systems of the kW range. Solar photovoltaic cells are still very expensive, and because of the erratic nature of winds, wind energy cannot be integrated locally on any significant scale unless coordinated deployments of these technologies are initiated in the region. These will enable a gradual introduction of renewable energy technologies and their integration within the local economy, namely the supplying of urban center and their mine processing industries. Provided an effect of scale can be achieved, these could pave the way to major renewable energy developments. Indeed, the saturation of the larger North African electricity grids to further wind developments – essentially due to grid stability problems – will quickly highlight a need to develop a more comprehensive and integrated approach.

The example of Denmark, as one of the world's wind energy leader is quite edifying in that regard. While relying on a highly interconnected grid, the country has not managed to cover more than 25% of its domestic electricity consumption through wind before encountering major grid stability problems. The country has frozen its domestic wind development activity for many years focusing its wind turbine manufacturing business – one of Denmark's main industrial employers – towards the servicing of export markets. The export of expensive European made wind turbines to lucrative markets (such as in the USA) may not be meant to provide a solution to Africa's electricity access challenges. Although 25% of Denmark's domestic electricity consumption may be quite significant, the same proportion (if achievable...) in the Saharan or sub-Saharan context will translate into very little quantities of wind turbines installed. Furthermore, the limited numbers of large wind turbines and their remoteness will make their maintenance extremely difficult to handle. With about 120 MW of total installed capacities, decentralized and distributed over territories that are twice the size of France, countries like Mauritania, Mali, Niger and Chad to name a few, will hardly make it possible for any imported wind energy technology from Europe to become commercially viable. Hence, developing alternative wind energy technologies to feed smaller electricity markets could be essential for tackling the region's decentralized energy access issues and enable the development of a local, viable wind energy industry which could be essential for tackling the region's economic and security challenges.

4. Wind Power, a Social Energy Economy

In the Saharan isolated urban environments, local diesel units or thermal power plants to provide basic electricity services (if the demand is large enough) are expensive and, maintenance and logistics are challenging issues. In these areas, distances between cities that represent marginal electricity markets in terms of sizes can stretch over 1,000 km. Developing alternative wind energy solutions to feed smaller electricity markets is essential for tackling the region's decentralized energy access issues and enabling the development of a local, viable wind energy industry. Unless mechanisms are developed for integrating renewable energy technologies such as wind power, that is relatively easy to manufacture for small capacities and inexpensive to generate, no basic electricity service can be secured on a sustainable basis.

Initially encouraged to provide employment in the relatively poor North Sea regions of Germany, the wind energy industry has emerged in the last 10 years, as a major business providing competitive prices of electricity even when operated under marginal European wind conditions. As the trade winds that blow along the Atlantic coast from Morocco to Senegal represent one of the largest and most productive wind potentials available on earth, this may open a realm of possibilities for a sustainable development of the region. These countries dispose of a vast wind energy source, and as they face similar social pressure from domestic and sub-Saharan migrant populations fleeing deteriorating environmental conditions, fostering collaboration on integrating clean and more sustainable energy technologies for tackling energy access on a regional base appears to be quite relevant. As mentioned previously, the import of expensive European made wind turbines may not provide a solution to Africa's electricity access challenges. However, since this region is located on the edge of one of the largest electricity grids (that of the EU), local capacity buildup would enable its large renewable energy potential to produce significant amounts of cheap wind energy that could ultimately end up supplying larger electricity markets as well. This however, will require an effect of scale. Developing initial mechanisms to progressively firm these intermittent energy sources locally is an imperative first step as this lies on the critical path of major alternative, sustainable energy developments.

5. Access to Basic Services Water, Electricity, Communications and Security

Although providing access to energy and basic services in a secure environment remains the fundamental responsibility of authorities and governments in these regions, least cost solutions and adequate support systems have to be provided for local populations distributed over very large areas. Conducting applied research on renewable energy technologies within Morocco and Mauritania's research institutions with local industries is critical as they may foster collaboration and regional synergies among developing countries that face common security threats in their loosely controlled remote areas. Indeed, areas of great economic importance are currently lost due to security considerations, particularly in the Sahel region where states rarely dispose of material means to secure their vast territories. It is therefore important that scientific communities integrate the security costs that this lack of alternatives represents to their own economies.

Energy access solutions and applications are relevant to communication infrastructures and permanent power supply systems in remote sites as well. Mobile phone networks and basic security infrastructures do rely on permanent power systems that have to be deployed within broad areas. The development of these infrastructure services and systems contribute to the prevention of security related problems which ultimately falls within the responsibility of sovereign states and governments. Indeed, in the Sahara desert, Mauritania is twice the size of a country like France as are Mali, Niger, Chad and other Saharan countries further to the east. This makes any logistics to service infrastructures for access to electricity, water or communication very challenging due to logistics. Utilizing wind or any other intermittent renewable energy source for distributed electricity supply solutions, or local electricity integrated applications, as the production of chlorine for water treatment, oxygen and even hydrogen as alternative fuels could also be explored. These could be relevant in local mine processing operations and will require adequate training and a capacity build-up in the field of scientific education. This is a fundamental step for harnessing renewable energies and their associated technologies. High tech storage of renewables in the form of hydrogen that can power everything from electronics to life support systems or even vehicles, open perspectives for integrated mine processing applications as well. Mobilizing academia in fulfilling these objectives may be appropriate, since complex hydrogen related technologies are likely to have greater importance in the future. These would enable researchers to identify local synergies and specific applications where these technologies may be relevant.

Providing access in exposing researchers, Engineers and Ph.D. students to these technologies may open a realm of opportunities for them, as well as for their countries. Besides preventing any technological gaps to widen in time, fields of specialization and excellence can be developed regionally, provided a targeted support and appropriate focus can be put on such installations.

The need to develop storage mechanisms is fundamental. If basic electricity service cannot be secured, local economies cannot thrive. Exactly this lack of local economic future leads to employment and population instability which represents a common security concern to both NATO and the countries of the region.

Building regional scientific capacities, and developing a common vision that can generate economic growth in integrating an environmentally friendly and sustainable energy industry (wind energy has 25% growth rates worldwide focused in Europe, the USA and China) could in the long term, become an alternative in fixing migrant population, and contribute to their social integration.

Hence, Involving local scientific communities, industries and end user groups to participate into an applied research program aimed at developing exploitable energy systems to integrate widely available renewable energies is essential in addressing this problematic. Coordinated by Sahara Wind Inc., the NATO Science for Peace SfP-982620 project is for that matter deploying applied research platforms within Morocco and Mauritania's main educational centers around which a far ranging, comprehensive strategy aimed at integrating intermittent sources of renewable energies in the weak grid infrastructure of the Saharan/Sahel region has been envisioned.

Perspectives of a hydrogen energy economy will enable North Africa's scientific communities to take a comprehensive look at energy systems and adopt a holistic, integrated approach to energy technologies which are linked to development issues that have been driven thus far mostly by external market forces providing unsuited ready-made solutions. Indeed, experiences in North Africa have clearly shown that efforts aimed at introducing (new) wind energy technologies in these developing countries amounts ultimately to the simple import of turn-key equipment through concessionary sources of financing and export credit packages. These policies have done very little in terms of local impact for a technology that could have been promising in terms of economic returns, in addressing energy access, energy security, and the creation of an accessible integrated industrial activity.

6.   The EU Neighborhood Policy, the Mediterranean Solar Plan

The Mediterranean Solar Plan, is one of six projects of the Union For the Mediterranean (UPM – Union Pour la Méditerranée). The UPM, launched with 43 Member States, by the President of the French Republic under the French Presidency of the European Union on July 13, 2008, aims at implementing common projects with the countries of the two shores of the Mediterranean. To this end, the UPM seeks to foster a new cooperation and development policy in the whole region of the Mediterranean basin. The Mediterranean Solar Plan (PSM – Plan Solaire Méditerranéen) is the flagship project of the UPM. Its goals are the development of new energy production capacities using renewable energy, in particular solar technologies, on the southern side of the Mediterranean, in order to satisfy demand at local level and to export part of the production towards Europe, as well as the implementation of major progress in reducing energy demand and increasing energy efficiency in the whole region. The concrete target is the construction, by 2020, of 20 GW of new electricity production capacities using low carbon technologies.

7. The Sahara Wind Project

Under the preparation phase of the Mediterranean Solar Plan, whose protagonists the Sahara Wind Energy Development Project helped inspire, and in order to support this political initiative, a pilot project has been submitted under this framework. The deployment of such pilot scale project is aimed at enabling the testing, configuration and adaptation of wind turbine technologies which could subsequently be manufactured locally and gradually deployed in the region as part of a larger project. In a first stage, taking into account the current Moroccan regulation, the Sahara Wind Energy Development Project's installed power capacity will be limited to 50 MW. Then, in subsequent clusters, the wind farm capacities will be progressively developed and extended in order to tap the whole 100 km^2 of land which has been reserved for the project in the region, for a capacity of 500 MW. The initial building phase of the large-scale Sahara Wind Project "PIMS#3292 Morocco: Sahara Wind Phase I/Tarfaya (400–500 MW) on – Grid Wind Electricity in a Liberalized Market ", submitted to the United Nations Development Program (UNDP) and the World Bank's Global Environment Facility (GEF), with the support of several other multilateral institutions would thereby be achieved. This multilateral backed framework would allow for a quick development of the additional capacities while guaranteeing effective, transparent regulations for the project's long term implementation. This would pave the way to the phased deployment of the High Voltage Direct Current Infrastructure that would link the region's vast wind resource to major regional load centers. At the final stage of the project, 5,000 MW will be installed in the region of Tarfaya (500 MW for the first 100 km^2 land). An interconnected 33/90/225 kV wind farm network will be developed; HVDC system will be used for long distance transmission lines whose technical terms of references – for assessing optimal transfer capacities – have already been established with ONE, the Moroccan State owned electric utility company operating the local grid infrastructures. These on-going developments will be further developed and expanded upon insuring the sustained wind energy developments in the region.

Since efficient power transfers over long distances are currently only available for large-scale integrated projects worldwide, the prospect of adding a hydrogen production component in the initial building phase of the Sahara Wind Project is likely to provide an alternative that could enhance the prospects and phasing of this large-scale project. As its original concept developer, Sahara Wind Inc. highlighted the production of hydrogen as an integrated and complementary component of the Sahara Wind Project that enhances prospective uses, perspectives and overall energy efficiencies of large scale renewable energy systems located in remote areas and operating in weaker grid infrastructures. These activities, along with the wind resource assessment of the region are currently being funded by the North Atlantic Treaty Organization under its Science for Peace and security program (under NATO SfP-982620), and the United Nation's Industrial Development Organization under UNIDO assignment post TF/INT/03/002/11-68.

8.   Other Alternatives or Lack of Renewable Energy Alternatives

8.1. ELECTRICITY FROM FOSSIL FUELS

Among European Union countries, both Spain and Portugal have in recent years boasted the highest electrical consumption growth rates. Taking these in consideration together with the even higher growth rates of North African countries, where yearly growth rates of up to 8% have been reached, as in Morocco (Hajroun), the need for deploying rapidly additional electric generation capacities becomes quite obvious. Among the options that are sought, and because of their lower costs to the kW/h generated, large coal fired power plants represent the main alternatives that are currently envisioned. Although no coal deposits are available in the region, coal fired power plants relying on imported coal purchased on the international market, remain the most competitive alternative. Within this context, and while North African countries are not bound by Emission limitations under the current Kyoto agreement, it is important to mention that Spain and Portugal have witnessed the highest growth rates within the EU's country greenhouse gas emissions since 1990. Reducing the carbon footprint on the Iberian Peninsula is a particularly important issue as these countries will be among the ones that will need to benefit from of the EU-wide carbon trading/compensation mechanisms.

Regarding the use of natural gas as a fuel, the GME natural gas pipeline supplies most of Spain and Portugal from the Algerian gas fields (Fig. 2). As this country represents their main natural gas supplier via its pipeline networks, some concerns over energy dependencies have been raised. To this extent, it is interesting to note that the Iberian Peninsula disposes of more Liquefied Natural Gas (LNG) terminals then are available in the rest of the EU (Barcélo, 2009).

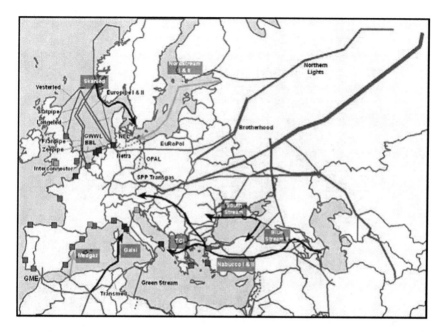

Figure 2. European Gas Networks with LNG Terminals (Source: EU-DG TREN)

The mains concern over electricity generation from natural gas remains its high fuel costs, which makes its use more relevant in matching peak power demands. It is important to notice that Spain and Portugal have also made significant investments in the wind energy sector, which provides a growing share of the region's demand. Some studies have for that matter illustrated that the purchase price premium provided to wind generated electricity, was lower than the actual cost of gas imports that would be needed to generate the same amount of electricity produced. Hence, once could conclude that the combination between gas turbines and wind generated electricity appears to be quite a complementary electricity mix, as the burning of natural gas resources can be matched as fuel savings from wind energy within the realm of the system's global stability. This is particularly relevant, when the operational advantages of both systems in terms of flexibility are taken in consideration.

8.2. NUCLEAR ENERGY OPTION

In Spain, an incident occurred in the final construction stages of a nuclear power plant located close to the Ria de Gernika estuary, in the Basque region (Fig. 3). As a result of a bomb attack of a terrorist action, the concrete reactor dome of the plant shown in the picture had been partly damaged. This incident effectively halted the construction of the plant, bringing the Spanish nuclear energy program effectively to a standstill.

Figure 3. Abandoned Nuclear Plant in Spain's Basque country (Photo: Sahara Wind Inc.)

It may be relevant to note that this event occurred in the mid-1980s when oil prices hit record lows. The more recent peaking of oil prices have not had any impact on the position of the Spanish government to remain opposed to the implementation of a nuclear energy program, preferring instead to support the development of a green energy industry. The green economy is often referred to by the current government as a promising example for the future. Indeed, Spain, along with Portugal has been very active in both the manufacturing and the deployment of renewable energy technologies domestically and abroad. The social dimension of the renewable energy industry may in fact represent one the key arguments playing against the nuclear energy option. Indeed, and as the sector grows, the employment levels from renewable energy industries tend to gain in significance, hence popular support, whereas their geographic distributions are spread in areas where economic activities are most needed. Desolated country sides are indeed one of the main areas benefiting from new wind turbine installations. These generate local jobs, and most importantly tax revenues that remain within the region where the systems have been installed. As jobs created by the nuclear industry tend to be generally concentrated in larger industrial centers where nuclear power plants parts are designed and built, renewable energies provide employments and tax revenues that are geographically much better distributed. Once installed and operational, a nuclear power plant requires fairly limited human attendance, in comparison to the power it generates. This, off course translates into very little employment positions at the plant level. As a comparison, in Germany the Nuclear Industry, which provides over 20% of the country's electricity supply, employs some 30,000 person. As a comparison, the renewable energy sector has, only in the year 2008, created 30,000 new jobs. It is important to mention that in the German context renewable energies account for less than 10% of the overall electricity supply and some 280,000 people are currently employed in the renewable energy sector in Germany alone, according to that country's Ministry of Economy (Fachblatt-Aktion Neue Energie Deutschland, 2009). The figures in Spain are likely to boast similar ratios and one can easily understand why the new Obama administration in the US has singled out renewable energies and related technologies in its stimulus package by providing it with US$78 billion of funding, with the objective to generate thousands of "new green jobs". It is important to mention that the Chinese government is putting the finishing touches to a US$440 billion incentive package designed to boost amongst others the use of solar and wind power in the country (Chan). Hence, the social dimension and growth of the renewables sector will definitely play a key role in the energy options of countries that have initiated industrial programs toward the building of their renewable energy economies.

In the case of North Africa, the Nuclear Energy option will provide very little added value in terms of industrial and economic impact as the high costs of a nuclear power plant is likely to be spent on a turnkey plant imported from abroad. The fact that these plants are hardly duplicable, due to the exiguity of the local electricity markets (smaller grids) will make any industrial integration hardly justifiable whereas stringent and mandatory procedural safety costs for running a single plant cannot be distributed over to other plants. It is also important to mention that base load generated nuclear power, does not match very well with intermittent sources of renewable energies, where more expensive fossils fuels can be saved on an operational basis. Funding for such type of plants is not likely to be easily available due to high upfront investments, and the type of guarantees required from local operators. Although this may less likely to be an issue for countries like Algeria which disposes of large foreign currency reserves from its gas revenues, Morocco may find some difficulties in financing such a program. Seen the high upfront capital costs and the delays associated to the implementation of a nuclear energy program before nuclear power plants can effectively be built and put on-line, alternative options may have an influence on such long term decisions. The renewable energy options followed by larger electricity markets on a regional basis, as on the Iberian Peninsula, will make a nuclear option very difficult to justify in North Africa. The ease for instance, with which a renewable energy industry can be built or transferred in the North African economic context on the basis of what has been achieved in Spain some 13 years ago, when the country did not dispose of any wind energy industry for instance, is likely to favor this alternative. This will be furthermore justified as the renewable energy resources such as wind and solar energies are capable of providing much higher productivities on the North African continent then what is currently being harvested on the Iberian Peninsula.

9. Conclusions

The potential of these resources are vastly higher, as the main limitations relative to their access will remain a weaker grid infrastructure network. The lower, although growing electricity demands versus the long distances from the vast Saharan Trade Wind resources to North Africa's load centers will constitute a major challenge. The access to these renewable energy potentials will in fact require the use of different technologies currently available only for much higher energy transfer capabilities that the size of North African load centers simply cannot absorb.

High Voltage Direct Current (HVDC) transmission lines for instance, can allow vast amounts of electricity (in the GW range) to be transported over long distances at minimal losses. If we were to connect the North African renewable energy resources into a Euro-Mediterranean electricity market as envisioned through our 5,000 MW Sahara Wind Energy Development Project, then a significant share of Europe's wind energy production would be complemented by these large productive sites, making wind energy more affordable.

Comparative advantages rather than a mere displacement of European wind energy productions would be achieved, as the High Voltage Direct Current (HVDC) transmission technologies used would contribute to stabilize surrounding grids on both ends, enabling them to integrate more wind energy. The advantages of integrating wind resources on a continental basis become even more obvious, as the seasonal distribution of winds in terms of peak power production are quite complementary. While winter highs are characteristics of European wind energy generations, the Saharan Trade Winds have their peak production in the summer season. This is particularly relevant as in Southern Europe; the tourism driven economy induces higher electricity consumptions at this time a year, which would, in such cases, be matched by a carbon-free renewable source of wind energy made available at competitive prices.

These perspectives highlighted by the 5,000 MW Sahara Wind Energy Development Project have been presented at the European Parliament in 2002, as they would also contribute to improve the economic prospects of marginal desert regions that currently dispose of very limited endogenous development possibilities.

References

Lahlou M., 2009. Migration and Development in ECOWAS countries: What role for the Maghreb? Regional Challenges of West African Migration. African and European Perspectives – West African Studies. OECD publications, April 2009, pp 99–122, www.oecd.org/csao/migrations

Hajroun M., 2008. Ministère de l'Energie de l'eau et de l'environnement, Direction de l'Electricité et des Energies Renouvelables – Le secteur de l'Électricité au Maroc: L'expérience Marocaine –Africa Power Forum, presentation June 12–13, Marrakech, Morocco.

Fachblatt-Aktion Neue Energie Deutschland., 2009. EUROSOLAR. Europäische Vereinigung für Erneuerbare Energien Kaiser-Friedrich-Straße 11 D-53113 Bonn. April 2009, pp 1–8, (http://www. neue-energie-deutschland.de/)

Barcélo J., 2009. (personal quote and supporting information) Analysis and External Relations. Operador del Mercado Ibérico de Energía – Polo Español, S.A. (OMEL) Madrid, Spain.

Chan Y., 2009. China ready to roll with $440bn green energy plan, New incentives expected to expand use of solar and wind power, BusinessGreen, 29 May 2009. http://www.businessgreen.com

RENEWABLE ENERGY POTENTIALS IN SERBIA

WITH PARTICULAR REGARD TO FOREST

AND AGRICULTURAL BIOMASS

MILAN PROTIC, DRAGAN MITIC, DEJAN VASOVIC[*],
AND MIOMIR STANKOVIC
*University of Nis, Faculty of occupational safety,
Nis, Serbia*

Abstract   In Serbian National Energy Strategy it is clearly stated that one of the main directives in upcoming years will be substitution of conventional fossil fuels with renewable sources. This opens up prospects for development of new business ideas for utilization of Serbian biomass resources which will undoubtedly contribute towards lessening domestic dependence on imported fossil fuels and consequently their environmental impact. Wood and agricultural biomass play key role in this as a two most important biomass resources in Serbia. The paper depicts the current situation and proposes options for future action.

Keywords: Energy efficiency, energy stability, biomass

1. Introduction

Biomass includes all kinds of materials that were directly or indirectly derived not too long ago from contemporary photosynthesis reactions, such as vegetal matter and its derivatives: wood fuel, wood-derived fuels, fuel crops, agricultural and agro-industrial by-products, and animal by-products (Van Loo and Koppejan 2008). It is a widespread source of renewable energy and the only source of renewable carbon. Additionally, it can be regarded as promising option for the large-scale economical production of

---

[*] To whom correspondence should be addressed: Dejan Vasovic, Environmental protection Dept. University of Nis, Faculty of occupational safety, Carnojevica 10a, 18000 Nis, Serbia; E-mail: djnvasovic@gmail.com

F. Barbir and S. Ulgiati (eds.), *Energy Options Impact on Regional Security*,
DOI 10.1007/978-90-481-9565-7_17, © Springer Science + Business Media B.V. 2010

renewable transportation fuels and chemicals. Worldwide production of terrestrial biomass has been estimated to be on the order of 200×10^{12} kg (220 billion tons) annually, which is approximately five times the energy content of the total worldwide crude oil consumption (via heat of combustion analysis) (Himmel 2008). Globally, interest in biomass projects has gradual but strong ascending trend with funding which totaled at least $4.8 billion in 2007 across 127 projects (Boyle et al. 2008).

Similarly, in Serbia interests for biomass are after years of stagnation in permanent rise. This urges for identification of locally available biomass resources as well as appropriate assessment of alternative economically acceptable biomass utilization technologies. To assist in identifying and prioritizing the indigenous biomass resources, the potential raw materials were grouped into two categories of special interest. These are:

- Wood biomass
- Agricultural biomass

2. Wood Biomass

Wood (forest) biomass is defined as the accumulated mass above and belowground of the wood, bark and leaves of living and dead woody shrub and tree species (Young 1980). One of the outstanding features of wood is that it is a readily renewable resource. Ash content in wood is very low. It is free of sulfur and other polluting and corrosive elements.

The challenge for energy production from wood is twofold. First to increase its share of energy supply in the developed countries; and secondly to retain and to grow its share of energy supply in the developing countries while introducing modern bioenergy production technologies (Walker 2006).

Wood raw materials are used selectively in forest industries. Even in the most favorable market conditions, industrial use of forest biomass is generally limited to stems that meet given dimensional and quality requirements (Richardson 2002). Remnants that are not used in future processes are referred to as residues. Most common classification divides forest biomass residues in *forest residues* and *wood processing industry residues*.

2.1. FOREST RESIDUES

In commercial harvesting operations, low-quality stems and other tree components, such as tree branches, tops of trunks, stumps and leaves are left at the site as forest residues.

2.2. WOOD PROCESSING INDUSTRY RESIDUES

Forest based industries fail to completely utilize the wood raw material at the mill, thus producing industrial process residues. In addition to primary process residues such as bark, sawdust, slabs and cores from the primary forest industries, downstream industries such as furniture manufacturing produce secondary process residues that are typically characterized by low moisture contents (Richardson 2002).

3. Serbian Wood Biomass Potentials

Forests in Serbia cover 2.36 million hectare, that is 27.3% of country area. Total forests area per inhabitant is 0.25 ha. Wood reserves are estimated at 235 million cubic metre of wood, using convenient methodological procedures. Pure beech forests represent 27.5%, oak forests 24.4%, other deciduous forests 8.5%, coniferous forests only 5.9% and rest are mixed deciduous–coniferous forests (Vukmirovic 2008).

According to the Republic of Serbia Spatial Plan afforestation will be significantly supported. It is planned to enlarge forests from 27.3% to 31.5% by 2010 and 41.4% by 2050 (Perisic 1996).

Enlargement is planed on the forest estates as well as on the estates which are incapable for agricultural production. Disposition of forests by the districts in the total area of Serbia is given in the Fig. 1.

3.1. FORESTRY OWNERSHIP STRUCTURE

The state owns 56.2% of the total forest area. The remaining 43.8% is under private ownership. The state owned forests in Serbia occupy area of about 1.021 million hectare. These forests are divided between seven public enterprises (PE): PE Srbijasume, PE Vojvodinasume and five PE National Parks.

PE "Vojvodinasume" is a new public enterprise, which was created by a law enacted in 2002, and became operational in early 2003. It is responsible solely for the management of state forests in Vojvodina. PE "Srbijasume" manages the rest of Serbian state forests.

The average size of private forests holdings is small (about 0.5 ha). Private forests tend to contain timber of poor quality, and are largely unproductive. Due to their small size and low productivity, owners can't afford to pay for professional management of their forests. The timber produced in these forests is used mainly for domestic heating.

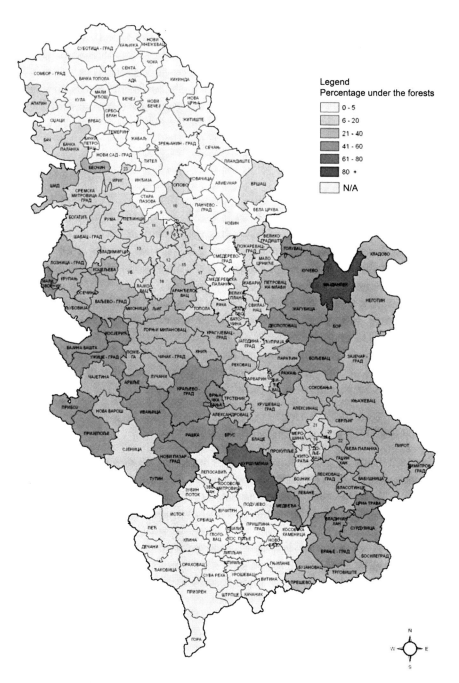

Figure 1. Forest area in the total area of municipalities in Serbia

Despite the fact that PE Vojvodinasume manages less than 20% of total forest area, it excels in sustainable forest management. This is primary based on their sound approach to artificial afforestation, especially with hybrid poplar species.

TABLE 1. Forests by area in 2005 (in 1,000 ha)

	Total	Pure stands	Broadleaved	Conifers	Mixed stand	Broadleaved	Conifers	Broadleaved conifers
State forests								
Republic of Serbia	1021.4	746.7	626.7	120	274.6	194.3	36	44.2
Central Serbia	858.5	652.8	533.9	118.9	205.6	131.1	30	44.1
Vojvodina	162.9	93.9	92.7	1.1	0.069	63.2	5.6	0.1
Private forests								
Republic of Serbia	963	671	608.8	62.1	292	269.5	13.3	9.1
Central Serbia	962.9	671	608.8	62.1	292	269.5	13.3	9.1
Vojvodina	0.1	0.1	0.1	0.001	–	–	–	–

Forest ownership structure as the clarification about district names are illustrated in the Fig. 2.

3.2. WOOD RESIDUES IN SERBIAN FORESTS AND WOOD PROCESSING INDUSTRY

Two forest assortments are produced from gross felled trees: round wood and stacked wood. In felling procedure part of the wood remains on the forest floor and is not used in further processes. That residue can be estimated at 12–18% of gross wood volume. Beside this residue from 100 m^3 of timber 15 m^3 remains on branches and approx. 25–30% on stumps and roots.

Average annual forest falling in Serbia can be estimated at 5.5 million cubic metre. According to the above calculations there are more than 3 million cubic metre of unused forest residues. Energetic value of unused forest residues is estimated at 43,000 TJ/year.

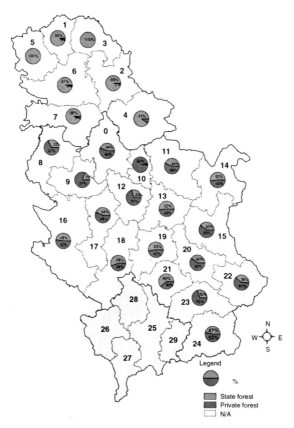

Figure 2. State/private forest ratio by administrative districts in Serbia

0. The city of Belgrade
1. North Backa administrative district
2. Central Banat administrative district
3. North Banat administrative district
4. South Banat administrative district
5. West Backa administrative district
6. South Backa administrative district
7. Srem administrative district
8. Macva administrative district
9. Kolubara administrative district
10. Podunavlje administrative district
11. Branicevo administrative district
12. Sumadija administrative district
13. Pomoravlje administrative district
14. Bor administrative district

15. Zajecar administrative district
16. Zlatibor administrative district
17. Morava administrative district
18. Raska administrative district
19. Rasina administrative district
20. Nisava administrative district
21. Toplica administrative district
22. Pirot administrative district
23. Jablanica administrative district
24. Pcinje administrative district
25. Kosovo administrative district
26. Pec administrative district
27. Prizren administrative district
28. Kosovska Mitrovica adm. district
29. Kosovo and Pomoravlje adm. district

TABLE 2. Structure of potential forest residues in Serbia

	Residues characteristics		Content		Note
	Moisture (%)	Dimensions	(%)	(m³)	
Foliage	30–60	–	–	144,000	Neglected
Stumps and roots	40–60	Bulky	47.3	1,255,000	Remain in forest
Unmerchantable branches	40–60	Bulky	25.7	750,000	Partially used
Offcuts	40–60	Bulky and tiny	27.0	900,000	Partially used
Total			100.0	3,049,000	

Various techno-economical factors influence the purpose and scale of potential residues utilization. Most important are price of the residues, transportation costs and terrain type.

It is possible to utilize virtually 100% forest residues in plain, artificially grown forests estates. However, in natural mountainous counties utilized quantities are substantially reduced since transportation and lack of appropriate forest roads appear to be the dominant problems. These problems intrinsically influence feedstock price.

Forest residues obtained from sound forest management do not deplete the resource base. On the contrary, the utilization of such residues can enhance and increase future productivity of forests.

Wood processing industry also generates considerable amounts of wood residues that can be used as energy source. Wood residues can be in the form of solid matter (offcuts, shavings, chips) or in the form of dust from sawing. The Table 3 depicts estimates for the generation of residues for different types of processes. The estimates should be used with care since actual values will vary widely depending on local conditions, wood species and moisture content.

TABLE 3. Residues from various mechanical wood processing industry

Process	Residue type	Residue rate (%)
Saw-milling	Solid	38
Saw-milling	Sawdust	12
Plywood	Solid	45
Plywood	Dust	5
Particle board	Dust	10
Fiber board	Dust	10

Available technical capacitates for wood processing in Serbia are used almost completely after great recession in previous years. Average (on last tree year basis) annual trimmed wood usage is estimated to be as it is presented in Table 4.

TABLE 4. Average annual trimmed wood usage in Serbia in m^3

	Beech	Oak	Conifer	Poplar	Total
Serbia	464,712	69,730	117,882	660,194	1,312,518

According to the statistical reports forest industry in Serbia comprise 2,758 registered wood processing enterprises of which more than 90% are micro and medium sized firms in private ownership. Estimated number of employees in 2005 was 45,000 of which one third works in primer production and rest in final wood processing sector.

The forest industry in Serbia is in the process of transition. The first and foremost step in this process is privatization of forest harvesting and wood-processing sector. Government efforts and incentives are directed towards the creation of a positive investment climate for direct foreign investments in the wood processing sector, and the establishment of joint ventures between foreign and domestic companies.

4. Agricultural Biomass

Agricultural biomass refers to as lignocellulosic biomass that remains in the field after harvest of agricultural crops as well as crops planted exclusively for energy utilization. According to this, agricultural biomass can be divided into following classes:

- Agricultural residues
- Energy crops

4.1. AGRICULTURAL RESIDUES

Agricultural residues are by-products remaining after harvesting of agricultural crops. Residues can be divided into two main classes:

- Residues from herbaceous cultivation
- Residues from arboriculture

Residues from herbaceous cultivation include wheat, rye, barley, oats and rice straw; maize leaves, stalks, and cobs; soybean stubble; sunflower and rice husks, and bagasse.

Residues from arboriculture include pruning and residues from vine, olive and fructiferous trees.

Due to high collecting costs for most of the agricultural residues (collecting costs are substantially higher for agriculture than for herbaceous cultivation residues), they are not yet widely used for energy purposes. One of the main challenges for the emerging lignocellulosic biomass processing industry will be how to produce, harvest, store and deliver large quantities of feedstock to plants (refineries) in an economically feasible way.

4.2. ENERGY CROPS

The term energy crops can be used both for biomass crops that simply provide high output of biomass per hectare for low inputs, and for those that provide specific products that can be converted into other bio-fuels.

Inevitable, in the longer term, dedicated energy crops will have a much grater role to play. Energy crops can be produced as dedicated plants or in combination with non energy crops. This is new concept for farmers which have to be accepted if large scale energy crops are to become integral part of farming practices. However in USA some extraordinary results have already been obtained with herbaceous perennials such as switchgrass.

4.3. AGRICULTURAL BIOMASS POTENTIALS IN SERBIA

Land and climate conditions in Serbia are highly favorable to the development of agriculture. The plains of Vojvodina, Pomoravlje, Posavina, Tamnava, Krusevac and Leskovac offer favorable conditions for mechanized field crop planting and vegetable production.

Agricultural land covers 5,734,000 ha of which 4,867,000 ha are arable land (0.46 ha per capita). Farmland comprises 70% of the total surface area of Serbia (Vukmirovic 2008).

The Serbian agriculture and food processing industry plays an important role in the contribution to the Serbian GDP, employment and exports. Most of the agricultural production and processing is located on the territory of Vojvodina province and the northern part of central Serbia (FAUB 2005).

Despite great potentials, many difficulties hinder the process of development. Inherited ownership structure of husbandry, social structure of population, place, function and role of agriculture in rural and economy development as well as lack of clear and sound agricultural policy seems to be the dominant problems. From the total number of households in Serbia, 30.89% have agricultural farm. In rural areas from the total amount of households 60.96% are agricultural husbandries. Farms are small and parceled out. From the total number of agricultural husbandries in Serbia, 27.5% have land lesser than 2 ha and only 5.52% bigger than 10 ha. Process of parcels aggregation, modernization and introduction of sound agriculture practices started only recently. However, results are already visible (Table 5).

TABLE 5. Production of important cereals and industrial crops in Serbia (2003–2005)

	Area harvested (ha)	Production (t)	Yield per ha (t)
	Wheat		
2003	611,633	1,364,787	2.2
2004	636,289	2,758,017	4.3
2005	563,269	2,007,060	3.6
	Rye		
2003	6,057	8,225	1.4
2004	6,107	14,902	2.4
2005	7,168	15,778	2.2
	Maize		
2003	1,199,871	3,817,338	3.2
2004	1,199,921	6,569,414	5.5
2005	1,220,174	7,085,366	5.8
	Barley		
2003	109,626	194,371	1.8
2004	109,862	407,411	3.7
2005	104,917	310,850	3.0
	Sunflower		
2003	199,361	353,784	1.8
2004	188,696	437,602	2.3
2005	197,843	350,762	1.8
	Soybean		
2003	131,403	225,963	1.7
2004	117,270	317,836	2.7
2005	130,936	368,023	2.8
	Rape seed		
2003	3,212	3,809	1.2
2004	1,896	4,531	2.4
2005	1,730	3,333	1.9

Principal crops produced in Serbia are summarized in Table 5. It is very difficult to estimate amount of agricultural residues but as a starting point seed/plant residue ratio was used (Table 6).

TABLE 6. Seed/plant residues ratio

Field crop	Seed:plant ration
Wheat	1:1
Barley	1:0.8
Rye	1:1.1
Maize	1:1.1
Sunflower	1:2.5
Soybean	1:2
Rape seed	1:3

Based on prior considerations, potential energy content of unused agricultural residues can be estimated at 20,000 TJ/year. Unfortunately, just negligible part is used for energy utilization (most commonly direct combustion of wheat straw and maize cobs, stalks and leaves). Residues (especially straw) are mainly used as animal feedstock or incinerated in field.

Energy crops planting is confined to rape seed which is used for simultaneous production of biodisel and vegetable oil in first Serbian biorefinery with capacity of 30,000 t/year.

Utilization of agricultural residues in producing energy is economically feasible only if producing plant is located near harvesting field. Main obstacles for wider utilization of agricultural residues in Serbia lie in great expenses of collecting procedure, transport and storage. Since these problems are stumbling blocks even on a global scale, some time will be needed before agricultural residues become economically suitable for energy production.

5. Identification of Potential Biomass Utilization Technologies

In this section alternative biomass utilization technologies were reviewed. Only technologies that are capable for processing biomass resources attainable in Serbian were analyzed. Biomass utilization technologies can be divided into diverse categories based on the type of technology application being proposed and the type of biomass being utilized. For this particular study, biomass utilization technologies were categorized by product application as:

- Bioenergy technologies
- Bioenergy products technologies
- Biofuel technologies

In the following text overview is given for every corresponding technology. Overview was restricted to the most important features. Also, in order to form the basis for unbiased assessment of proposed technologies, information about investment cost for every technology was provided.

5.1. BIOENERGY TECHNOLOGIES

Bioenergy (biomass energy) technologies utilize biomass for the production of heat, steam and electricity. The potential of biomass energy is substantial but lack of proper policy is main obstacle of its proper utilization and development. This is particularly the case with agricultural and wood residues, which are unexploited resources.

5.1.1. *Direct Combustion*

Direct combustion is by far the most common and traditional way of producing heat from biomass. Combustion technologies convert biomass fuels into several forms of useful energy e.g. hot air, hot water, steam and electricity. The simplest combustion technology is furnace that burns biomass in combustion chamber. Biomass fired boilers are today convenient appliances that convert biomass to electricity, mechanical energy or heat.

Commercial and industrial combustion plants can burn many types of biomass ranging from woody biomass to municipal solid waste. Large scale combustion systems use mostly low quality fuels, while high quality fuels are more frequently used in small application systems. Today various industrial combustion systems exist which can be defined as fluidized bed combustion, fixed-bed combustion, and dust combustion.

Combustion technology still needs to be optimized. Fundamental breakthroughs are not expected, but rather small improvements considering cost reduction, increase of fuel flexibility, and reducing environmental impact.

Investment cost: 300 €/kW$_{th}$[1] for 40 MW$_{th}$.[2] production facilities (Ruchser 2000).

5.1.2. *Cogeneration*

Cogeneration is the process of producing two useful forms of energy, normally electricity and heat, using the same fuel source. In developing countries

---

[1] W$_{th}$ stands for Watt thermal.

[2] For direct combustion, pelleting and briquetting facilities investment costs are lead down to 40 MW$_{th}$ facility while for rest of the technologies investment cost were deduced to 400 MW$_t$ facility.

natural gas is mostly used as fuel source. Cogeneration plants can also be based on renewable resources like biomass. Energy experts foresee increasing role of cogeneration in power supply since dissemination is substantially stimulated by the changes taking place in the electricity sector. This technology for certain offers the potential for a much cleaner environment.

Cogeneration plants on large scale are an established technology that is cost competitive with conventional power. Challenges are directed towards micro cogeneration plant as a direct replacement of conventional gas fired boiler. Most promising features of micro cogeneration plants are fuel flexibility control. This technology is still in emerging phase and first units are on commercial trial.

Investment cost: 400–500 €/kW$_{th}$ for 40 MW$_{th}$ production facilities (Ruchser 2000).

5.1.3. Gasification

Gasification is one of the most important ongoing energy research areas in biomass for power generation as it is the main alternative to direct combustion. Gasification is the process of decomposition of solid and liquid organic materials into a combustible gas by controlling the amount of oxygen available. The product gas consists of carbon monoxide, carbon dioxide, hydrogen, methane, trace amounts of hydrocarbons, water nitrogen and various contaminants such as char particles, ash and tars. The importance of this technology relies on the fact that it can take advantage of advanced turbine designs and heat recovery steam generators to achieve high energy efficiency. Gasification processes are relatively insensitive to input material but feedstock with higher moisture content will generally give a gas with lower heating value. Although gasification has not reached a commercial status yet, it is widely accepted that, if fully commercialized, it will bring many environmental and economic benefits. This technology is close to commercialization with over 90 installations and over 60 manufactures around the world (data from 2002).

Investment cost: 550 €/kW$_{th}$ for 400 MW$_{th}$ production facilities (Sims 2002).

5.1.4. Pyrolisis

Pyrolysis is process of decomposition of complex organic molecules in the absence of oxygen to produce tree main energy products in different quantities: gas, oil and coke. The main advantage of pyrolysis over gasification is exactly in wider range of products that can potentially be obtained. From that, pyrolisis is more profitable technology. Pyrolisis technology has been studied in the past decade in many countries and findings are promising. Any form of biomass can be used, but cellulose gives the highest yields. Liquid oils obtained from pyrolysis have been tested for short periods on gas turbines and engines with some initial success, but long term data is still lacking.

Investment cost: 350–1,000 €/kW$_{th}$ for 400 MW$_{th}$ production facility (cost depends on production process) (Sims 2002).

5.2. BIOENERGY PRODUCTS TECHNOLOGIES

Bioenergy products technologies are classified in separate category to distinguish them from bioenergy and biofuels production. Bioenergy product can be regarded as value added bio based manufactured products.

5.2.1. *Carbonization*

Carbonization is technology for biomass conversion to charcoal for use as a boiler fuel. Another approach is referred to as torrefaction, low temperature carbonization process that produces a substitute product for conventional charcoal in some applications. These technologies can create a various bio carbons (e.g. charcoal and carbonized charcoal) for retail use. Contrary to the general view, charcoal consumption has increased in recent years and is becoming an important source of energy as people from rural and urban areas of developing countries shift from wood to charcoal use.

Investment cost: 600 €/kW$_{th}$ for 400 MW$_{th}$ production facility (Sims 2002).

5.2.2. *Pelleting*

Pelleting is process of densification of raw material into a steady and mechanically stable form. Pellets standardized size, allows the manufacturers of boilers, even at the low output range, to implement fully automated combustion and heating system. The compaction of pellets under the great pressure enables the wood pellets to remain stable even under mechanical loads during transport and filling, until they come to be burned. Even if the production of wood pellets with its pressing and drying processes is energy intensive, the energy requirements are well below 2% of the energy content of the end product. At this rate, wood pellets are significantly better than fossil energy sources, for which 10–12% of their own energy content is required for their refinement. Almost all types of agricultural and woody biomass are in principle suitable raw materials for pellet production. In order to keep costs for drying and grinding low, dry sawdust and wood shavings are used predominantly. The production of pellets from bark, straw and crops is usually more appropriate for large scale systems. Research indicates that wood pellet production is in principle economically possible both in small scale (up to 100 t/year) as well as in large-scale plants (10,000 t/year) – with the risk of production being uneconomical higher in smaller plants (less than 1 t/h).

Investment cost: 95 €/kW$_{th}$ for 40 MW$_{th}$ production plant (Mitic 2008).

5.2.3. *Briquetting*

Briquetting is process alike to pelleting with final product similar to split logs. In most cases briquettes have circular cross section with diameter of between 5 and 10 cm and of any length depending on the briquette technology used. Until recently the briquettes were considered the pellets poor relation because, since it doesn't have the same characteristics of availability such as ease of distribution (deliverable by tanker) and a wide range of combustion systems on offer even for individual homes. In most cases raw material is simply compressed without the addition of glues or additives but there is possibility to use different binding materials and additives to form briquettes with predefined features.

Investment cost: 130 €/kW$_{th}$ for 40 MW$_{th}$ production plant (Mitic 2008).

5.3. BIOFUEL TECHNOLOGIES

Biofuels are another category of potential products that can be produced from biomass. Biofuels are typically considered to be liquid transportation fuels generated from biomass. Since this study is confined to Serbian biomass resources, ethanol and biodiesel seems to have most economical potential.

5.3.1. *Ethanol*

Fuel ethanol can be produced from starch or cellulose. Starch hydrolysis for ethanol production is a mature and proven technology. Bioethanol can be produced from various kinds of biomass but only few crops contain simple sugars, which can be easily separated and made available to the yeast in the fermentation process.

The conversion process of lignocellulosic biomass to ethanol only differs from the process described above with respect to the break down of the raw material to fermentable sugar. This hydrolysis process is more difficult than the hydrolysis of starch. Current research and development activities are mainly focused on this issue. Technology is not available on a commercial scale yet and scaling up still proves difficult and commercially unattractive.

Investment cost: 290–400 €/kW$_{th}$ for 400 MW$_{th}$ production facility (first given price is for ethanol produced from sugar-containing crops and the second for ethanol production from lignocellulosic raw materials) (Hamelinck 2001).

5.3.2. *Biodiesel*

Biodiesel is another type of biofuel that can be produced from biomass. It is produced from vegetable oils, which can be derived from oil crops. Vegetable oils have been used as a fuel for a long time already. The conversion of biomass into vegetable oils for automotive fuel applications is similar to the production of vegetable oils for the food industry, which is a well-established process. In most cases the use of biodiesel does not require any adjustments to the engine. Moreover, for the most part, it is possible to use a fuel blend and the existing injection system.

Investment cost: 350 €/kW$_{th}$ for 400 MW$_{th}$ production facilities (Mitic 2008).

6. Closing Remarks

Selection of most suitable biomass utilization technology largely depends not only on technological but on series of socio-economic, environmental and political conditions. In addition, biomass and bioenergy is intrinsically interlaced with land use and labor. This implies that bioenergy programs have to be thoughtfully designed in order to ensure that varying needs are met, for example:

• Satisfying/improving basic needs
• Providing income opportunities
• Making good and effective use of land resources
• Promoting health needs and environmental protection (Heinloth 2006)

Government regulatory role (particularly through subsidizing) in these issues is decisive. Although, in some cases opposite is true. The notable exception is Brazil, where extensive investment in the sugar cane industry helped its well-established bioethanol industry to flourish. It is one of the few countries capable of generating biofuels economically and efficiently, independent of government incentives (Boyle 2008).

Present situation in Serbia renewable energy sector reveals that some pioneering efforts have already been made. This originally pertains to projects grounded on mature and proven technologies (co-combustion, pelleting, briquetting and biodiesel production technologies). First results indicate that even without concrete government support these production facilities could work profitably. However for wider acceptance of renewable energy technology and appropriate market development concrete government incentives are needed.

In order to achieve better, meaning acceptable and sustainable energy policy, Serbian government is obligated to promote clear political will, arrange the energy sector (in terms of national laws harmonization) and encourage further researches and commercial utilizations. Incentives related to renewable sources usage are more than welcomed. It is expected from government to provide concrete financial impetus for renewable energy project and according to recent official announcements this is hopefully going to happen in near future.

Finally, we must be aware that further activities regarding the decrease of fossil fuels consumption will not be possible without renewable resources promotion.

References

Boyle, R., et al., 2008, Global trends in Sustainable Energy Investment 2008, United Nations Environment Programme and New Energy Finance Ltd., pp. 35–37

FAUB – Faculty of Agriculture, University of Belgrade, 2005, FP6 Project – CEEC AGRI POLICY – Agro Economic Policy Analysis of the New Member States, the Candidate States and the Countries of the Western Balkan, First 6-monthly report

Hamelinck, C., Faaij, A., 2001 Future prospects for production of methanol and hydrogen from biomass, Utrecht University, Copernicus Institute, Utrecht

Heinloth, K. (ed.), 2006, Energy Technologies – Subvolume C: Renewable Energy, Springer, New York, p. 358

Himmel, M. (ed.), 2008, Biomass Recalcitrance – Deconstructing the Plant Cell Wall for Bioenergy, Blackwell Publishing, London, p. 7

Mitic, D., 2008, Energy, Faculty of Mechanical Engineering, Nis, pp. 97–110 (in Serbian)

Perisic, D. et al., 1996, Spatial Plan of Republic of Serbia, Službeni glasnik, Belgrade (in Serbian)

Richardson, J. et al. (ed.), 2002, Bioenergy from Sustainable Forestry: Guiding Principles and Practice, Kluwer, The Netherlands, pp. 18–20

Ruchser, M., 2000, Leitfaden für die Errichtung von Bioenergieanlagen, EFO, Bonn

Sims, R.H., Gigler, J., 2002, The brilliance of bioenergy – small projects using biomass, Renewable Energy World 5 (1) pp. 56–63

Van Loo, S. and Koppejan, J. (ed.), 2008, The Handbook of Biomass Combustion and Co-firing. Earthscan, London, p. 1

Vukmirovic, D., 2008, Statistical Yearbook of Serbia, Statistical Office of the Republic of Serbia, Belgrade

Walker, J., 2006, Primary Wood Processing – Principles and Practice, Springer, The Netherlands, pp. 536–540

Young, H.E., 1980, Biomass utilization and management implications. In: Weyerhaeuser Science Symposium 3, Forest-to-Mill Challenges of the Future, Tacoma, Washington, pp. 65–80

DEVELOPMENT OF STRATEGIC SCENARIOS AND OPTIMAL STRUCTURES FOR SUSTAINABLE PRODUCTION OF BIOFUELS IN SLOVENIA

HELLA TOKOS, DAMJAN KRAJNC,
ZORKA NOVAK PINTARIČ[*]
*Laboratory for Process Systems Engineering and
Sustainable Development, Faculty of Chemistry and Chemical
Engineering, University of Maribor, Maribor, Slovenia*

Abstract   The introduction of biofuels in Slovenia is lagging behind the reference values in EU Directive on the promotion of biofuels and other renewable fuels for transport. In order to define the real possibility of achieving political objectives, a survey for Slovenia was implemented, in order to prepare development scenarios for the achievement of the agricultural, industrial and commercial potential of biofuels production in Slovenia. This contribution presents analysis of production processes and technologies of biofuel production. Mathematical model for selection of optimal biofuel production technology depending on raw material availability and economic criteria is briefly presented. Partial results related to the transformation of the closed sugar factory as one of the suitable biofuel production locations into bioethanol plant are also presented.

Keywords: Biofuels, strategic development, bioethanol, mathematical modeling, sustainable development of biofuels, polygeneration

[*] To whom correspondence should be addressed: Zorka Novak, Laboratory for Process Systems Engineering and Sustainable Development, Faculty of Chemistry and Chemical Engineering, University of Maribor, Smetanova 17, SI–2000 Maribor, Slovenia,   E-mail: zorka.novak@uni-mb.si

F. Barbir and S. Ulgiati (eds.), *Energy Options Impact on Regional Security*,
DOI 10.1007/978-90-481-9565-7_18, © Springer Science + Business Media B.V. 2010

1. Introduction

The development of biofuels in the European Union mainly aims at partially replacing diesel and gasoline, in order to abide by the commitments regarding climate change, ensure a durable security of energy supply, and promote use of renewable energy. The transportation sector, road transport in particular, produces more and more greenhouse gas emissions. The transportation market is dependent on oil at 98% (EUBIA 2007). According to the Directorate-General for Energy and Transport of the European Commission (EC 2000), the energy dependence of the EU on its external suppliers could reach 70% by 2030 (of which 90% oil) if no action is taken. At a time when the impending "peak oil" is at the heart of all the discussions, it is urgent to develop and promote sustainable alternatives to conventional fuels.

Because of their renewable nature and their properties similar to those of conventional fuels, biodiesel and bioethanol are some of promising alternatives in the short term. The Directive 2003/30/EC on the promotion of the use of biofuels or other renewable fuels for transport sets a minimum percentage of biofuels to replace use of diesel or petrol for transport purposes in each Member State.

Between the different EU member states, significant differences occur, in terms of production volumes and in the ratio between biodiesel and bioethanol production. The production of biofuels is currently concentrated in a limited number of member states: Germany, France, Spain, Italy and Sweden cover more than 80% of total production.

The introduction of biofuels in Slovenia, and the objectives in this area are lagging behind the reference values in EU Directive on the promotion of biofuels and other renewable fuels for transport, amounting to 2% by the end of 2005 and 5% by the end of 2010. Thus, Slovenia should ensure 36,200 t of biodiesel and 42,170 t of bioethanol by 2010. Slovenian aims to achieve objectives of the EU will require the introduction of new approaches to the production of biofuels. In order to define the real possibility of achieving political objectives, it is essential to perform a multi-criteria systemic analysis of the economic, environmental, energy and social aspects of biofuel production in Slovenia. Such a survey for Slovenia has not been implemented yet, although it is of strategic importance for the preparation of development scenarios for the achievement of the agricultural, industrial and commercial potential of biofuels production in Slovenia.

2. Development of Strategic Scenarios and Optimal Structures for Sustainable Production of Biofuels in Slovenia

In order to determine the strategy of Slovenia in the field of biofuel production, an examination of the actual capacity contribution of biofuels from various sources of raw materials, the impact on climate change, the effects on biodiversity, the importance of plant residues on the sustainable management of land and water resources, degradation of food structure etc. is needed. Thus, the research project related to these issues was initiated, aiming at answering the question whether the Slovenian goals regarding fulfilling requirements of the EU directive are achievable or not, and by which circumstances (environmental impacts, economic, and social effects).

The ongoing project focuses on the development of strategic scenarios and optimal structures of a sustainable production of biofuels in Slovenia. The main goal is to evaluate land availability and suitability for biomass-to-biofuel production, to analyze production processes and technologies of biofuel production (i.e. available technologies for biofuels production and the processes in the development phase, which will be potentially available in the commercial context until 2015), and to assess environmental impact of biofuels production. The project aims to determine optimal structures of complete biofuels production chains, to perform sensitivity analysis of parameters influencing optimal structure, and to support political decision-making regarding biofuels in Slovenia.

The project aims to provide the political authorities with the establishment of a framework that can be achieved in meeting the biofuel energy needs in Slovenia in 2020. The research is to determine whether the plan on 10% biofuel use in Slovenia by 2020 is feasible and, if it is not possible, what should be adapted to plan for Slovenia. The results of research in the project will directly benefit the ministries in formulating and designing appropriate policies regarding intensification or introduction of technologically advanced methods of biofuels production. Results will be for the benefit of the investors who will acquire a more accurate insight into the possibilities and opportunities of biofuels production in Slovenia.

2.1. SUGAR BEET AS A RAW MATERIAL FOR SUGAR AND/OR BIOETHANOL PRODUCTION IN SLOVENIA

In the first part of the project, existing and possible locations for biofuels production have been investigated. Currently there are two planned biodiesel plants in Slovenia, which shall start operating in 2010. Also, a sugar plant which was closed due to EU sugar reform could serve as a perfect location for bioethanol production. According to the closure of this sugar factory and planned conversion to bioethanol production, technical analysis and economic opportunities were analyzed in the project.

The Slovenian sugar industry was recently undergoing a period of change, largely influenced by European sugar production reform. Owing to the surplus of sugar on the World Market, the European Union (EU) decided to reduce economic support for refined sugar by about a third, in order to prevent export of excess sugar to non-EU markets. Arising from these changes, beet growing and sugar production became unviable for one and only Slovenian sugar plants, which was compensated for full dismantling of facility.

However, sugar beet is an efficient energy plant, especially suitable for the production of bioethanol and biogas. With a high content of organic mass and a high share of sucrose, it enables the achievement of very high energy yields per unit area (Strube-Dieckmann 2007). High yield per unit area is also notable from the perspective of increasing competition for arable land, and between field crops.

The purpose of research work in the project was to examine the contribution that mathematical modeling with mixed integer non-linear programming can offer to any decision-making about different co-production strategies for sugar and bioethanol, according to changing market situations. An optimization-based conceptual process design that relies on superstructure optimization was used to simultaneously search for an economically and environmentally-optimal strategy for co-producing sugar and bioethanol. Two scenarios were considered, the first assuming that a facility for sugar production already exists and serves as a possible candidate pertaining to bioethanol production, whilst the second assumed a non-existing capacity for sugar and bioethanol production.

Raw, thin, thick juice, run-off syrup, molasses, or their mixtures, are all suitable as feedstock for bioethanol production using a biochemical process based on fermentation. Our approach was an optimization-based conceptual process design that relies on the use of optimization for identifying the best configuration. For this purpose, a so-called superstructure was defined (Fig.1), which considers a significant number of variations in the topology of sugar and bioethanol processes technological configurations considered. On the basis of results from the computer simulations of both production processes, all the individual process steps were modeled and connected in all possible ways into a superstructure of possible production strategies (i.e.

producing only sugar, producing only bioethanol, and producing sugar and bioethanol in various proportions). The following primary strategies for utilizing the beet resource have been explored in the superstructure:

- Diverting raw juice after the extraction step to the ethanol plant ($Y_{raw} = 1$) or to the next sugar processing step ($Y_{raw} = 0$)
- Selection of a processing route for thin juice after juice purification ($Y_{thin} = 1$, when diverting thin juice to the ethanol plant, else $Y_{thin} = 0$)
- Selection of thick juice utilization after the evaporation step ($Y_{thick} = 1$, when diverting juice to the ethanol plant, else $Y_{thick} = 0$) and selection of mass fraction of thick juice to be diverted to juice storage tanks ($0 < w_{thick\_store} < 1$)
- Determination of a crystallization process with an optimal number of steps, and diverting white sugar run-off syrup A, raw sugar run-off syrup B, and/or molasses to the ethanol plant, by selecting the one-, two- and three-stage crystallization scheme ($Y_{sec\_A} = 1$, $Y_{sec\_AB} = 1$, or $Y_{sec\_ABC} = 1$, respectively)

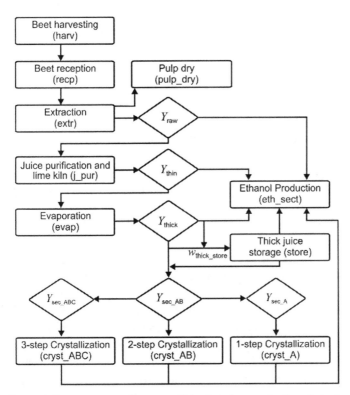

Figure 1. Superstructure of sugar and bioethanol co-production strategies

2.1.1. *Discussion on results*

Superstructure optimization for the *first scenario* with investment needed only for annexing an ethanol facility to an existing sugar plant revealed that, under current market conditions, the best solution is to keep focusing on sugar production by using a two-stage crystallization scheme and divert raw sugar run-off syrup B to ethanol production. The results of the *second scenario* optimization (considering investment in both processes) showed that diverting thick sugar juice directly to bioethanol production gained the best economic results followed by the second best strategy of diverting raw juice to the ethanol plant. This could explain why investments in bioethanol plants are currently more intensive than investments in new sugar plants, across the EU.

The optimal strategy for the second scenario could change, however, if after 2014 EU sugar prices were to be subjected to a further round of substantial price cuts, to bring EU sugar prices more in-line with world market prices, and if world oil prices were to rise considerably. In the long term, however, assuming substantial increases in oil prices, the production of bioethanol from sugar beet should become increasingly viable.

Even if anticipated EU sugar prices in 2014 are likely to remain substantially above world-market price levels, the margin between the optimal strategies for sugar and/or ethanol production is very slim. The results of our study reveal that annexing an ethanol production plant to an existing sugar manufacturing facility is very likely to be an attractive consideration for sugar producers over the coming years, as they will gain higher economic returns. However, the economics of sugar/bioethanol co-production from sugar beet or molasses also depends on other factors, such as the volume of feedstock supply, alternative markets and prices for beet and molasses, local and national demand for bioethanol etc. These factors also need to be included in a more rigorous modeling of sugar and bioethanol co-production strategies. The optimization model, used in our study, allows easy integration of these additional constrains.

This case study clearly proves the usefulness of process synthesis and mathematical modeling when applied to the simultaneous selection of optimal sugar and bioethanol co-production strategy. Information from this study might help to set development strategy, not only for sugar plants, but could also help to create a sustainable long-term policy perspective for the EU sugar sector in the years following deregulation of the market. In our opinion, the EU sugar reform should consider such strategies more carefully, in order to gain better outcomes of the reforms. However, the competition between bioethanol and sugar production will not be the only question faced by the EU sugar industry over the years. Decreasing prices of both commodities will probably force the sector to also investigate new pathways to fully utilize the potential of sugar beet as a raw material (Halasz et al. 2007). Thus, extending the proposed model to include other possible production processes (e.g. biogas and base chemicals), raw materials, logistics, production capacities, environmental impact etc. presents some of the challenges ahead.

2.2. OVERWIEW OF BIOFUEL PRODUCTION TECHNOLOGIES

Important task in the project is a review of current technologies for biofuels production and technologies ready for commercialization in the midterm (2010–2025). These technologies have been screened and are briefly presented in the following.

2.2.1. *Current technologies*

Current biofuel production technologies could be grouped in starch to ethanol fermentation, conversion of fatty acid to methyl esters (transesterification), and sugars to ethanol fermentation. Technologies based on starch fermentation to ethanol are dry and wet mill fermentation. In case of dry mill fermentation, the grains (corn or sorghum) are ground into flour and the starch is converted into sugar with enzymes in a cooker, which is followed by fermentation to ethanol. By-product of the dry-mill process is the distillers' grain, which includes fiber, oil, protein components of the grain, and the non-fermented starch. In the wet milling process, the grain is separated into components (germ, gluten, fiber and starch) by grinding and screening before fermentation, yielding a number of valuable by-products as corn gluten feed, corn gluten mill and corn oil. In case of transesterification vegetable oils and fats are filtered (in order to remove water and contaminants) and converted via base or acid catalyzed transesterification to biodiesel and glycerin, as by-product, which need to be separated. The conversion of sucrose into ethanol is easier in compare to starchy materials because previous hydrolysis of the feedstock is not required since this disaccharide can be broken down by the yeast cells. The process is very similar to the dry milling process, except that no cooking is required. By-product of this production technology is the dry pulp.

2.2.2. *Technologies ready for commercialization in period 2010–2025*

Technologies in development are based on conversion of lignocellulosic biomass to biofuel. These technologies are more complex in comparison to current technologies leading to higher ethanol production costs. However, the fact that many lignocellulosic materials are by-products of agricultural activities, industrial residues or domestic waste (they are not part of human food resources) offers possibilities for biofuel production at large scale. It is considered that lignocellulosic biomass will become the main feedstock for biofuel production in near future.

One of the possible technologies converts cellulose and hemicellulose to sugar by hydrolysis, which follow after pretreatment. Several pretreatment options is studied e.g. chemical pretreatment with diluted acid, physical pretreatment by steam explosion or liquid hot water, and biological pretreatment by fungi. The produced sugar is converted to ethanol by fermentation. Solid residuals from purification, after the fermentation, could be used in cogeneration system, which produce steam for pretreatment and "green electricity".

Several technologies are based on syngas production by biomass gasification. The produced syngas could be conditioned, compressed and fermented to ethanol. The tail gas from fermentation could be also used in cogeneration system. Another option is to use the purified syngas for diesel production (low temperature reactor, 220°C) or gasoline production (high temperature reactor, 325°C) by Fischer Tropsch synthesis. In addition, the syngas could be converted to mixture of alcohols by thermo-chemical conversion, where the ethanol needs to be separated from the mixture.

The lignocellulosic biomass could be converted to bio-oil by fast pyrolysis, by co-processing with fossil fuels in petroleum refinery, or to diesel and other hydrocarbons via hydro-treating methods.

2.3. DEVELOPMENT OF MATHEMATICAL MODEL FOR SELECTION OF OPTIMAL BIOFUEL PRODUCTION TECHNOLOGY

Due to high number of possible biofuel production technologies, it is important to establish a system which is capable of selecting the economically and environmentally the most suitable technology for given circumstances. In this section, a short description of mathematical model for selection of optimal biofuel production technology is given, depending on raw material availability and economic criteria. The model at this stage of project does not include environmental criteria and it is based on superstructural approach with the net present value as an objective function. The superstructure is shown in Fig. 2.

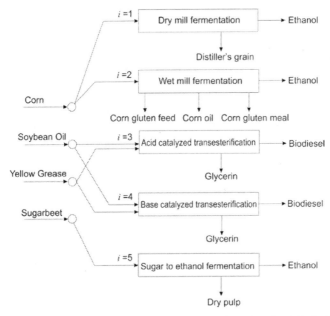

Figure 2. Superstructure of mathematical model for selection of optimal biofuel production technology

According to the superstructure, the model enables selection between the current technologies: dry mill fermentation ($i = 1$), wet mill fermentation ($i = 2$), acid catalyzed transesterification ($i = 3$), base catalyzed transesterification ($i = 4$) and sugar to ethanol fermentation ($i = 5$). In the case of processes $i = 4, 5$ the model was selecting between raw materials for biodiesel production (soybean oil or yellow grease). The model was tested by assuming equal growing area available for all raw materials. In this way, the model shows the influence of raw material yield, raw material cost, by-product selling price, environmental tax on CO_2 emission, operating and capital cost on the selection of optimal biofuel production strategy. The results of mathematical modeling are shown in Table 1. The optimal strategy for biofuels production would be selection of wet mill fermentation, which resulted in the highest net present value. Revenue from selling by-products (corn gluten feed, corn oil and corn gluten meal) makes wet mill fermentation favorable than dry mill fermentation. Sugar to ethanol fermentation also shows high net present value, but the investment is high because of the basic formulation e.g. on the same available area the sugar beet yield is higher which results in increased plant capacity and ethanol production. Processes based on transesterification are economically not attractive because of lower raw material yield and revenues from by-product. Based on these results, if the objective function enables only economic evaluation of production processes, the optimal solution will depend mostly from raw material yield and revenue from by-products.

TABLE 1. Results of mathematical model for optimal biofuel production selection

Process	Feedstock	Yield (kg/ha)	Biofuel production (m³/t)	By-product (kg/m³)	Net present value (G€)	Investment (M€)
Dry mill fermentation	Corn	9,200	0.378	802.85	23,333	2.24
Wet mill fermentation	Corn	9,200	0.338	146.36 38.21 20.19	23,339	28.3
Acid catalyzed transester.	Soybean oil	5,940	0.9766	95.86	2,289	8.3
	Yellow grease	3,200	0.9426		0,047	5.6
Base catalyzed transester.	Soybean oil	5,940	0.9766	95.86	0,062	19.2
	Yellow grease	3,200	0.9426		1,214	12.6
Sugar to ethanol fermentation	Sugar beet	48,500	0.938	500	20,269	53

In this stage of the model development the potential of "green energy" production (by-product in case of hydrolysis and fermentation of lingo-celluloses to ethanol) was also analyzed. Poly-generation increases the efficiency of energy use in production, decreases or eliminates the electricity use from the public network, reduces CO_2 and SO_2 emissions to the environment etc. With the aim to select the optimal co- or trigeneration system based on process heating, cooling and electricity demand a mathe-matical model was developed. The superstructure of alternative poly-generation systems (Fig. 3) includes a cogeneration with a back-pressure steam turbine at three pressure levels ($i = 1, 2, 3$), and a cogeneration system with back-pressure steam turbine at three pressure levels, and higher heat production during heating season ($i = 4, 5, 6$). In this way, the influence of inlet pressure and steam flow rate on electricity production in the cogeneration system with a steam turbine is taken into account. The model also includes a cogeneration system with a gas turbine ($i = 7$) and a trigeneration system with a back-pressure steam turbine at three inlet pressure levels (i = 8, 9, 10). It should be noted that steam production in options i =1–3 and 7–10 should exactly fulfill the current demand of heat in the factory, while the additional production of heat is optimized in the options i = 4, 5, 6 assuming regular demand of external consumers.

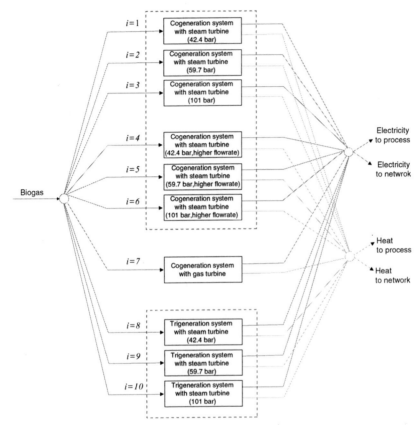

Figure 3. Superstructure of mathematical model for selection of optimal polygeneration system

The model is tested for a brewery, which produces biogas from wastewater (high COD value) with anaerobic fermentation. The results are shown in Table 2. The optimal solution selected by the model is the cogeneration system with back-pressure steam turbine at the pressure level of 42.4 bar (i = 4). During the heating season, heat production should increase above the current level by 50%. Electricity production would cover 25% of the company's requirements. The net present value is 1,089,395 € and the payback period 2.2 years. The disadvantage of this solution is that the brewery would become dependent on smaller consumers of surplus heat.

TABLE 2. Results for polygeneration

Polygeneration system, i	$i = 1$	$i = 2$	$i = 3$	$i = 4$	$i = 5$
Pressure level (bar)	42.4	59.7	101.0	42.4	59.7
Annual electricity production (kW h)	2,821,548	3,400,360	4,557,984	3,381,689	4,075,409
Savings by electricity production (€/a)	182,498	219,935	249,810	218,728	263,597
Saving by additional heat energy production (€/a)	0	0	0	175,328	175,328
Investment (€)	408,112	491,832	659,272	612,168	737,748
Maintenance cost (€/a)	8,464	10,201	13,674	10,145	12,226
Additional fuel consumption cost (€/a)	104,509	125,948	168,827	125,257	150,952
Savings by relief on CO_2 emission tax (€/a)	83,031	77,181	65,478	77,369	70,357
Payback period (a)	3.1	3.5	4.1	2.2	2.5
Net present value (€)	396,944	372,919	324,867	1,089,395	1,041,670

Polygeneration system, i	$i = 6$	$i = 7$	$i = 8$	$i = 9$	$i = 10$
Pressure level (bar)	101.0	11.4	42.4	59.7	101.0
Annual electricity production (kW h)	5,462,847	13,568,370	3,046,712	3,672,592	4,921,808
Savings by electricity production (€/a)	353,337	877,490	197,061	237,543	318,342
Saving by additional heat energy production (€/a)	175,328	0	0	0	0
Investment (€)	988,908	1,471,908	440,680	531,208	711,896
Maintenance cost (€/a)	16,388	67,842	9,140	11,018	14,765
Additional fuel consumption cost (€/a)	202,343	670,092	112,849	136,032	182,303
Savings by relief on CO_2 emission tax (€/a)	56,334	0	80,755	74,429	61,802
Payback period (a)	3.1	—	3.2	3.6	4.3
Net present value (€)	946,220	<0	387,598	361,619	309,766

3. Conclusions

Slovenia experiences several barriers in introducing biofuels production, such as economic barriers (high feedstock prices, production costs, sales prices), technical barriers (need for research and development of second generation biofuels, vehicles available to use bio fuels, distribution of fuels), supply of feedstock (quality of feedstock, available biomass) and other barriers such as competition for feedstock with heat and power.

It is evident that biofuels represent only small part of possible solutions to these barriers. Slovenia should go for many types of renewable energy, focusing on dispersed, small-scale conversion of biomass and bio-waste to transport fuel for local use. A holistic analysis is still needed, taking into account complex relations between food and feed production, biofuels production and environmental impacts.

Thus, in the ongoing project our efforts shall be devoted to estimation of crops' yields future development, land availability for biomass to biofuels, optimal regions/locations for biofuel production, and detailed holistic analysis of environmental, economic and social impacts. All described models represent only some segments in our effort to determine optimal structures for sustainable production of biofuels in Slovenia. They have to be incorporated in a common model, in order to arrive to the best strategy based on both, economic and environmental criteria.

References

Directive 2003/30/EC of the European Parliament and of the Council of 8 May 2003 on the Promotion of the Use of Biofuels or other Renewable Fuels for Transport.

EUBIA – European Biomass Industry Association, 2007. Biofuels for transport. Available at: http://www.eubia.org/ (accessed 11.08.2009).

European Commission (EC), 2000. Green Paper Towards a European strategy for the security of energy supply. COM, 2000, 769 final, Brussels.

Strube-Dieckmann, 2007, December 5. Bioenergy from sugar beet. Retrieved September 29, 2008, from Strube-Dieckmann Web site: http://www.strube-dieckmann.com/inhalte/download/BioenergyfromSBNov2007.pdf.

Halasz, L., Gwehenberger, G., and Narodoslawsky, M., 2007. Process synthesis for the sugar sector – computer based insights in industrial development. Computer Aided Chemical Engineering, 24, 431–436.

BIOMASS FOR ENERGY AND THE IMPACTS ON FOOD SECURITY

SANDERINE NONHEBEL[*]
Center for Energy and Environmental Studies
University of Groningen, The Netherlands

Abstract   In climate policies in the developed world the use of biomass as an energy source plays an important role. Indications exist that these policies are affecting global food security. In this chapter we compare the global demands for food, feed and energy in the near future. We distinguish between developing countries, transition countries and the developed countries. The first group of countries needs extra food for their growing population, the second one needs extra feed, since the increased incomes among their population lead to increased demand for animal products. The developed countries require biomass to reduce the CO_2 emissions of their energy use. On global scale the extra needs for biomass as a fuel (1,100 MT) turn out to be larger than the needs for food and feed (900 MT each). This huge demand for biomass from the energy system is likely to result in large instabilities on the international agricultural markets.

Keywords: Food supply and demand, energy supply and demand, feed, biomass for fuel

1. Introduction

Last year (2008) world market food prices reached highest levels ever, leading to increased costs for food and food security problems in the third world. FAO reported food riots in 33 countries and estimated that 75 million people became food insecure over and above the 850 million in the

---
[*] To whom correspondence should be addressed: Sanderine Nonhebel, Center for Energy and Environmental Studies University of Groningen, The Netherlands; E-mail: S.Nonhebel@rug.nl

F. Barbir and S. Ulgiati (eds.), *Energy Options Impact on Regional Security*,
DOI 10.1007/978-90-481-9565-7_19, © Springer Science + Business Media B.V. 2010

years before. Later that year food prices dropped again and presently food prices are relatively low (FAO 2009). Frequently mentioned reasons for the last year high prices pointed to shortages on the world market due to lower yields in main exporting countries, increased demand for livestock feed in Asia and increased demand for biofuels in the US and EU (Trostle 2008).

The increased demand for biofuels in OECD countries is the result of Climate Policies developed in these countries (FAO 2008). The use of biomass as energy source is one of the options to reduce CO_2 emissions to atmosphere. In many national climate policy plans the use of biomass as energy source plays an important role. The EU directive on renewable energy, for example, mentions that 10% of all transport fuels should be from biomass by the year 2020 (FAO 2008). Presently these biofuels (ethanol and biodiesel) are obtained from food crops like maize, wheat and rapeseed. The contribution of biofuels to the total transport fuels is very small (less than 1%) but its production volume is increasing very fast: 30% per year (OECD/FAO 2009).

The fact that organizations like FAO and OECD mention the use of maize as feedstock for biofuels in the US as one of the causes for the large disturbances in the global food market last year makes it interesting to analyze the magnitude of needs for crops for fuel in relation to the need for crops for food and livestock feed on a global scale. When the present very small use already affects food security situation in Africa and Asia the energy policies aiming at 10% biofuels in 2020 might have disastrous effects on global food security.

In this paper we determine the demands for food, feed and fuel in the coming decades and compare the order of magnitude of these demands. We use the findings to discuss the events on global food markets last year with respect to high food prices. With respect to global demands for food three different factors are of importance: the size of the global population (more people need more food), the type of diet of this population (luxurious diets require more resources than basic menus) and the use of cereals as source for biofuels. These factors show different patterns in time and space. To understand the processes behind these demands the needs for food, feed and energy are first discussed in historical perspective on the global level, then the differences between various countries are analysed. With this knowledge a very simple model is constructed to determine present and future needs for food, feed and fuel. The model results are compared with the present situation on the global cereal market, and consequences of present Climate Policies in OECD countries for food security in the developing countries will be discussed.

2. Food Demand and Supply System

Human diets show an enormous variation (Menzel and D'Aluisio 2005; Popkin 2002). This has to do with the availability of food crops. Not all crops grow everywhere. In the cool climates wheat and potato are the important staple crops, while in the subtropical climates maize is the most important one and in the tropics rice is the major source for food.

Next to this vegetables and fruits consumed also show large variation: carrots, unions, cabbage and apples in cool climates, tomatoes, peppers and oranges in subtropical climates and mangos, pineapple in the tropics. Further differences over the year exist, often climates only allow one harvest per year, while food has to be available whole year long and has to be stored. Staple foods as wheat, potatoes and beans can easily be stored over long periods (months), fresh fruits, however should be consumed within days after harvest. This means that in harvest season food is available in abundance both in volume as in variety, during the year this surplus declines and can even end up in shortage just before the new harvests are available.

With the increased possibilities for (cheap) transport tomatoes and tropical fruits became available for the inhabitants of the cool climate regions, but their historical menus are based on the crops grown locally. The cheap transport also makes out of season consumption possible. On the other side of the globe harvesting seasons occur in other parts of the year and apples from New Zealand are transported to Europe to solve the out of season shortage.

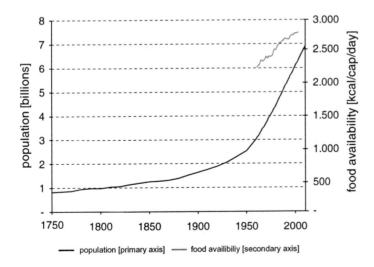

Figure 1. Global population (Evans 2001) and food availability (FAO statistics)

A part of the menu consists out of animal origin products (dairy, meat, fish, eggs), again large variation is found. In some areas hardly any animal products are consumed, while in other regions animal products are the largest share of the menu (90% of the food from Inuits in Northern Canada originates from animals, Receveur et al. 1997). Again the availability of animals in the neighborhood plays an important role. People on islands usually eat a lot of fish, nomads travelling around with herds do have large dairy consumption etc. The main driver for the global demand for food is the size of the population, increase of the population leads to an increased demand for food. Figure 1 shows the global population from 1750 onwards. In 1750 the global population was less than one billion people, it doubled in the next 200 years up to 1950 and from than on steep increase was observed, it nearly tripled in 50 years. In these 250 years food production sector was able to feed this increasing population. Up to the two billion people the increased need for food was fulfilled by increasing the area under cultivation. Later on the green revolution with the improved crop varieties, application of fertilizer and pesticides etc, led to large increases of the crop yields per hectare and area under agriculture remained the same (Evans 2001). Presently large difference between individual countries can be found with respect to population growth. To give an indication of the differences Fig. 2 shows the population growth for a large variety of countries from very poor countries in Africa to the rich countries in Europe. The GDP is used as indication for the development of the countries. The rich countries are on the right hand of the graph and the poor countries on the left hand.

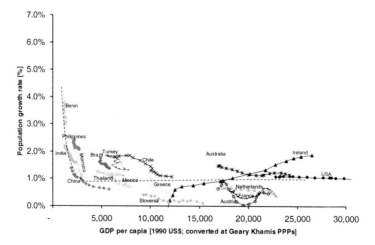

Figure 2. Population growth as function of the GDP per capita The line shows the general pattern used later in this chapter

The graph shows that in the developing countries population growth is around the 2–3% per year, it declines rapidly with increasing GDP and levels off around 1% in the rich western countries. It should be noted that population growth is plotted here, not birth rates. The increase in Ireland over the last 15 years is mainly caused by Irish returning back to Ireland, after having left for the US in the economical crisis in the 1980s.

2.1. THE CONSUMPTION OF MEAT

Next to the size of the population the consumption pattern of this population plays a role. Luxurious consumption patterns (with meat, dairy, exotic fruits and vegetables, beverages like coffee, tea, beer and wine) require more resources than consumption patterns mainly based on staple foods (solely rice). Presently over 35% of the cereals produced in the world is fed to livestock (USDA 2007). Meat consumption plays an important role in global demands for cereals.

Figure 3 shows the global meat consumption from 1960 onwards. In the last 50 years the meat consumption increased from 70 to 250 MTon. The steep increase was mainly due to increased consumption in the developing world, since meat consumption in the rich countries was more or less stable (data FAO statistics).

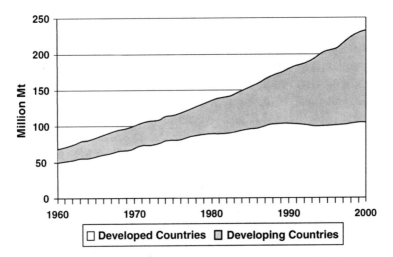

Figure 3. Trends in global meat consumption (FAO statistics)

The feature of stabilizing meat consumption in Western Europe and USA can also be recognized in Fig. 4. This graph shows the changes in consumption of animal origin food products (meat, milk egg, fish) over the last 15 years for the same countries as studied in Fig. 2. The dotted line shows the relation between GDP and consumption of animal products is will be used later on in this chapter. The consumption is expressed as percentage of the total calories consumed.

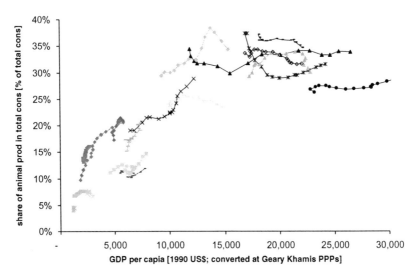

Figure 4. The consumption of animal origin food products in relation to the GDP per capita (1990–2005)

It shows that in the developing countries meat consumption is low (5–10%) with rising GDP values the consumption increases fast to 35% of the calories. At GDP values of about $10,000 per capita, a saturation level is reached and meat consumption levels remain at this 35%. In the high income countries like USA and Western Europe consumption of animal products actually did not increase over the last 15 years.

It should be noted that the graph expresses all products from animal origin (also dairy and eggs). This is the explanation for the fact that consumption in France and The Netherlands is far higher than the consumption in USA. Meat consumption is the USA is highest of the world but they hardy consume dairy products, as a result countries with a large share of dairy in their menu end up higher in this graph.

The largest changes in meat consumption occur in the fast growing economies in Asia (India and China). These countries have GDP values above 2,000 per person, and due to the fast economic growth, their GDP is increasing fast. Figure 4 shows that meat consumption per person in China doubled over the recent 15 years.

3. Energy Demand and Supply System

Figure 5 shows the development of global energy use since the mid-nineteenth century, both the total consumption as the main energy sources are shown. Until 1900 the consumption was more or less constant at 20 EJ year^{-1}. After the industrial revolution the total energy use increased to nearly 500 EJ year^{-1}. Besides this enormous increase in total consumption also a change in energy sources is evident. Before the industrial revolution energy consumption was mainly supplied by wood and in present days it is a mix of several energy sources. These changes in energy sources came along with large changes in society. Before the industrial revolution the total consumption was low and energy conversion was limited to burning of fuel wood for heating and cooking. Humans, horses, windmills and watermills provided mechanical power.

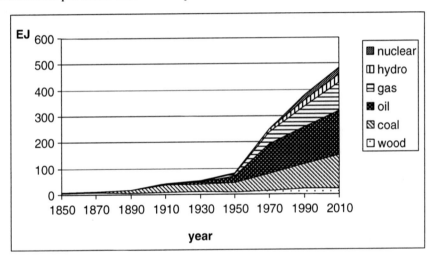

Figure 5. Historical trends in global energy consumption and energy carriers used (Adapted from Grubler and Nakicenovic 1997 using data from IEA 2009a)

The introduction of the steam engine powered by coal caused a large change in the energy system. It was the first conversion from fossil energy into work. Further since coal can be transported and stored, the work could be done on the site where it was required. This in contrast to windmills and watermills where the work had to be done on the site where the energy was. This led to a new organisation of the production: the factories. The development of the mobile steam engine had large impacts for the transportation system. At the beginning of the twentieth century global energy use had increased to 50 EJ year^{-1} and coal had become the most important energy carrier (Fig. 5).

The introduction of electricity as energy carrier (that can supply light, heat and work) and the development of the internal combustion engine triggered the second transition: the introduction of oil as a carrier and later on the introduction of natural gas. At the beginning of the twenty-first century total energy use equals 450 EJ year^{-1} and this amount is supplied by a variety of sources: natural gas, oil, coal, hydropower, nuclear power and wood.

The changes from 1850 onwards showed an increase in total use and a change in sources. These changes in energy sources came together with technological, economic and institutional changes in societies and these changes are interrelated. The situation in individual countries deviates from the global average. Differences are found in the total amount of energy used, the energy carriers in use and the timing of the change from one carrier to the other.

Major differences can be observed between the industrialised countries and the developing countries. Industrialised countries use 200 GJ capita^{-1} year^{-1}, while in the rest of the world it is only 35 GJ capita^{-1} year^{-1} (IEA 2009a). Next to the total amount the carriers used differ. In the developing countries biomass (wood), is still an important energy source (38%), while in the industrialised world it only accounts for 3% of the energy supply (Fig. 6). With respect to the use of hydropower no difference is found in both the industrialised countries as in the developing countries 5% of the energy is generated with hydro.

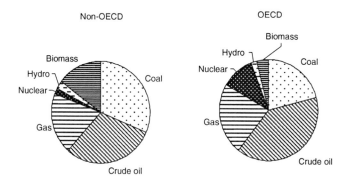

Figure 6. Differences between OECD and non-OECD countries with respect to energy carriers used

Between industrialised countries also large differences exist with respect to energy carriers used. The types of energy sources used in the energy systems of individual countries depend on the availability of natural resources. (IEA, statistics 2009b). Presently coal, oil and gas are account for over 90% of the energy use (Fig. 5). These energy carriers are fossil, they originate from plant material produced in ancient times and combustion of these energy sources lead to emissions of CO_2. Historical data on CO_2 concentrations show fast increasing global levels from 1950 onwards (Fig. 7).

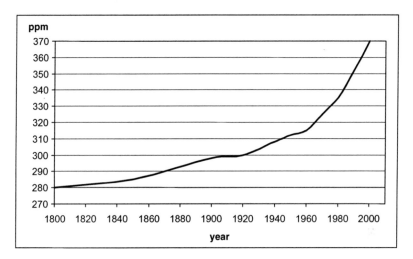

Figure 7. Historical trends in CO_2 concentrations

As mentioned earlier large differences exist in energy use between developing countries and developed world. Energy use per capita for the same group of countries is shown in Fig. 8 as in Figs. 2 and 4. With respect to energy use a more or less linear relation is found between GDP and energy use. The rich countries are requiring nearly five times as much energy as the poor countries. But also large differences between countries at the same level of welfare exits. A person in the USA is using 350 GJ year^{-1}, while a person in Ireland, at the same welfare level only needs 200 GJ year^{-1}.

The rising levels of CO_2 are effecting global climate, in OECD countries policy plans aiming at reduction of the national CO_2 emissions exist. One of the options to reduce emissions is the use of biomass as an energy source. The CO_2 emitted when using biomass as energy source equals the CO_2 captured earlier that year in the photosynthetic process of the crops, on an annual basis no extra CO_2 is emitted to the atmosphere. Both in the EU and in US policy programs exist aiming at 10% biomass as energy source by the year 2020 (FAO 2008; Steinbeck 2007). These programs include the use of biomass as feedstock for electricity generation but also use of biofuels (ethanol and biodiesel) as transport fuel. Ethanol and/or biodiesel are obtained from food crops like wheat, maize, sugarcane, rapeseed, oil palm etc. and they tend to compete with food production. The so-called second generation biofuels use other types of biomass as feedstock. However, techniques to produce second generation biofuels are still under development and not available on the market yet. Therefore it is not likely that they will play a role in the coming years. This implies that in nearby future, biofuels are produced in the old fashioned way, using food crops as feedstock.

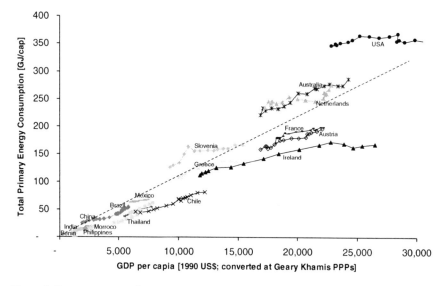

Figure 8. Energy consumption per capita as function of GDP. Line shows relation used later on in this chapter

4. Determining the Need for Food, Feed and Fuel in the Coming Decades

With respect to the needs for food, feed and fuel in the coming decades three different processes can be distinguished. In the first place there is population growth, this growth occurs in the countries with GDP levels below $2,000 per person year. These countries accommodate three billion people. Assuming a populating growth rate of 2.5% this population of three billion people will increase with two billion people in 20 years. In countries with GDP values above $2,000 per capita per year population growth is much smaller.

Secondly there is the change in consumption patterns. The major changes take place in countries with GDP levels between $2,000 and $10.000 per capita per year. In these countries people change from a basic menu mainly consisting out of staple foods (cereals, beans and potatoes) to luxurious menus including animal origin products, (exotic) vegetables and fruits, beverages etc. (Poleman and Thomas 1995). Around $10,000 per person per year a saturation level seems to be reached.

The GDP values of the emerging economies (China, India etc. two billion inhabitants) lay between $2,000 and $10,000 per person. Since the economic growth in these countries is fast (6% per year) that they will pass through this traject from $2,000 to $10,000 GDP within 20 years. So in the coming decades two billion people can be expected to change from a basic into a luxurious diet.

Finally there is the impact of using biomass as energy source. The rich countries (one billion people), are implementing programs aiming at 10% biomass as energy source by 2020. In these countries people use about 200 GJ per person per year, so 20 GJ per person should be generated from biomass.

Food consumption patterns consist out of hundreds of different food items, originating from a large number of crops showing large variations over the globe. The same accounts for the animal products in the food, different livestock types are used, consuming different types of feed and providing different types of products (meat, milk, cheese, eggs). The amount of food needed is therefore difficult to determine in detail. Comparable situation exists for biomass used as energy source. There are different ways to convert biomass into a useful energy carrier: it can be used as feedstock for electricity plants, as feedstock in co-digestion systems (generating biogas) and it can be converted into a biofuels like ethanol or biodiesel. All these conversion routes show different efficiencies (energy obtained from the biomass material), the characteristics of the biomass used (moisture content, fat and carbohydrate composition) determine the most profitable route. It is obvious that we cannot include all this variation in our analysis. Therefore we simplify the food system into just one crop: wheat. So we assume that all our needs for food are fulfilled with wheat, that animals providing the animal products in the diet are fed with wheat and that wheat is used as biomass for energy generation. This simplifying the food consumption pattern into one crop is a commonly applied methodology in food security studies (Penning de Vries et al. 1997). Often the so called grain equivalent is used and tables and calculation procedures exist to convert all kinds of food items into these grain equivalents (Luijten 1995). In this paper we use the methodology developed in food security studies and extend it with the biomass needed for energy.

4.1. THE NEED FOR FOOD IN LOW INCOME COUNTRIES

The human body requires about 2,500 kcal of food per day. On an annual basis this involves about 2.2 GJ. The most simple way to fulfill the metabolic requirements involves the consumption of staple products like wheat, maize, beans and potatoes. This is the menu of the people in the developing countries (three times rice per day), but also the diet of the people in the western world some 100 years ago. One kilogram of wheat provides about 3,000 kcal, so that on an annual basis 300 kg is needed to fulfill the basic demand for food of human being. This 300 kg is the absolute minimum a person needs to stay alive. It resembles the menus provided to people in refugee camps. Even in the poorest countries of Africa (Burkino Faso, FAO statistics) menus show more variation (small amounts of meat, vegetables etc.). To account for this variation the diets in the poor regions of the world are assumed to require 450 kg of wheat.

The population growth takes place in the poor countries where menus are based on staples and hardly any energy is used (Figs. 2, 4, and 7). So in the coming decades the extra food needed as a result of the increasing population involves 450 kg wheat per person for two billion new people.

4.2. THE NEED FOR FEED IN TRANSITION COUNTRIES

To obtain animal products livestock must be kept and during its live this livestock needs feed. The amount of biomass required to obtain 1 kg of meat (feed conversion factor) depends on the type of livestock used. Chicken have low feed conversion factors (around 2) ruminants like cows and sheep do have rather high ones (10) and pigs are in between (4 kg of wheat required for the production of 1 kg of pork (Elferink et al. 2007). This large amount of feed required to produce animal products makes that the more luxurious diets in the western world require far more agricultural resources than the sober diets in the third world. The nutritive value of meat shows large variation, mainly due to the fact the fat percentage differs. In here we assume that 400 kg of wheat is used for feed, this would results in 100 kg meat (pork) per person per year, comparable to meat consumption levels in the rich countries (FAO statistics, Fig. 3). So a change in diet from basic to luxurious will require 400 kg of wheat extra. In the coming decades two billion people are expected to change to luxurious diets and 800 billion kilogram of wheat will be required on a global scale.

4.3. NEED FOR FUEL IN THE RICH COUNTRIES

Modern societies use different energy carriers (gas, oil, coal, electricity etc.) and use different technologies to convert these energy carriers into heat and/or power (combustion engines etc). To be of use for present modern societies, biomass has to be converted into a carrier that fits in the present energy systems. Fermentation of wheat, resulting in ethanol that can be used in gasoline car engines is such an conversion. Another example involves the use of wood as feedstock for electricity generation in coal plants. The last one is the most efficient (unit of biomass per unit of energy). The heating value of dry biomass is 18 MJ/kg. When all conversion losses are ignored 1,100 kg biomass is needed to obtain 20 GJ of energy. The extra need for fuel involves 1,100 kg wheat per person for one billion people.

5. Comparison Between Needs for Food, Feed and Fuel in the Coming Decades

With respect to the needs for biomass in the coming decades three different processes can be recognized: an increase in population in the low income countries leading to and increased demand for food. An increased consumption of luxurious food products in the fast growing economies requiring large amounts of livestock feed and finally an increased need for biomass as a feedstock for energy generation in the high income countries. Table 1 shows the results. In the present situation there are five billion poor people, mainly eating staple crops and one billion people living in the rich world having a luxurious diet with a lot of food products from animal origin. It requires 450 kg of wheat per person per year for a very basic diet and 400 kg of wheat extra for a luxurious diet. Multiplying food per person with the total population results in a total need for wheat of 3,100 billion kilogram on a global scale.

In the coming decades the following events will happen. In the first place two billion people, presently living in the fast growing economies (China, India etc.) are likely to achieve some welfare and change from a basic menu to a luxurious one (transition). This implies that for two billion people 400 kg of wheat per year is needed to fulfill their needs for animal products. In the rich countries 10% of energy supply will be fulfilled with biomass, requiring 1,100 kg of wheat per person. Finally in the poor world, population will increase from three billion to five billion due to high population growth rate (2.5%). In this part of the world two billion people extra have to be fed. As a result of all these changes 2,800 MT wheat extra will be needed.

When comparing the data for food, feed and fuel in Table 1 it is striking that the requirements for fuel overrule all other needs. To fulfill 10% of rich world energy requirements with biomass more wheat (1,100 kg) is needed than for food and feed together (850 kg). Next the amount of wheat needed for fuel in the rich countries (1,100 billion kilogram) is far higher than the amount of feed needed to supply the two billion people in the transition countries with a luxurious diet (800 billion kilogram) and also higher than the amount needed to supply the two billion extra people in the poor countries with a basic diet.

TABLE 1. Overview of the results obtained using assumptions discussed in text

	Present				20 Years				
	Needs per person (kg)		Global needs (Mton)		Needs per person (kg)				Global needs (Mton)
	Food	Feed	Pop.			Food	Feed	Fuel	Pop.	Total
Devel.					Devel.					
Poor	450		5	2,250	Poor	450			5	2,250
					Transition	450	400		2	1,700
Rich	450	400	1	850	Rich	450	400	1,100	1	1,950
Total				3,100						5,900

The quantities of wheat required for food, feed and fuel presently and in the coming decades, in poor, transition and rich economies, using assumptions discussed in the text.

6. Discussion Methodology

The calculation done above is very simple and imply large simplification of the real world. In the first place it is assumed that no losses occur in the food chain, so that all wheat produced is actually consumed. Between the production and the consumption large losses emerge, some publications even mention 30% as losses occurring. The amount of wheat needed to produce animal products (4 kg wheat for 1 kg animal product) is derived from intensive livestock production systems of pigs. More extensive systems require more (Elferink 2009). Further only wheat required for meat is calculated. A change to a luxurious diet involves more changes than just the addition of meat. Dairy products, beverages, vegetables, frying fats etc. also require large amounts of resources, in some food security studies the luxurious diets are estimated to require five times as much grain equivalents as the basic diets. So the amount of wheat to produce the luxurious diets is underestimated.

With respect to biofuels the conversion into an energy carrier is ignored. When biomass is used as feed stock for electricity plants the using 1 GJ of energy from biomass saves 1 GJ of coal or other feed stock. However when wheat is used to produce ethanol which is used as biofuel for cars the losses in the process are large. One ton of wheat (18 GJ) results in 350 l of ethanol (FAO 2008). The energy content of ethanol is 33 MJ/l, so 350 l of ethanol equals 11 GJ and involves an energy loss of 40%!

So in general estimates done in here provide an underestimation of the amounts of wheat required. A more detailed analysis is likely to result in more appropriate estimates. However, the simplicity of the model presented here makes it far easier to understand what processes play an important role, and how do they affect the magnitude of the needs for food, feed and fuel in the coming years. The absolute values of the categories are likely to change but the order of differences shown between categories is very likely to remain.

Although the model is very simple, when the model results for the present situation (Table 1) are compared with the food available at present the resemblance is striking. The present global cereal production is 2,000 Mton (FAO stat 2009), these cereals supply 70% of the global need for food (FAO stats), the remainder is fulfilled with beans and root crops. In this paper all consumption is assumed to be in cereals, the value of 3,100 Mton is therefore in accordance with the present global food consumption.

In the recent OECD/FAO outlook on agriculture (OECD 2009) also comparison was made between requirements for cereals for food, feed and fuel in nearby future. They apply a different methodology but their estimates show same order of magnitude as calculated here.

So it can be concluded that this very simple model provides good insights in present and future needs for food, feed and fuel on global scale. In the following sections the results of the model will be used to discuss the food situation in coming decades.

7. Consequences Results Found for Global Food Security

In the coming 2 decades the global needs for wheat nearly double from 3,100 to 5,900 MT. The production of food has to double to fulfill this needs. Up to the present the food production system was well able to meet the increasing demands as is shown in Fig. 1. There the production per capita is shown over the last 40 years. In 1960 the food production equaled

2,200 kcal per person, while 50 years later the production increased to 2,700 kcal per person. Within this timespan the global population increased from three billion to over six billion people! To obtain some insight in the growth of the demand in comparison with recent history the needs for 1985 are calculated based on the model presented earlier. In 1985 the total population was five billion people of which 0.8 billion where consuming luxurious diets. This situation requires 2,570 MT wheat (five times 450 +0.8 times 400). Figure 9 shows the results from 1985, 2009 and 2030. The last 20 years the global needs increased with 27%, and in the coming decades it has to increase with over 90%.

So the claims that emerge from the expected needs for food, feed and fuel are huge in comparison to changes observed in the past. Assuming that the food production system will easily adapt to these forthcoming needs is therefore not realistic. And shortages on global markets can be expected.

To discuss the consequences of the food production system not being able to meet the needs it is essential to convert the model results to the real

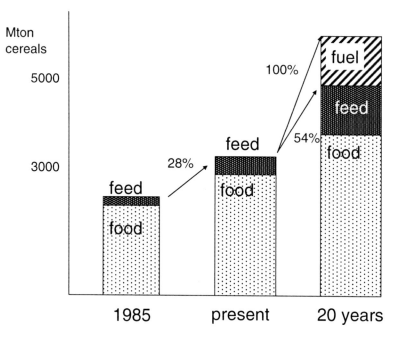

Figure 9. Comparison of the needs for food (green), feed (brown) and fuel (red) on global scale, in 1985, present and in 20 years, based on the assumption in text

world. This means that instead of mentioning poor, transition and rich countries we change to real people countries and continents.

7.1. ANALYSIS IN THE NEEDS PER CONTINENT

Three different populations are distinguished: the three billion people in the low income countries (GDP below 2,000 per person). Most countries in Sub-Sahara-Africa are in this group, and many countries in Asia (Cambodia, Bangladesh). These people live on very basic diets mainly consisting out of staple foods. They spent about 70% of there household income on food (World Bank 2008). The needs for these people with respect to biomass only involve wheat for food. Their demands include 450 kg. Since their economic growth is very small, their consumption per person is not expected to increase.

The second group of people involves the two billion people in the transition countries ($2,000–10,000). For these people menus are changing from basic to luxurious. They spent about 50% of their household income on food. Typical countries are China and India. Due to their shift to more luxurious diets their needs for food will increase from 450 kg per person to 850 kg per person.

Finally the rich countries (GDP above 10,000 per person) are recognized. Their food consumption are luxurious with lots of animal products etc. requiring 850 kg wheat per person. Their consumption patterns have reached the saturation levels with respect to animal products and no changes with respect to the wheat required for food is expected there. They spent 20–30% of their household income on food. Typical countries are USA, Canada and countries in Western Europe. About one billion people live in these countries.

These countries however, have energy programs that aim at obtaining 10% of their energy from biomass, which equals 1,100 kg wheat per person. When we compare the needs between the three populations it is striking that a person mainly living on staple food requires 450 kg, a person on a luxurious diet 850 and a person on a luxurious diet that obtain a small part (10%) of its energy from biomass 1,900 kg (this is nearly five times as much a person in Africa!!). So through using biomass as energy source the claim on the food system doubles! The fact that a lot of biomass is needed to provide energy is known. Densely populated countries are aware of the fact that they will never be able to produce this biomass within their own borders. Often the necessity of importing biomass from other nations is mentioned in the climate policy documents.

8. Global Cereal Production and World Cereal Markets

The global cereal production is in order of 2,200 MT, wheat (630 MT), maize (1,060 MT) and rice (44 MT) are the most important cereals (OECD 2009). Presently wheat is used for food and feed, maize mainly for feed and rice mainly for food.

Figure 10 shows the cereal production in various continents and the imports and exports. The production is highest in Asia and lowest in Africa, due to the differences in population. It is remarkable that continents are producing more or less enough food for their own population.

Only a small fraction (285 MT (10%)) of the total production is actually sold on the world markets. North America and Europe and exporting, while Africa and Asia are importing (Fig. 11).

The fact that local markets are far larger than the global market, makes it interesting to analyze the consequences of the future needs per continent. The present production of Asia and Africa is 1,250 MT and they import 100 MT from the global market. In the coming decade their needs for food increase from the present 1,350 to 2,050 MT (50%), due to increase of population and changing consumption patterns. In the last 20 years production in Africa increased with over 70%, in Asia the increase was smaller: 40%. Based on the achievements in history the increase in production of 50% in the coming decades seems possible.

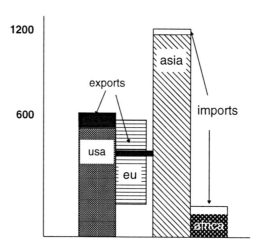

Figure 10. The present situation with respects to cereal production and consumption, imports and exports for four continents

Another picture emerges when we study Europe and the US. Presently they are net exporters they produce more food (1,050) than they can consume (950) and export 100 MT. Their production have to double in 20 years to meet the needs for energy (1,100). However, in the last 20 years the cereal production in these continents increased with only 10%, so in these continents major system changes are needed to meet the increased requirements. When the systems don't adapt in time a shortage on local markets will occur and both continents will change from net exporters into net importers.

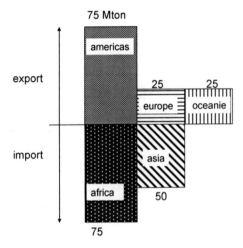

Figure 11. The quantities of cereals imported and exported on the world markets in Mton

However, presently the exports from both EU as USA play vital roles in food security in many low income countries. 75–80% of grains available on the world market is imported by developing countries and used for food and feed. The situation in Africa is worst: this continent is actually importing 30% of its demand for wheat and maize. For individual countries even larger values are found: Egypt is importing 50% of its demand (OECD/FAO 2009).

9. Conclusions

In present Climate Policy plans the availability of biomass as feedstock for energy generation is not a point of attention. Sometimes a remark is made that imports will be needed to achieve the policy goals. In this chapter is shown that even very moderate policy goals aiming at 10% energy from biomass, already leads to demands for biomass for energy that are far larger than the present demands for food. Even very small amounts of biomass as energy source may course large instabilities on the global food markets and large fluctuations in price. Several low income countries are dependent on imports from the global market to fulfill the needs for food of their populations. These populations spend about 60% of their income on food. When global market prices rise these people are affected first. This implies that the present energy policies in the OECD countries carry the risk of generating food insecurity in low income countries.

References

Elferink, E.V., Nonhebel, S., Moll, H.C., 2008. Feeding livestock food residue and the consequences for the environmental impact of meat. Journal of Cleaner Production, 16, 1227–1233.

Elferink, E.V., 2009. Meat, Milk and Eggs: Analysis of the Animal Food Environment Relations. Ph.d. Thesis, Groningen University, The Netherlands.

Evans L.T., 2001. Feeding the ten billion. Cambridge University Press, Cambridge, 247 pp.

FAO, 2008. The state of food and agriculture 2008, biofuels, prospects, risks and opportunities.

FAO, 2009. The state of agricultural commodity markets 2009. FAO, Rome, 63 pp.

Grubler, A., Nakicenovic, N., 1997. Decarbonizing the global energy system Technological Forecasting and Social Change 53, 97–110

Hall, D.O., RosilloCale, F., Williams, R.H., Woods, J., 1993. Biomas. for energy: supply prospects. In: Johansson, T.B., Kelly, H., Reddy, A.K.N., and Williams, R.H. (eds), Renewable Energy, Sources for Fuels and Electricity. Island Press, Washington, pp. 593–651.

IEA, 2009a, Annual Energy Review 2008 Report No. DOE/EIA-0384(2008).

IEA, 2009b. Key world energy statistics. IEA, Paris, France.

Luijten, J.C., 1995. Sustainable World Food Production and Environment. AB-DLO Report no. 37. ABDLO, Wageningen, The Netherlands.

Menzel, P., D'Aluisio, 2005. Hungry planet: what the world eats. Material World Books, California, USA, 287 pp.

OECD-FAO, 2008. Agricultural-outlook 2009–2018, OECD, Paris, France, 279 pp.

Penning de Vries, F.W.T., Rabbinge, R., Groot J.J.R, 1997. Potential and attainable food production and food security in different regions. Philosophical Transactions of Royal Society of London, 352, 917–928.

Poleman, T.T. and Thomas, L.T., 1995. Income and dietary change. Food Policy 20 (2), 149–159.

Popkin, B.M., 2002. The dynamics of the dietary transition in the developing world. In: Caballlero, B. and Popkin B.M. (eds), The nutrition transition: diet and disease in the developing world, Academic, Amsterdam, pp. 111–121.

Receveur, O., Boulay, M., Kuhnlein, H.V., 1997. Decreasing traditional food use affects diet quality for adult Dene/Me'tis in 16 communities of the Canadian Northwest territories. Journal of Nutrition, 127, 2179–2186.

Speedy, A.W., 2003. Global Production and consumption of animal source foods. Journal of Nutrition, 133, 4048S–4053S.

Steinbeck, R., 2007. Biofuels at what cost? International Institute for Sustainable development, Geneve, Switzerland. 72 pp.

Trostle R., 2008. Global Agricultural supply and demand: factors contributing to the recent increase in food commodity prices, Economic research service/ USDA. WRS-0801.

World Bank, 2008. World Development indicators 2008, Worldbank, New York, USA

United States Department of Agriculture (USDA) Foreign Agricultural Service (FAS). 2007. *Production, Supply & Distribution Online Database*. USDA: Washington, D.C. Available online at http://www.fas.usda.gov/psdonline/.

CHANGING ENERGY PRODUCTION, EMERGING TECHNOLOGIES AND REGIONAL SECURITY

BURKHARD AUFFERMANN AND FRANCESCA ALLIEVI[*]
Turku School of Economics, Finland Futures Research Centre, Tampere, Finland

Abstract   The aim of this foresight study is to bring together the issues involved in energy production and emerging technologies, as well as the related threats at the regional level. Scenarios are developed on the basis of a STEEPV workshop.

Keywords: Futures studies, foresight, energy scenario, energy security

1. Introduction

This foresight study focuses on the issues of evolving energy production and emerging technologies, and the related threats at the regional level. These include for example security/safety hazards through inadequate use, and threats from terrorism or organized crime. The time scale taken into consideration for the scenarios development is of 20 years, thus reaching year 2030.

   The first step of the analysis is the assessment of the current energy security situation in the chosen areas of interest: this is carried out through the selection of suitable indicators and their calculation for the period considered. Potentially threatening new technologies and fields of techno-science research are identified, evolving security threat scenarios drafted and suitable policies suggested.

*To whom correspondence should be addressed: Francesca Allievi, Turku School of Economics, Finland Futures Research Centre, Pinninkatu 47, 33500 Tampere, Finland; E-mail: Francesca.Allievi@tse.fi

F. Barbir and S. Ulgiati (eds.), *Energy Options Impact on Regional Security*,
DOI 10.1007/978-90-481-9565-7_20, © Springer Science + Business Media B.V. 2010

Finally, the scenarios (Best Case, Worst Case and Business As Usual) are developed from the results obtained from the workshop carried out during the NATO Advanced Research Workshop which took place in Split in June 2009.

2. Issues to Be Dealt with

This paper is part of a new research project at the Finland Futures Research Centre, Turku School of Economics titled "Land use: Synergies and trade-offs between energy and food production? (LUST)". Climate change, population growth on a global scale and a rising energy demand are interlinked processes with conflicting dimensions. Bioenergy is often addressed as one of the main factors in the portfolio of solutions proposed to fight climate change. Given competing uses for land, including food production, housing, industry and leisure, the land use of energy systems becomes an important issue. The relationship between energy production and food production is increasingly competitive and the global population growth also sets new requirements for agricultural food production in terms of land use. Simultaneously, population growth goes hand in hand with economic growth and increasing energy demand. At the same time, diminishing reserves of fossil fuel, requirements set forth from the climate change mitigation point of view, and ongoing debate on the use of nuclear power indicate a need for increasing amount of renewable energy such as different bio fuels. Consequently, alternative ways of land use become a central problem related to renewable energy.

The objectives of the research are: (1) to develop a new integrated assessment methodology including both quantitative and qualitative methods in order to evaluate social, economic and environmental impacts of possible contradictions, synergies and trade-offs between energy and food production from a land-use perspective; (2) to test this methodology through a number of case studies representing different contexts; (3) to generate future images of land uses at the regional, national, and international level; and (4) to identify and recommend effective policies that would support food and energy security in the future.

In the research project, a tool able to provide both scientists and policy makers with an improved understanding of the prospects for future land-uses will be generated.

This paper will aid to the aforementioned project by drafting preliminary scenarios for future changing energy production, considering emerging technologies and security aspects. European energy security is depending on changing foreign trade and international relations, the changing (global) market situation, technological progress, and the challenges of global climate policies. Secondly, we illustrate our scenarios with a comparison of seven countries/groups of countries in the field of several central indicators related to energy policies/energy security.

3. Overview of Energy Issues Development Between 1960 and 2004

In order to contextualize the issues which have been brought up in the previous paragraph, some energy indicators have been calculated for the following countries: Croatia, Finland, Germany, Italy, Spain, USA and OECD Europe. The input data was taken from the IEA databases (Energy Balances of OECD Countries and Energy Balances of Non-OECD Countries).

3.1. SHARE OF ELECTRICITY LOSSES OVER ELECTRICITY FINAL CONSUMPTION

The first indicator calculated is the share of electricity losses over the final consumption of electricity, which helps to give a picture of how much is lost during the transportation of electricity in the national grids (Fig. 1).

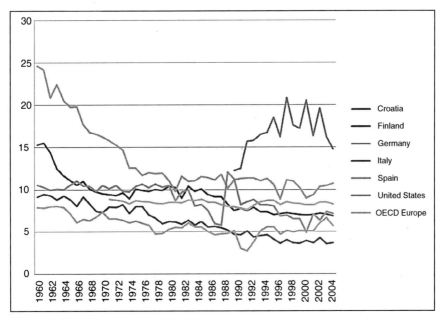

Figure 1. Share of electricity losses over final electricity consumption. Source: Energy Balances of OECD Countries (2007 edition, IEA) and Energy Balances of Non-OECD Countries (2007 edition, IEA)

Spain started with the highest share of electricity losses (24.7%) but rapid decrease to values which are just slightly above those of OECD Europe.

Finland has a steady decreasing pattern throughout the period considered: since the mid-1990s it is constantly the country with the lowest distribution losses.

On average the electricity distribution losses for the countries included in this analysis have diminished from 7.9% in 1990 to 6.8% in 2005.

3.2. ENERGY DEPENDENCY

The energy dependency indicator shows how much countries rely on imports to satisfy their energy needs (Fig. 2).

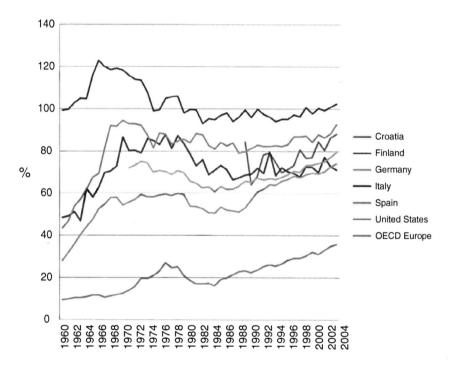

Figure 2. Energy dependency (calculated as imports/Total Final Consumption). Source: Energy Balances of OECD Countries (2007 edition, IEA) and Energy Balances of Non-OECD Countries (2007 edition, IEA)

All the countries follow a similar pattern: increase in energy dependency from 1960 to the beginning of 1970s, overall constant dependency until the end of the 1980s and again an increase until 2005s values.

3.3. TOTAL FINAL ENERGY CONSUMPTION PER CAPITA

With the calculation of this indicator it is possible to compare the amount of energy used per capita in the countries considered (Fig. 3). A more detailed description of the results is reported below.

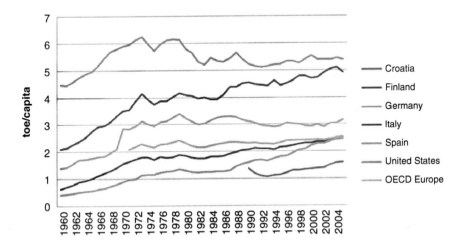

Figure 3. Total Final Consumption per capita. Source: Energy Balances of OECD Countries (2007 edition, IEA) and Energy Balances of Non-OECD Countries (2007 edition, IEA)

In general situation the Total Final Energy Consumption per capita has increased. In United States the TFC/capita has increased until the end of the 1970s, and decreased after that, maintaining rather constant values between 1996 and 2005. A similar situation is that of Germany, which also presents first an increase, then a slight decrease in TFC/capita values, and finally pretty constant values from the beginning of the 1990s on. The country that shows the steepest increase in TFC/capita is Finland, with values ranging from 2.1 toe/capita in 1960 to nearly 5 toe/capita in 2005.

3.4. TOTAL FINAL ENERGY CONSUMPTION OF THE RESIDENTIAL
    SECTOR PER INHABITANT

This indicator shows how much energy is used by the residential sector and in order to make it possible to compare different countries, it has been calculated as a "per capita" value.

In many of the countries considered the Total Final Energy Consumption of the residential sector has seen an increase from 1960 until the beginning of the 1970s, and has subsequently stabilized since the beginning of the

1990s. The situation is however a little different in the case of Spain, where the TFC/capita of the residential sector shows a constant, even if not too steep, increase throughout the period. Also Croatia presents increasing values of this indicator. On the contrary the United States present a decrease between the beginning of the 1970s and the beginning of the 1980s, and a subsequent stabilization of their values around 0.9 toe/capita.

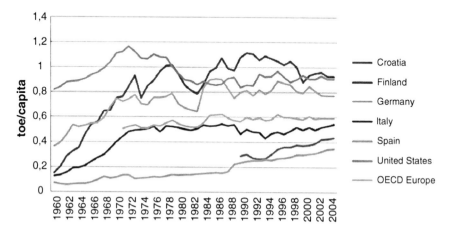

Figure 4. Total Final Consumption of the residential sector per capita. Source: Energy Balances of OECD Countries (2007 edition, IEA) and Energy Balances of Non-OECD Countries (2007 edition, IEA)

4.   Definition of Scenario

A scenario is "a hypothetical sequence of events constructed for the purpose of focusing attention on causal processes and decision points." (Hermann Kahn) Scenarios describe future developments of a specific subject matter considering alternative possibilities, in other words: alternative futures. Scenarios are not about foreseeing the future, but instead about concrete objectives and plausible visions, in this sense one could say that scenarios are stories about the future. Scenarios are no forecasts, they do not schedule any probabilities of future developments, but are interested in causalities of different factors basing on certain aims to be named. Epistemologically, the scenario method helps to identify and describe alternative paths of development and to focus on the related decisions. The starting point of Future Studies is the central assumption that the future is not foreseeable, and the future is not fixed in advance. In societies we are making decisions between different alternative futures (plural!), on the basis of the circumstances and the values involved. Thirdly, future is not to be seen as something we are going to achieve automatically but as the result of joint efforts. We shape our future. Scenarios are thus based on the collection, evaluation and interpretation of facts and trends. An overall picture of what is meant with the word "future" in the field of futures studies is reported below in Fig. 5.

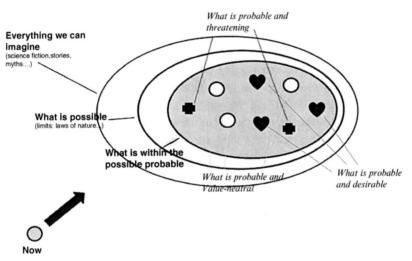

Figure 5. The concept of "future" in the field of futures studies

5. The Securitization of Energy

An important question of analysis is whether energy is seen as an issue of security and defence, or as an issue related to economics and politics. Such difference in perspective is programming the responses:

- In the case energy supply to reasonable prices is seen as an issue of national security, a comprehensive and broad range of diplomatic, foreign policy, trade policy, security and even military policy and measures is coming on to the agenda. Geopolitics and geo-economics will in this case define the agenda of political action.
- If instead the production, the supply and the distribution of energy are anticipated first of all as an economic issue, the answers are necessarily primarily in economics and general politics.
- A third value-based perspective might be found in the field of ecological thinking: Energy-use, distribution, and production would be analysed and political measure proposed in a perspective of security for societies and human beings considering issues like climate change, environmental degradation etc. in the first place.

Securing the energy supply for European societies with their energy intensive industries became a major political issue in the context of the politically motivated oil embargo of the OAPEC countries in autumn 1973 and the following drastic rise of the oil price. This oil crisis and the subsequent price shock addressed for the first time that the dependence of the industrialized countries on oil imports is not only an economic issue, but also a political one. The scientific and political debate about energy prices and strategies to guarantee sufficient energy to a reasonable price became over the last decades linked to the debates about the environmental consequences of the intensive use of fossil sources of energy like oil and gas. The environmental movements changed the political systems in most Western European countries, thus soon establishing themselves as political parties in most of these countries. But the Greens or Environmentalists were not only skeptical about unlimited economic growth and the use of fossil energy, but – first of all – they opposed to the use of nuclear power for the purpose of energy production. As a consequence, environmental thinking influenced politics over the next decades. This led to the establishment of environmental policies, ministries, environmental education and a rather broad consciousness about the limits of growth and the need to use natural resources carefully. Green parties joined governments in several countries, and environmental friendly or "green politics" became part of the established discourse. But this did not mean that hegemonic politics would have been changed drastically. In the policy of most European countries and the European Union, as well as in the policy design of US-President Obama, we can observe a mix of different discourses: security and (national) interest policies are linked to the issue of safeguarding the energy supply. Economic and political instruments are combined with instruments belonging traditionally to "hard security" issues: defense planning as well as planning of future police activities. They, again, are strongly linked to arguments belonging traditionally to the environmental discourse. This has changed only lately, thanks to two factors:

- The awareness raised by the Intergovernmental Panel on Climate Change, because the scientific evidence brought up by the first IPCC Assessment Report of 1990 unveiled the importance of climate change as a topic deserving a political platform among countries to tackle its consequences. It therefore played a decisive role in leading to the creation of the United Nations Framework Convention on Climate Change (UNFCCC), the key international treaty to reduce global warming and cope with the consequences of climate change. Since then the IPCC has delivered on a regular basis the most comprehensive scientific reports about climate change produced worldwide, the Assessment Reports. The

IPCC Second Assessment Report of 1995 provided key input in the way to the adoption of the Kyoto Protocol in 1997.

- The rising prices of oil and gas since the beginning of the Millennium, political instabilities and shortages in supply in recent years changed the picture and brought the issue of energy security on top of the agenda. This has resulted in an increasing competition on the energy markets between the traditionally established end users and the newly rising economies like China and India.

Just recently the treats to energy security had been placed on the agendas of the national security agendas besides security threats like terrorism, transnational crime, environmental catastrophes, AIDS/HIV and pandemics. After 9/11 (2001), the security strategies of many countries had been largely reduced to the threat posed by terrorism. This reduction had been softened during recent year and national security strategies are becoming more multidimensional again, including energy issues, too.

A central question is, whether energy security is seen as an issue of security and defense, or as an issue related to economics and politics. Such difference in perspective is programming the responses: In the case energy supply to reasonable prices is seen as an issue of national security, a comprehensive and broad range of diplomatic, foreign policy, trade policy, security and even military policy and measures is coming on to the agenda. Geopolitics and geo-economics will in this case define the agenda of political action. Once, the production, the supply and the distribution of energy are anticipated as first of all an economic issue (like "energy supply to a reasonable price"), the answers are necessarily first of all in economics and general politics. A third perspective (value-based?) might be found in the field of ecological thinking: Energy-use, distribution, and production would be analyzed and political measure proposed in a perspective of security for societies and human beings considering issues like climate change, environmental degradation etc. at the first place.

6. Workshop Methodology

In relation to the topics presented above, this analysis drafted scenarios based on the results obtained during the workshop carried out within the NATO Advance Research Workshop "Energy Options Impact on Regional Security" which took place in Split, Croatia, in June 2009.

The methodology used is the so called STEEPV analysis. STEEPV is the acronym of the six themes used to develop future scenarios: Society, Technology, Economics, Ecology, Politics, and Values. Moreover, the scenarios are based on the following assumption:

1. Changing foreign policies, including "energy foreign policies" and changing energy markets are the core of any future decision-making on issues related to energy production, supply and used technologies. This includes necessarily attempts for new multilateral agreements about regional and global solutions.
2. Global climate change calls for new climate policies and energy policies aiming at carbon dioxide neutral economies. As the International Energy Agency has formulated in Energy Technology Perspectives 2008: "A global revolution is needed in ways that energy is supplied and used...the global energy economy will need to be transformed over the coming decades".
3. Technological progress offers in future not only a more efficient production of energy but enables in the relative short period until 2030 essential changes towards the use of new technologies of energy production including alternative modes (photovoltaic, gasification, wind energy...).

Scenarios are built by identifying the "drivers" leading to certain conditions. In particular, three groups of drivers can be identified:

1. Foreseeable developments which hardly can be stopped (population developments, non-renewable resources, environmental problems, efficiency of the technologies to be used)
2. Assumptions about issues without clear indication of the direction of the developments (political topics, wars, different social behavior, social systems, market systems, different dynamics of structural adaptations)
3. Wild cards: extremely uncertain and improbable, and non-foreseeable developments, events and influences which are having drastic effects on the whole environment (examples are 9/11, implosion of the Soviet empire, Chernobyl disaster, Oil price shock, Sputnik shock, Black Friday)

After introducing the methodology, participants were divided into three groups corresponding to the three types of possible scenarios: worst case, best case and business as usual. These types are described in further detail below:

6.1. BEST CASE SCENARIO

The leading thought of a best case scenario is rather undisputable: cheap energy supply is sustainable, guaranteed in a long run; its production is not basing on oil, nearly carbon-free and limited burden for the environment; environmental risks are excluded not only in a short, but also in a long run (no new nuclear waste produced).

6.2. WORST CASE SCENARIO

A worst-case scenario has to draw the attention on issues and developments leading to the worst thinkable development. In the case of energy politics this means energy nationalism, wars in competition for energy access, a continuation of the unsustainable energy production resulting in the emission of carbon-dioxide. Nuclear power for energy production is still in use despite the lack of a 100% control of this technology and the long-term consequences of its use.

The purpose of such worst-case scenario is to draw the attention on necessary measures to gain some kind of a minimal security to avoid such development.

6.3. BUSINESS AS USUAL

This type of scenario implies that no big changes occur in relation to the six themes mentioned above during the timeframe considered. Therefore, the deriving scenario will not be much different from the present situation.

7.   Workshop Results

The workshop was carried out by reporting the STEEPV themes vertically on three different posters, on the top of which the writings "Best Case", "Worst Case" and "Business As Usual" were reported. Participants were then given post-its and pens and were asked to write their ideas for each of the six themes and within their corresponding group.

The resulting scenarios are summarized below.

7.1. WORKSHOP RESULTS FOR THE BUSINESS AS USUAL SCENARIO

- Society: more knowledge on energy saving; increasing social conflicts, poverty and conflicts between rich and poor; poor countries develop resulting in population and energy consumption increase; so far so good
- Technology: large improvements in housing sector/zero energy houses; technology development driven by investment – sustainability secondary; gradually increasing efficiency; renewables do not play a major role until oil is exhausted
- Economics: energy prices rising, renewables becoming more competitive; higher taxes on energy from fossil fuels
- Ecology: environment getting better; higher environmental quality demand

- Politics: more liberal; no communication between industry and politics
- Values: values moving towards environmentalist

7.2. WORKSHOP RESULTS FOR THE WORST CASE SCENARIO

- Society: "1989" scenario; population goes up to 25 billions; fortifications of natural borders; no immigration tolerance; brain drains from developing countries only; elderly are easily manipulated ; crisis of the democratic process; breaking down of society; second generation of immigrants →riots; aging society: conservative and nationalist parties taking power; politicians; individualization, not looking at how other people are living
- Technology: big energy companies and state owned companies preventing the development of alternative energies; co-existence of oil and alternatives gives an advantage to those using oil; no transformative technology breakthrough; renewable too labour and material consuming option for sustaining the economy; putting 5,000 MW of wind power in the Sahara and connecting with one cable to EU→1,500 km of unprotected cable in terrorist region; new sophisticated technologies become everyday reality; not reaching progress in energy storage technologies (batteries and hydrogen); huge investment in RES with no solution for the growing world demand
- Economics: energy getting out of market regulations: lobbies of power determining availability of energy; growth paradigm dominates the West and the South; educational level is gradually decreasing; bilateral deals of big money; dynamics of economy turn chaotic; state bankruptcies; no market; oligopolies
- Ecology: everybody gets a small box with a plug in it: free non-polluting energy for everybody...environmental collapse; very cold winter in the EU, Russia can not deliver gas, people burning the forest; environmental refugees; resource plundering, keep cost as low as possible and create crisis; no international cooperation for ecological concerns; nuclear attacks
- Politics: shift to totalitarian society; international conflicts; power production is concentrated in few hands (more than today); weak states increase; war for oil
- Values: Material values prevail; life style "protection" as value and dogma; "good intentions" values of people are manipulated to maintain the "status quo". Example biofuels, Development of Granfalloon distorting the discussion of reasonable alternatives; profit is the only value

7.3. WORKSHOP RESULTS FOR THE BEST CASE SCENARIO

- Society: more open-minded; shift from consumerism to environmentalism; slow down/downshifting; solving problems by taking care of each other, replacing competition; people are content with less stuff; less people; people become more environmental conscious
- Technology: local renewables, in combination with energy storage technologies; donkeys can fly, with or without technical devices; RES are widely used even if they are expensive; no new technologies for problems that require systems solutions; integrated long-range high voltage grid from EU/North Africa; CSS works; 100% energy from RES; nuclear fusion is widely used; GMOs have a positive impact on biofuels
- Economics: there are no more economists; renewable energy will become cheaper; heavy taxes on excessive use of energy; limiting power of corporations by introducing new actors (trusts) that represent individual rights and other species rights; end of capitalism as we know it; create money for public benefit (centrally and locally) no private benefit
- Ecology: climate change is stopped; RES will reverse degradation; appropriations of resources by humans will be limited; waters/land ownership to include guardianship
- Politics: Israel and Palestine live in peace in their own state; Pakistan and India solve the Kashmir conflict; all nuclear weapons are destroyed; strengthening of the EU parliament; politicians finally have got the capacity to make decisions; campaigning by progressive groups will be helpful
- Values: happiness first, community and collaboration values; society starts to value scientists instead of celebrities; "transformative philanthropy" by wealthiest; GDP switched to Gross Peaceful Product (weapons spending); "positive development" ideas of progress

8. Conclusions

The purpose of the workshop was not to draw full scenarios. Instead of telling full stories of what has to happen in the future, we decided to leave it on the level of collecting participants' views on some of the major issues related to future use of energy, the use of technologies, and related economic, ecological, political and value-oriented effects.

Considering the fact that most of the participants had some kind of energy expertise as their background, the results were rather surprising. Under the headline "business as usual" participants addressed rather optimistic views of future developments: strengthening of environmentalist thinking, more wide-spread consciousness among citizens about the use and possible saving of energy, including a certain increase in the use of renewable energy sources. Rather broad was the perception that such "business as usual scenario" will lead to increasing social conflict inside societies and globally between rich and poor countries.

The worst case scenario for the year 2030 addressed clearly a need for major changes in energy policies: crisis of democracy, rise of nationalisms, societies breaking down, failing states and/or totalitarian societies, emerging international conflicts etc. Environmentally, participants views tended to see rather catastrophic developments to be happen until environmental collapse and nuclear war due to such developments.

The material collected for a best case scenario again war surprising, as participants obviously could imagine, that new energy technologies might contribute radically to positive developments in society, politics and environment.

Acknowledgments

The authors would like to thank the participants to the workshop for the successful and productive work done together. Thanks also to the NATO Advance Research Workshop of 2009 organizers, without whom it wouldn't have been possible to develop the scenarios presented.

References

Bell, Wendell (1997) *Foundations of Futures Studies. Vol 1.* Transaction Publishers, New Brunswick. 365 p.

Brown, Lester R. ym. (1996) *Maailman tila 1996.* Gaudeamus, Tampere.

Elämänlaatu, osaaminen ja kilpailukyky. Tietoyhteiskunnan strategisen kehittämisen lähtökohdat ja päämäärät. SITRA 206: Helsinki 1998.

De Jouvenel, Bertrand (1967) *The Art of Conjecture.* Basic Books, New York

Godet, Michel – Roubelat, Fabrice (1996) Creating the Future: the Use and Misuse of Scenarios. *Long Range Planning* 29, 2, 164–171.

Godet, Michel (1994) *From Anticipation to Action, A Handbook of Strategic Prospective.* UNESCO Publishing, Paris.

Hergert, Roland (2007) Strategische Früherkennung. Wahrnehmung relevanter Umweltreize oder Wie ticken Unternehmen? Metropolis.

Ilmola, Leena/Osmo Kuusi (2006) Filters of weak signals hinder foresight: Monitoring weak signals efficiently in corporate decision-making, *Futures* 38, 908–924.

Kuusi, Osmo (1999) Expertise in the Future Use of Generic Technologies. Theoretical and Methodological Considerations Concerning the Delphi Method. Väitöskirjan käsikirjoitus, Helsinki.

Masini, Eleonora (1993) *Why Futures studies?* Grey Seal Books, London.

Niiniluoto, Ilkka (2009) Future Studies: Science or Art? Futura 1/2009, 59–64.

AUTHOR INDEX

SUBJECT INDEX

LaVergne, TN USA
23 September 2010
198115LV00009B/5/P